T0283658

Spectroscopy: Essential Topics and Diverse Applications

Spectroscopy: Essential Topics and Diverse Applications

Edited by **Jason Penn**

New York

Published by NY Research Press,
23 West, 55th Street, Suite 816,
New York, NY 10019, USA
www.nyresearchpress.com

Spectroscopy: Essential Topics and Diverse Applications
Edited by Jason Penn

International Standard Book Number: 978-1-63238-425-6 (Hardback)

Printed in the United States of America.

Contents

Preface

The purpose of the book is to provide a glimpse into the dynamics and to present opinions and studies of some of the scientists engaged in the development of new ideas in the field from very different standpoints. This book will prove useful to students and researchers owing to its high content quality.

This book talks about latest developments in technology, functions and other advancements in spectroscopy. Spectroscopy scientists from various domains had been encouraged to share and discuss information that has been included in this book. With inputs from experts in Chemistry, Biochemistry, Physics, Biology and Nanotechnology, this book is a treasured resource of knowledge for readers. This book will augment the understanding of researchers about complications of different spectroscopic approaches and urge professionals and beginners to contribute part of their future research in understanding relevant mechanisms and applications of chemistry, physics and material sciences. This book focuses on the following sections: atomic absorption spectroscopy, UV-VIS spectroscopy, FT-IR spectroscopy and fluorescence spectroscopy.

At the end, I would like to appreciate all the efforts made by the authors in completing their chapters professionally. I express my deepest gratitude to all of them for contributing to this book by sharing their valuable works. A special thanks to my family and friends for their constant support in this journey.

Editor

Section 1

Atomic Absorption Spectroscopy

1

Atomic Absorption Spectroscopy: Fundamentals and Applications in Medicine

José Manuel González-López[1], Elena María González-Romarís[2], Isabel Idoate-Cervantes[3] and Jesús Fernando Escanero[4]
[1]*Miguel Servet University Hospital, Clinical Biochemistry Service, Zaragoza*
[2]*Galician Health Service, Clinical Laboratory, Santiago de Compostela*
[3]*Navarra Hospital Complex, Clinical Laboratory, Pamplona*
[4]*University of Zaragoza, Faculty of Medicine, Department of Pharmacology and Physiology, Zaragoza*
Spain

1. Introduction

Spectroscopy measures and interprets phenomena of absorption, dispersion or emission of electromagnetic radiation that occur in atoms, molecules and other chemical species. Absorption or emission is related to the energy state changes of the interacting chemical species which characterise them, which is why spectroscopy may be used in qualitative and quantitative analysis.

The application of spectroscopy to chemical analysis means considering electromagnetic radiation as being made up of discrete particles or quanta called photons which move at the speed of light. The energy of the photon is related to the wavelength and the frequency by Plank's constant ($h = 6.62 \times 10^{-34}$ J second) and the speed of light in a vacuum ($c = 3 \times 10^8$ m/s) according to the following equation (Skoog et al., 2001):

$$E = h\nu = hc/\lambda$$

The interaction of radiation with matter is produced throughout the electromagnetic spectrum which ranges from cosmic rays with wavelengths of 10^{-9} nm to radio waves with lengths over 1000 Km. Between both extremes, and from the shortest upwards, can be found the following: gamma rays, X-rays, ultraviolet rays (far, mid and near), the visible portion of the spectrum, infrared rays and radio microwaves. All radiations are of the same nature and travel at the speed of light, being differentiated by the frequency, wavelength and the effects they produce on matter (Skoog et al., 2008).

The Bouguer-Lambert-Beer Law is fundamental in molecular absorption spectrophotometry. According to this law, absorbance is directly proportional to the trajectory of the radiation through the solution and to the concentration of the sample producing the absorption although there are limitations to its application. The Law is only applied to monochromatic radiation although it has been demonstrated experimentally that the deviations with polychromatic light are unappreciable (Skoog & West, 1984).

According to the Bouguer-Lambert-Beer Law:

$$A = abc$$

when b (trajectory of the radiation) is expressed in cm and c (concentration of the substance) in $g.L^{-1}$, the units of a (absorptivity) are $L.g^{-1}.cm^{-1}$, or

$$A = \varepsilon bc$$

when b is expressed in cm and c in $mol.L^{-1}$, a (absorptivity) is called molar absorptivity and it is represented by the symbol ε and its units are $L.mol^{-1}.cm^{-1}$

The absorption of light (A) = log Po/P (= Optical density or extinction)

where:

Po: Incident radiation
P: Transmitted radiation
Absorptivity, a, is A/bc (= Coefficient of extinction)
Molar absorptivity, ε, is A/bc (= Coefficient of molar extinction)

The Bouger-Lambert-Beer Law is fulfilled with limitations in molecular absorption spectrophotometry (Skoog et al., 2008).

In 1927, Werner Heisenberg proposed the principle of uncertainty, which has important and widespread implications for instrumental analysis. It is deduced from the principle of superposition, which establishes that, when two or more waves cross the same region of space, a displacement is produced equal to the sum of the displacements caused by the individual waves. This is applied to electromagnetic waves in which the displacements are the consequence of an electric field, as well as to various other types of waves in which atoms or molecules are displaced. The equation $\Delta t \times \Delta E = h$, expresses the uncertainty principle, signifying that, for finite periods, the measurement of the energy of a particle or system of particles (photons, electrons, neutrons, protons) will never be more precise than $h/\Delta t$, in which h is the Planck´s constant. For this reason, the energy of a particle may be known as a zero uncertainty only if it is observed for an infinite period (Skoog et al., 2001).

In 1953, the Australian Physicist Alan Walsh laid the foundations and demonstrated that atomic absorption spectrophotometry could be used as a procedure of analysis in the laboratory (Willard et al., 1991). The theoretical background on which most of the work in this field was based is due almost entirely to this author (Elwell & Gidley, 1966).

1.1 Fundamentals

The spectra of atomic absorption of an element are made up of a series of lines of resonance from the fundamental state to different excited states. The transition between the fundamental state and the first excited state is known as the first line of resonance, being that of greatest absorption, and is the one used for analysis.

The wavelength of the first line of resonance of all metals and some metaloids is greater tan 200 nm, while for most non-metals it is lower than 185 nm. The analysis for these cases requires modifications of the optical systems which increases the cost of atomic absorption instruments.

In atomic absorption spectrometry, no ordinary monochromator can give such a narrow band of radiation as the width of the peak of the line of atomic absorption. In these conditions the Beer Law is not followed and the sensitivity of the method is reduced. Walsh demonstrated that a hollow-cathode, made of the same material as the analyte, emits narrower lines than the corresponding lines of atomic absorption of the atoms of the analyte in flame, this being the base of the instruments of atomic absorption. The main disadvantage is the need for a different lamp source for each element to be analysed, but no alternative to this procedure improves the results obtained with individual lamps.

The energy source most frequently used in atomic absorption spectroscopy is the hollow-cathode lamp.

1.2 Types

The field of atomic absorption spectroscopy (AAS) includes: flame (FAAS) and electrothermal (EAAS or ETAAS) atomic absorption spectroscopy (Skoog et al., 2008). The base is the same in both cases: the energy put into the free atoms of the analyte makes its electrons change from their fundamental state to the excited state, the resulting absorbed radiation being detected. However the fundamental characteristic of the FAAS is the stage of atomization which is performed in the flame and which converts the analyte into free atoms, whereas in the EAAS the stage of atomization goes through successive phases of drying, calcination and carbonization, and it is not required to dissolve the sample in the convenient matrix as occurs with the FAAS (Skoog et al., 2008; Vercruysse, 1984).

1.2.1 Flame atomic absorption spectroscopy

Prior steps to the stage of atomization in flame are the treatment of the sample, dissolving it in a convenient matrix, and the stage of pneumatic nebulisation. In FAAS the stage of atomization is performed in flame. The temperature of the flame is determined by the fuel/oxidant coefficient. The optimum temperatures depend on the excitation and ionization potentials of the analyte.

The concentration of excited and non-excited atoms in the flame is determined by the fuel/oxidant coefficient and varies in the different regions of the flame (Willard et al., 1991).

1.2.2 Electrothermal atomic absorption spectroscopy

The electrothermal atomic absorption spectrophotometer has three part: the atomizing head, the power source and the controls for feeding in the inert gas. The atomizing head replaces the nebulising-burning part of the FAAS. The power source supplies the work current at the correct voltage of the atomizing head. The computer control of the atomizing chamber ensures reproducibility in the heating conditions, establishing a suitable profile of temperatures in the heating scale from environmental temperature to that of atomization so that the successive stages of drying, calcination and carbonization the sample must go through are those required. The working temperature and the duration of each stage of the electrothermal process must be carefully selected taking into account the nature of the analyte and the composition of the matrix of the sample. The control unit which measures and controls the flow of an inert gas within the atomizing head is designed to avoid the destruction of the graphite at high temperatures due to oxidation with the air.

One variant of the graphite oven is the carbon bar atomizer.

The main advantages of electrothermal atomic absorption are:

a. high sensitivity (absolute quantity of analyte of 10^{-8} to 10^{-11} g);
b. small volumes of liquid samples (5–100 µL);
c. the possibility of analysing solid samples directly without pretreatment and,
d. low noise level of the oven (Willard et al., 1991).

2. Applications of atomic absorption spectroscopy in medicine

Atomic absorption spectroscopy is a sensitive means for the quantification of some 70 elements and is of use in the analysis of biological samples (Skoog & West, 1984). FAAS allows the detection of Ag, Al, Au, Cd, Cu, Hg, Pb, Te, Sb and Sn with great sensitivity (Taylor et al., 2002). For most elements, the EAAS has lower detection limits than the FAAS. The incorporation of the new technology in the Laboratory of Clinical Biochemistry opened the possibility of approaches which had been unthinkable until then. For many of them it meant a reinforcement in the central position they held in hospital research. For those which incorporated spectroscopy it meant the possibility of new diagnostic, therapeutic and toxic controls.

In this chapter, a series of research studies are presented as example of the above mentioned. Thus, with respect to the Sr, refer to section 2.1, the first paper deals with the discrimination factor between Ca/Sr in absorption intestinal mechanisms. Afterwards, the different behavior of these metals in the binding to serum proteins is studied. And finally, the possibility of a hormonal regulation mechanism of serum levels of this element is evaluated, given the similarity of its biological behavior with the Ca.

The quantification of element bound to protein derived towards the direct applications for the study of medical problems such as the distribution of Zn in the acute and chronic overload and de Zn and Cu in the serum proteins in myocardial infarction.

Later, in section 2.2, a specific problem resulting from industrial development among other causes, threatening a part of the population -Pb poisoning- was tackled, analysing the serum and urine concentration of Pb and the hem biomarkers. This example is particularly useful not only because of what the technology meant for the diagnosis and control of this disorder but also because it has allowed to observe how the levels of this element in our city and, in general, in the West has declined over the years.

Finally, in section 2.3, an actual research is included: the design of a new strategy or approach possibility in the knowledge of the physiopathology of different neurological conditions based on the concentration of certain trace elements in the CSF, as well as of other parameters such as the cellularity, the proteins concentration, etc.

2.1 Strontium

This first group of papers serves as a model to analyze how this new technique allows evolving from the specific research problems (intestinal absorption, transport of element bound to proteins, etc.) to applications in medical pathology, such as the binding of Zn and Cu to plasma proteins after myocardial infarction.

2.1.1 Introduction

The disintegration of uranium and plutonium atoms in atomic explosions provokes the appearance of a series of elements with maxima in atomic weight of around 90 and 140. The isotopes of heavier atomic weight (140) fall in the area of the explosion while those of lower atomic weight (90) enter the troposphere and stratosphere. The particles which enter the troposphere spread out forming a gigantic belt around the area and are later deposited in local rain. Those others which reach the higher zones - the stratosphere - can be disseminated in over wide areas.

Atmospheric and tropospheric precipitation follow more or less quickly but the contaminants in the stratosphere may take many years before falling into the troposphere and being deposited in zones of greater rainfall.

Among the elements thrown into the troposphere and stratosphere are found those of the first peak of atomic weight (about 90), with two of the artificial isotopes of Sr, ^{89}Sr and ^{90}Sr, with different half lives. In particular, the half life of ^{90}Sr is 28.79 years.

The Sr deposited by the rain together with the Sr present in nature is absorbed by plants through the roots and that which is deposited on the leaves may also be absorbed. From here it enters the human organism, either directly by consuming the plants or, indirectly, by eating the animals which have eaten them.

Once the Sr has entered the organism it is carried in the blood to the cartilage and bone, choice sites for bonding. Far lower rates are found in other tissues.

In face of this threat, the analyses of animal milk for human consumption as markers of radioactive contamination and strategies to prevent the uptake (intestinal absorption) or to facilitate the elimination of ^{90}Sr from the bone once it has bonded are priority research into Sr domain (Escanero, 1974).

2.1.2 Development

2.1.2.1 A curiosity in biological barriers: discrimination between strontium and calcium

It is assumed axiomatically that biological organisms use Sr less effectively than Ca, which means that they discriminate against Sr in favour of Ca. This may be expressed in another manner by the concept "Strontium-Calcium Observed Ratio" (OR), the value of which is lower than 1. Comar et al. (1956) introduced the term to denote the overall discrimination observed in the movement of the two elements from one phase to another under steady-state conditions. The term OR denotes the comparative rates of Sr and Ca in balance between a sample and its precursor and is defined as:

$$OR= (Sr/Ca)\ sample/(Sr/Ca)\ precursor.$$

More precisely, the OR can be defined as the product of a number of 'discrimination factors' (DF), each of which is a measure of the extent to which the physiological process to which it refers contributes to the overall discrimination.

In this line it is shown one of the first studies which aimed to ascertain at what intestinal level the Sr/Ca discrimination takes place. Vitamin D_3, 25-hydroxy-cholecalciferol (25-0H-CC) and 1,25-dihydroxy-cholecalciferol (1,25 $(OH)_2$-CC) were administered to rats. The apparent and real intestinal absorption were analysed at those moments when their activity was maximum. Different concentrations of Sr were used and it was concluded that the discrimination occurred in the passive rather than in the vitamin-dependent transport (Escanero et al., 1976).

2.1.2.2 Hormonal regulation of strontium

From a general point of view, this research with Sr was designed to find the hormonal regulatory mechanisms in order to be able to act on them, provoking or forcing its elimination. This approach was based on several facts: a) The chemical similarity between Ca and Sr and their common participation in certain physiological processes lead one to think that both elements share a hormonal regulatory mechanism; b) The contribution of Chausmer et al. (1965) who had demonstrated the existence of specific action of calcitonin (CT) on the distribution of Zn, contrary to that exercised in the distribution of Ca, reducing the levels of this element in different tissues -thymus, testicles- seemed guiding; and c) Comar's idea (1967) that the plasmatic levels of Ca could participate in the regulation of Sr metabolism was particularly attractive. Taking these facts into account and with the possibility of a shared hormonal regulation for different trace elements (Zn and Sr among others), with specific properties for each one, the first steps in this approach were addressed to finding out more about serum Sr.

In the first study (Escanero, 1974), Comar's idea was proved since, at the same time as the concentration of Sr in bovine blood increased, so did that of Ca in total proteins, the contrary phenomena occurring with inorganic phosphorus and Ca in albumin.

In a later study (Alda & Escanero, 1985), the association constant and the maximal binding capacity for Ca, Mg and Sr to human serum proteins taken as a whole were determined. For Sr, a maximal binding capacity of 0.128 mmol/g of proteins and the association constant (K_{prs}) of 49.9 ± 16 M^{-1} were obtained. These values were not far from those of Ca (0.19 mmol/g of proteins and 55.7 ± 18 M^{-1}).

Later, Córdova et al. (1990) studied the regulating hormones of Ca and the effects of the administration of parathormone (PTH) and CT in rats and after thyroparathyroidectomy (TPTX) were analyzed. The PTH excess and defect (TPTX treated with CT + T_4) showed plasmatic increases in Sr. However, CT excess provokes decreases while the defect (administration of PTH + T_4 to TPTX rats) causes increases. Consequently, CT may be the hormone that plays a regulating role in the plasmatic Sr concentrations.

In this line, a study (Escanero and Córdova, 1991) was conducted in order to known the effect of the administration of glucagon on the serum levels of the alkaline earth metals since the phosphocalcic response to glucagon was already known in different animal species and it had been reported that, in mammals, glucagon triggered the release of CT by the thyroid gland. The study was completed with the analysis of the changes induced in these metals by the administration of CT. The effect (reduction) was observed two hours after administration and in daily administration the effect peaks on the 3rd day and CT significantly reduces the serum levels of these metals up to the 3rd day of treatment, just when the glucagon effect is highest.

2.1.2.3 Continuing with serum proteins

While the alkaline metals hardly bond in proteins and the alkaline earth metals do so in a proportion of 50% or slightly less, the trace elements do so almost completely. Zinc is deposited in a proportion of more than 99% and the proteins involved with albumin and the alpha$_2$ macroglobulin (α_2MG). The α_2MG bound tightly to the metal is responsible for 30-40%. The aminoacids are only responsible for about 2% of bonded Zn (Giroux & Henkin, 1972). The serum levels of Zn vary in different pathological symptomatologies and in relation to the amount of exercise. Several studies have shown these variations are related to the percentage of the element bound to albumin and another one established this relation between total serum Zn and α_2MG-bound Zn, in athletes after exercise (Castellano et al., 1988). This last study was aimed at analysing the variations in the serum levels of Zn after acute and chronic overload of this element and verifying whether these variations may be correlated to changes in the percentage of the element bound to albumin, α_2MG or both.

After a single intragastric administration of 0.5 mL of a solution containing 1000 ppm of Zn, the levels of this metal increased significantly ($p<0.01$) in serum 30 min after the beginning of the experiment, reaching maximum values at one hour and returning to normal levels 24 h later. It should be noted, however, that 8 h after administration the increases were no significant ($p>0.05$). The concentration of Zn bonded to albumin varied in parallel to the total serum levels of Zn. The values of bonded Zn in the globulins α_2MG also varied but the values returned to basal concentrations 4 h after the beginning of the experiment.

The chronic overload was performed with different groups who underwent a daily intragastric dosis of 0.5 mL of the same solution used for the acute overload. Different group of animals were sacrificed on days 7, 14, and 30 after the beginning of the experiment, with the aim of collecting values at different times of the overload. The control group and one other were used for recuperation and were kept for ten days with no treatment after day 30. Chronic overload of Zn caused significant increases in the serum levels of Zn throughout all the experiments. The highest values were found on the 14th day. The amount of Zn bonded to albumin varied in parallel with the total serum Zn; however, the concentrations of Zn bonded to globulins (α_2MG) showed a significant decrease ($p<0.05$) on the seventh day, increasing significantly ($p<0.01$) on the 14th day and the 30th day and returning to normal values 10 days after Zn overload was interrupted.

The results of the acute overload suggest a correlation with the proportion of element bonded to the albumin and those of the chronic overload showed a rapid response of the albumin while the increases of element bonded to the α_2MG responded slowly and remained constant until the end of the experiment. However, at this time the total serum Zn and the Zn bonded to albumin decreased.

These results suggest that albumin may play a new physiological role by adjusting its binding capacity to the serum Zn levels.

In the same line the levels and distribution of serum Cu and Zn were studied in patients diagnosed with acute myocardial infarction from the day of admission to the Cardiovascular Intensive Care Unit until the 10th day following the attack (Gómez et al., 2000). The results obtained showed that Cu increased significantly after the 5th day after the myocardial infarction, while Zn decreased significantly ($p<0.01$) with relation to the control group after the 1st day, the lowest values being found on the 3rd day after the attack.

Later, the total serum Cu showed an excellent correlation with the Cu bonded to the albumin and to the globulins (ceruloplasmin), as well as with the concentration of both fractions of serum proteins. In contrast, the total serum Zn only presented this correlation with the Zn bonded to the albumin, but not with the Zn bonded to the globulin or the albumin concentration.

These findings suggest the existence of some kind of relationship between the two fractions of the element bonded to proteins, which is probably different for each metal.

A further step was taken when wanting to analyse the possible role of albumin in the uptake of Zn by erythrocytes (Gálvez et al., 2001). Zinc is incorporated to erythrocytes by several mechanisms: i) passive transport, ii) anionic exchanger, iii) amino acid transport and iv) especially in the efflux, the Zn^{2+}-Ca^{2+} exchanger. In accordance with the free-ligand hypothesis only the free fraction could be used for the erythrocyte uptake. The results showed a significantly higher uptake ($p<0.01$) of Zn in the absence than in the presence of albumin for equimolar concentrations in both cases. However, the uptake of Zn in an albumin-free medium with a similar free-Zn concentration to Zn ultrafiltrable (20%) to another with albumin, a significantly ($p<0.01$) greater Zn uptake was observed in the latter. The DIDS (4-4´-diidothiocyanatostilbene-2.2´-disulphonic acid), that inhibits an important fraction of the Zn bonded to the anion carrier, also triggered a greater inhibition in the uptake of Zn when the albumin was present. Consequently, it was suggested that albumin must be directly or indirectly involved in Zn capture, facilitating the processes of passive transport and anionic exchanger.

Other properties of the uptake of Zn by erythrocytes were published in previous reports (Galvez et al., 1992, 1996a, 1996b):

- high dependence on temperature (Zn uptake was almost eliminated at 4° C)
- dependence on the concentrations of external Na^+ and K^+ and
- the apparent dissociation constant for the fast uptake step (15 minutes) is 0.46 µM for a medium without albumin and 0.121 µM with human albumin. For physiological concentrations of Zn the value was 15,3 µM (unpublished data).

2.2 Study of the effects of lead in hem biosynthesis

These studies are shown as an example of integration of the analysis provided for Pb by FAAS and EAAS with the results of the biomarkers of hem obtained with other laboratory techniques.

2.2.1 Introduction

Lead produces interferences in the hem biosynthesis pathway (Cambell et al., 1977), inhibiting the enzymatic effects of ALA-Dehydratase (ALA-D, EC 4.2.1.24) in cytosol and coproporphyrinogen oxydase (EC 1.3.3.3) and ferrochelatase (EC 4.99.1.1), both in mitochondria (González-López, 1992).

In the exposed organism, Pb produces affection of the target organs and critical effects which are characteristic of the disease known as saturnism since the days of Ancient Rome. Lead poisoning is diagnosed in the clinic and is shown by the analysis of Pb in blood. However, the concentration of Pb in blood depends on the metabolic condition of the individual (pH of the internal medium, bone activity, etc.) as well as on interactions with other metals such as Ca, Fe, Zn, Cu and Mg, among others. In order to ascertain the intensity and degree of affection of the Pb intoxication, it is necessary the study the so-called biomarkers of Pb exposure and poisoning. The most frequently analysed are the enzyme ALA-D in erythrocytes, protoporphyrin IX in erythrocytes, 5-aminolevulinic acid (5-ALA) in urine and the coproporphyrins in urine (Meredith et al., 1979).

The extreme sensitivity of ALA-D to divalent Pb ions has resulted in the measurement of its activity, as an indirect measurement of Pb in human blood (Berlin et al., 1977). Of all the enzymes involved in the hem biosynthesis pathway, it is the one which has been most studied due to the inhibiting effect that Pb has on its activity and the practical importance of the measurement of the enzymatic activity of ALA-D is considered to be of interest as a bioanalytic marker of environmental exposure to Pb. This has been assisted by the development of a method which has been standardised at the proposal of the executive council of the European Union. Hemberg & Nikkanen (1972) have published an extensive report on the biological meaning of ALA-D inhibition and its use as an exposure test.

As well as because of the effect of Pb, the activity of ALA-D may be also reduced by the effect of ethanol in alcoholics and by carbon monoxide in smokers although, in both cases, the reduction of activity is slight. In this line, porphyria of Doss is a recessive autonomous hereditary disease produced by the alteration of the gen which codifies the synthesis of ALA-D located in allele q34 of chromosome 9. It is a rarely presented porphyria, characterised by a deficit of ALA-D. In the homozygote form, a great reduction of ALA-D activity is observed in erythrocytes (2% of the control mean value), while in the heterozygotes, the enzymatic activity of ALA-D is reduced to 50%, being asymptomatic but especially sensitive to the toxic effect of Pb even with scarcely increased levels of Pb. The improvement of environmental and working conditions as well as the use of unleaded petrol, has led to the reduction of the concentration of Pb in blood in the population, a phenomenon observed over the last twenty years (González-López, 1992).

Taking into account the precedent facts, this section will analyse the effects of Pb poisoning in the biomarkers of hem studied in order to evaluate the recovery in the post-treatment with $CaNa_2EDTA$ as chelating agent.

2.2.2 Development: Subjects and methods

Subjects: The study of the biomarkers of Pb poisoning was performed in 377 adults (30-65 years old) and in 36 healthy children aged between 6 and 14. The adults were distributed as follows: 325 healthy (control group) and 52 cases of Pb poisoning: 24 severe, 15 slight and 13 treated with chelating agents. The children were another control group.

For inclusion in the group of healthy people, control groups, were required to:

a. No symptoms of lead poisoning and other diseases,
b. No changes in routine biochemical parameters and
c. Normal values of biomarkers characteristic of lead poisoning.

The group of the 13 patients treated with chelating agents were integrated as patients with severe poisoning and they were treated with $CaNa_2$-EDTA at doses of 50-75 mg/Kg body weight per day for five days, not exceeding the amount of 500 mg. It may be given an additional set after two days of interruption. This treatment requires hospitalization and clinical management with special attention to controlling renal function due to its nephrotoxicity.

Methods: The parameters analyzed were: Pb in blood and urine and the biomarkers of hem characteristic of Pb poisoning, ALA-D and protoporphyrin IX in blood and 5-ALA and coproporphyrins in urine.

The methods used were as follows:

Lead was analyzed in heparinized whole blood and 24-hour urine collection in container without additives. The first results of Pb were performed in both blood and urine by FAAS, using a modification of the method of Hessel (1968) by extraction into n-butyl acetate of the complex formed by the Pb with dithiocarbamate ammonium pyrrolidine (Pb-APDC).

Subsequently, the latest determinations were obtained by EAAS with Zeeman correction spectrometry, given the recent introduction of this technique to the laboratory. Blood was used diluted to 0.2% nitric acid (Pearson & Slavin, 1993).

Biomarkers: ALA-D of erythrocytes (in total blood with heparine) was determined applying the method of the European Standards Committee (Sibar Diagnostici) (Berlin & Schaller, 1974; Schaller & Berlin, 1984).

Erythrocyte-free Protoporphyrin IX (in total blood with heparine) was performed applying the method of Piomelli (1977), (Sibar Diagnostici).

At present, both products are manufactured by Immuno Pharmacology Research (IPR) Diagnostics.

ALA/PBG in urine was determined by means of column chromatography. For analysis a modification of the method of Mauzerall & Granick (1956), manufactured by Bio-Rad, was applied.

Determination of porhyrins in urine was performed after the separation of uro- and coproporphyrins by means of ethyl acetate which extracts the coproporphyrins from the aqueous phase and later absorption of uroporphyrins from the aqueous phase with activated Al_2O_3. Finally, the uroporphyrins are extracted from the Al_2O_3 activated and the coproporphyrins from the ethyl acetate by means of HCl 1.5 N, performing a fluorimetric reading of the extracts obtained (Schwartz et al., 1960).

In cases of the increased excretion of porphyrins, it is important to analyze the porphyrin biosynthesis pathway, applying the following analytic methods:

Analysis of carboxylic acids of free porphyrins in urine, by means of HPLC/FD. The chemical structure of the porphyrins presents the natural property of being fluorescent compounds, which makes them detectable by spectrofluorimetry.

The application of high pressure liquid chromatography (HPLC), constitutes a valuable resource in research applied to the study of the porphyrin biosynthesis pathway.

The analysis of the carboxylic acids of free porphyrins does not require any prior treatment of the samples (urine or faeces) to be chromatographed. They are only passed through a 22 µm Millex-GS (Millipore) filter. If the quantity of porphyrins contained in the sample is very high, it will be diluted with the eluyent.

The method used allows us to obtain a type of chromatogram in which the separation can be observed and the identification performed of the carboxylic acids of free porphyrins, from the octacarboxylic porphyrins (8-COOH, uroporphyrins) followed by the heptacarboxylic porphyrins (7-COOH), hexacarboxylic (6-COOH) and pentacarboxylic (5-COOH) and the tetracarboxylic porphyrins (4-COOH, coproporphyrins). The dicarboxylic form (2-COOH, protoporfirin) is not excreted in urine (Meyer et al., 1980a; Meyer, 1985).

As standard the Porphyrin acids chromatographic marker kit is used (Porphyrin products, INC., CMK-IA).

Analysis of type I and II isomers of uroporphyrins and coproporphyrins in urine by means of HPLC/FD. According to the disposition the substitutes may adopt around the tetrapyrrol ring of the porphyrin molecule, there are only four possible types of uroporphyrinogens (I, II, III and IV), but in nature only the type I and III uroporphyrinogens exist and, consequently, decarboxylation will only produce coproporphyrinogens I and III.

The use of high pressure liquid chromatography (HPLC) with fluorimetric detection (FD) allows to separate and identify the isomer forms of type I and III uro- and coproporphyrins as metabolytes derived from the oxidation of the corresponding porphyrinogens, produced in the process of natural metaloporphyrin biosynthesis. With this aim, the application method described by Jacob et al. (1985) was used.

With this method, a chromotagram is obtained in which the separation and identification can be observed of the isomers of uroporphyrins I and III and coproporphyrins I and III which are eluated and detected in this order.

The following standards are used: Uroporphyrin fluorescence standard, Uroporphyrin I (Porphyrin products, INC. UFS-I), Uroporphyrin III octamethyl ester (Sigma), Coproporphyrin I (Sigma), Coproporphyrin fluorescence standard, Coproporphyrin III (Porphyrin products, INC. CFS-3).

Analysis of protoporphyrin-Zn of erythrocytes in blood.

The alteration of the enzymatic activity of ferrochelatase due to the inhibition of this enzyme by the effect of Pb produces an increase in the protoporphyrin IX concentration in erythrocytes. This increase of protoporphyrin IX produces and accumulation of Zn-protoporphyrin I due to the complexation formation of this porphyrin with Zn^{2+}.

The method described by Meyer et al. (1980) was used for the analysis of porphyrins in erythrocytes. They developed a procedure for the separation of porphyrins from erythrocytes in blood with HPLC/FD in reverse phase by formation of the ionic pair (Meyer et al., 1982).

The standards used were Coproporphyrin fluorescente Standard (Porphyrin products, INC. CFS-3, Logan, UTA, USA), Protoporphyrin fluorescent Standard (Porphyrin products, INC. PFS-9, Logan, UTA, USA) and Mesoporphyrin IX (Porphyrin products, INC. M 566-9, Logan, UTA, USA).

2.2.3 Development: Results

The first group of values (results not published) were obtained in an early study and show the concentrations of Pb in blood and urine, as well as the values of various biomarkers of the porphyrin biosynthesis pathway. Moreover, it was also included the results of a second study from some years later, presenting the normality values of Pb in blood as an update, observing the difference of Pb concentration with respect to the earlier study.

I.A Control group. The results of the control group for Pb in blood and in 24 hours urine are shown in Table I. In Table 2 are presented the values of ALA-D and protoporphyrin IX in erythrocytes for the same population and in Table III those of the hemoglobin concentration and protoporphirin IX/g of haemoglobin coefficient. In table 4 are indicated the values of 5-ALA and porphobilinogen (PBG) in 24 hours urine and in the last (table 5) are presented the results found in urine of 24h of uroporphyrins and coproporphyrins.

	Pb in blood **µg/dL** **x ± SD**	**Pb in 24 hours urine** **µg/dL** **x ± SD**
Men (n = 104)	17.72 ± 6.01	45.10 ± 39.85
Women (n = 61)	14.00 ± 3.86	35.78 ± 27.26
Adults (n = 165)	16.54 ± 5.49	40.96 ± 35.01
Children (n = 36)	14.58 ± 2.79	-

Table 1. Values (x ± SD) of Pb in blood and in 24 hours urine

	ALA-D **U. of CEE/mL erythrocytes*** **x ± SD**	**Protoporphyrin IX** **µg/dL of blood** **x ± SD**
Men (n = 104)	44.70 ± 13.94	28.92 ± 11.50
Women (n = 61)	50.41 ± 17.27	26.80 ± 9.66
Adults (n = 165)	46.81 ± 15.45	28.13 ± 10.87
Children (n = 36)	64.42 ± 13.39	28.53 ± 9.28

(*): Units of the European Standards Committee

Table 2. Values (x ± SD) of ALA-D and protoporphyrine IX in erythrocytes

	Hemoglobin **g/dL of blood** **x ± SD**	**Protoporphyrin IX** **µg/g Hb** **x ± SD**
Men (n = 104)	15.27 ± 1.54	1.92 ± 0.84
Women (n = 61)	13.39 ± 1.16	2.02 ± 0.78
Adults (n = 165)	14.57 ± 1.68	1.96 ± 0.82
Children (n = 36)	12.94 ± 0.71	2.19 ± 0.65

Table 3. Values of Hb in blood and protoporphyrin IX/g of Hb coefficient

	5-ALA mg/24 hours x ± SD	PBG mg/24 hours x ± SD
Men (n = 223)	2.96 ± 1.30	0.41 ± 0.42
Women (n = 102)	2.42 ± 1.17	0.35 ± 0.35
Adults (n = 325)	2.79 ± 1.28	0.39 ± 0.40
Children (n = 36)	2.84 ± 0.91	0.46 ± 0.43

Table 4. Normal values of 5-ALA and PBG in urine of 24 hours

	Uroporphyrins µg/24 hours x ± SD	Coproporphyrins µg/24 hours x ± SD
Men (n = 223)	11.09 ± 6.00	101.95 ± 55.46
Women (n = 102)	9.08 ± 4.95	64.34 ± 35.07
Adults (n = 325)	10.46 ± 5.76	90.15 ± 52.88
Children (n = 36)	4.18 ± 2.98	56.59 ± 34.03

Table 5. Normal values of uroporphyrins and coproporphyrins in 24 hours urine

The values of Pb in blood and urine (Table 1) for the different population groups studied are within the range of those reported by other authors of the time the study was conducted (Carton, 1985; Carton, 1988). Because of that environmental improvements and labor have reduced Pb concentrations in the environment, the serum concentration of Pb have been also reduced (Trasobares, 2010).

With regard to the values of the biomarkers of Pb poisoning in blood analyzed (Tables 2, 3, 4 and 5), all of them are within the range reported by other authors (Goldberg, 1972; Tomokuni, K. & Ogata, 1976; Campbell et al., 1977; Goldberg et al. 1978; Granick et al., 1978; Meredith et al. 1979; Sakai et al., 1982; Barbosa et al., 2005). Although the standard deviation values can be considered high for some parameters, this should not be attributed to the methodology used given the biological variability that is observed in the study population.

There have also been included the values for Hb in the blood and the ratio of protoporphyrin IX/g of Hb. This last value increases in the Pb poisoning (protoporphyrin increased and decreased hemoglobin) while in the protoporphyria the Hb did not decrease and consequently the ratio does not increase as much as in the Pb poisoning. Likewise the values of 5-ALA, PBG and porphyrins (uro-and copro-) are included in the study as they provide a more complete picture of potential changes in Pb poisoning.

I.B Pb poisoning. In table 6 and 7 are presented the statistical tests of comparison of means observed in large samples with independent data and their degree of significance, performed for each biomarker of Pb poisoning analysed in each group studied with respect to the control group. Specifically, in table 6 are presented the results in blood of Pb, ALA-D, protoporphyrine IX and Protoporphyrine IX/g Hb and in table 7 the results in urine (24 hours) of Pb, 5-ALA, PBG, uroporphyrins and coproporphyrins.

	BLOOD			
	Pb µg/dL x ± SD	ALA-D U/mL x ± SD	Protoporphyrine IX µg/dL x ± SD	Protop. IX/Hb µg/g Hb x ± SD
Control group (n = 165)	16.54 ± 5.49	46.81 ± 15.45	28.13 ± 10.87	1.96 ± 0.82
Severe pois.	101.04±58.03[b]	12.00 ± 5.56[b]	198.49 ± 89.51[b]	17.24 ± 8.74[b]
Slight pois.	41.60±15.86[b]	26.47 ± 20.50[b]	26.57 ± 92.15[b]	9.65 ± 7.35[b]
Post Treatment	40.92±18.74[b]	37.08 ± 14.20[a]	116.94 ± 106.21[b]	8.70 ± 8.26[b]

Concentration of Hb (g/dL blood) in each group studied:
Control group: 14.57±1.68;
Severe poisoning: 11.46±1.98b
Slight poisoning: 13.59±1.46a;
Post treatment: 14.12±1.17

Degree of statistical significance: a ($p < 0.050$); b ($p < 0.001$);

Table 6. Values (x ± SD) of Pb, ALA-D, protoporphyrine IX and Protoporphyrine IX/g Hb in blood of the indicated group

The above table shows that in both, severe and slight Pb poisoning, the increases in the concentration of blood Pb are associated with significant inhibition of ALA-D activity (Campbell et al., 1977; Goldberg et al., 1978; Sakai et al., 1982) and with significant increases of protoporphyrin IX concentrations (Goldberg et al., 1978; Meredith et al., 1979). After treatment the values still remain significantly altered, indicating that the patients need a new series of treatment, because of the guidelines therapy is performed during five days with breaks in which they carry out checks on biomarkers. The fact of treating patients with different series of treatment would explain the dispersion of the results.

	URINE				
	Pb µg/24h x ±SD	5-ALA mg/24h x ±SD	PBG mg/24h x ±SD	Uroporphyrins µg/24h x ±SD	Coproporphyrins µg/24h x ±SD
Control group (n = 325)	41±35	2.79±1.28	0.39±0.60	10.46±5.76	90.15±52.88
Severe pois.	361±189[b]	47.3±21.8[b]	2.21±1.84[a]	26.4±18.8[b]	1612.9±682.8[b]
Slight pois.	114±81[b]	4.89±1.96[b]	0.49±0.62	5.61±2.25	87.7±54.2
Post treatment	197±167[b]	5.43±3.52[b]	0.42±0.62	9.67±6.50	148.9±119.1[b]

Degree of statistical significance: [a] ($p<0.002$); [b] ($p<0.001$).

Table 7. Values (x ±SD) of Pb, 5-ALA, PBG, uroporphyrins and coproporphyrins in 24 hours urine

The identification and quantification of high levels of porphyrins in erythrocytes, mainly protoporphyrin IX and its chelated form, Zn-protoporphyrin, are essentials in the diagnosis of Pb poisoning and in erythropoietic porphyrias (Meyer et al., 1980b).

The results of the above table present significant differences in the Pb urinary elimination in Pb poisoning both severe as slight in relation to the control group. Moreover, increases in the elimination of all parameters (biomarkers) analyzed in severe poisoning, being the most important those of the 5-ALA and coproporphyrins. In contrast, in slight poisoning has only significantly increased the elimination of 5-ALA. After treatment of severe cases there is a significant decrease in excretion of 5-ALA and coproporphyrins, keeping the levels still increased. Of the above comments it can be seen that the most effective urinary biomarkers are the urinary elimination of 5-ALA and coproporphyrins (type isomeric III).

These results agree, in the literature review performed, with the studies published by numerous authors and refered in the Doctoral Thesis of González-López (1992) and others (Sakai, 2000; Gurer-Orhan et al., 2004). All of them showed the influence of Pb on the heme biosynthesis pathway and the effects produced in some of the enzymes which take part in the biosynthesis of porphyrins, phenomena demonstrated in clinical research, *in vitro* studies and in experiments on animals.

In the diagnosis and evolutionary control of Pb poisoning, the study of the biomarkers of the hem biosynthesis pathway is very efficient, even more so than that demonstrated by the concentration of Pb in blood and urine due to the susceptibility of these to hormonal influences as well as metabolic ones such as the pH of the internal medium and the activity of bone turnover. According to this, Pb in blood means the degree of uptake; Pb in urine, the degree of elimination; Pb in urine provoked by EDTA or some other chelating agent, the degree of accumulation; ALA-D, the degree of exposure and is directly related to Pb in blood; erythrocyte free protoporphyrin IX, especially Zn-protoporphyrin IX, the degree of intake and chronic evolution; and the coproporphyrins (coproporphyrin III) the severity of the poisoning.

II. In the second study was performed a review of the values of Pb in blood in the present. The values were obtained by EASS in a graphite furnace with correction of the Zeeman effect, in a sample of 156 individuals. The results obtained are shown in Table 8.

	Pb in blood µg/dL x ± SD
Men (n = 83)	3.51 ± 2.16
Women (n = 73)	2.29 ± 1.64
Adults (men + women) (n = 156)	2.94 ± 2.02

Table 8. Normal values (x ± SD) of healthy individuals of Pb concentration in blood

These results indicate a decrease of the concentration obtained in 1989 which was 13.17 ± 3.47 µg/dL, attributed to the improvement of environmental and working conditions as well as to the suppression of Pb as antiknock agent in petrol (Izquierdo-Álvarez et al., 1985; Trasobares, 2010).

In the review of results for 2008 (results not published), the values for ALA-D in the adult sample were 60.59 ± 16.49 (x ± SD), higher than those obtained in 1990: 46.81 ± 15.45 (x ± SD), which is logical if we take into account the fact that the activity of ALA-D in erythrocytes has a negative or inverse correlation with respect to the concentration of Pb in blood. In the 1990 study, Pearson's coefficient of linear correlation obtained was: $r = -0.568$ ($p<0.001$).

2.3 Study of some elements in cerebrospinal fluid: physiopathological evaluation

This study, still in progress, can serve as an example of using the atomic absorption spectrometry technique for the assistance, together with other techniques, in the clinical diagnosis of some diseases.

2.3.1 Introduction

The cerebrospinal fluid (CSF) fills the subarachnoid space between the arachnoid membranes and the pia mater called leptomeninges which protect the Central Nervous System -CNS- (encephalon and spinal cord). Seventy percent of the CSF is formed in secretory structures called choroid plexi and the remaining 30% is produced from the cerebral capillaries (Carpenter, 1985).

Although the composition of CSF is similar to a plasma filtrate, there are differences which indicate that the CSF formed is produced both by a process of filtration and by active secretion, an osmotic balance being observed between CSF and plasma. A similarity can be seen in the composition of CSF and the extracellular liquid of the nervous system, indicating an easy interchange between both compartments. The CSF and in the cerebral interstice are separated from the blood circulation by the hematoencephalic and hematocephalorachideal barriers which prevent the free passing of substances. This is why they are considered to be functional elements of protection of the nerve cells (Nolte, 1994). In comparison with plasma, the CSF contains a greater concentration of Na, Cl and Mg and a lower one of glucose, proteins, amino acids, uric acid, K, bicarbonate, Ca and phosphate (Guyton, 1990). These differences indicate that the CSF is produced by a mechanism of active secretion and varies according to the location of CSF extraction with regard to the structures it bathes.

Water passes from the stroma to the CSF following the concentration gradient produced by the ATPase-dependent carrier proteins, Cl, Ca and Mg (Nolte, 1994). Cellular metabolytes also enter the extracellular liquid from neurones and glial cells.

The CSF maintains an appropriate chemical environment for neurotransmission and removes metabolic products and substances which are harmful for the CNS.

2.3.2 Development

In this research, it has been studied the concentration of Ca, Mg, Zn, Cu, Fe and Mn in CSF in order to analyze their influence on the pathogeny of some neuropathies (González-Romarís et al., 2011).

The mineral chemical elements and the trace elements were analysed in the CSF extracted from 37 people (17 men and 20 women, between 27 and 73 years of age) who were considered to be healthy after performing a clinical and analytical study. They made up the control group.

In addition, analysis was made of the CSF of 136 individuals from the Services of Neurology, Neurosurgery and Emergency. The CSF was extracted by lumbar puncture in all cases.

The analysis of Ca, Mg, Zn, Cu, Fe and Mn was performed by flame atomic absorption using the corresponding hollow cathode for each metal. The wavelengths for the reading of the absorbance corresponding to each cation analysed were the indicated by the manufacturer.

The values (mean and standard deviation) obtained in the control group for each metal analysed were as follows: Ca (mg/dL), 4.95 ± 0.70; Mg (mg/dL), 2.74 ± 0.10; Zn (µg/dL), 17.40 ± 7.50; Cu (µg/dL), 15.70 ± 4.50; Fe (µg/dL), 13.10 ± 3.60; and Mn (µg/dL), 2.50 ± 0.70. These values agree with the findings published by other authors (Hazell, 1997; Kapaki et al., 1997; Levine et al., 1996).

With regard to the results obtained in the pathological CSF, significant increases were found ($p<0.05$) in the concentrations of Ca, Cu, Fe, Zn and Mn in the groups classified with cell and protein increase in CSF in comparison to the control group. It was also see that the significant increase of the Ca, Zn and Cu concentrations is greater in those groups which present a higher concentration of proteins, while the increase of Mn corresponds to the increase of cell count. With regard to magnesium, it was seen that the significant reduction of its concentration in relation to the control group corresponds equally both if the cell count or the protein concentration was increased.

Interest in analysing these metals in CSF is directed to explaining the pathogenesis of some dysfunctions of the CNS. Clinical and experimental studies reviewed in the literature confirm the influence that these metals have on the pathogeny of some CNS dysfunctions and diverse neuropathies.

In this way the Ca and Mg ions play an important role in the action of glutamate, which is one of the most important neurotransmitter of vertebrates in the brain. The receptor of glutamate N-methyl-D-aspartate (NMDA) can only be activated in certain conditions of depolarisation of the membrane (Johnson & Ascher, 1988). The Mg ion blocks the channel, not being permeable to the Ca ions. When the receptor of glutamate are activated, the receptor reduces its affinity for Mg and the channel becomes permeable, permitting the entry of Ca ions to the neurone, a phenomenon which has been related to memory and the learning process (Hammond & Tritsch, 1990; Thomson, 1986). A reduction has been found in the glutamate and Mg concentration in the CSF in schizophrenic patients (Levine et al., 1996).

Some studies have found an increase of Cu in serum and CSF and an increase of Mn in the spinal cord in Amyotrophic Lateral Sclerosis (ALS), which suggests that this metal plays a role in the pathogeny of this disease (Kapaki et al., 1997).

The clinical association of Pb poisoning and ALS with an increase of Pb in blood and bone has been reported (Kamel et al., 2006). It has been suggested that patients with polymorphism in the gene of the ALA-D enzyme might be more at risk of presenting ALS in exposure to Pb (Kamel et al., 2003). Other studies have found no association between exposure to metals and ALS (McGuire et al., 1997; Bergomi et al., 2002).

It has been suggested that, regarding Mn, the binding of this metal in the basal ganglions of the brain may contribute in the pathogeny of the symptomatology of hepatic encephalopathy (Kuliseusky & Puyols, 1992; Weissenborn, 1995; Noremberg, 1998).

Moreover, it has been demonstrated that Mn reduces the uptake of glutamate in cultivated astrocytes. The great capacity of astrocytes to accumulate Mn suggests that its uptake by these cells may play an important role in the development of Alzheimer's type II astrocytosis (Hazell, 1997; Aschner & Gannon, 1992).

In conclusion, this research line, still in progress, can be highly promising to clarify the pathogenesis of some brain conditions.

3. Conclusions

After incorporating the atomic absorption spectrophotometry (spectroscopy) to the hospital laboratories has been observed that the medical research has improved in these laboratories. In this chapter the highlighted technique has been presented through a few research examples with different metals.

The conclusions from the research with the metal studies have been the following:

a. Strontium:
   - The intestinal absorption discrimination between Ca and Sr takes place in the passive transport mechanism and not in the active one or of vitamin D dependent mechanism.
   - With respect to the hormonal regulation of the plasmatic Sr, the CT is the only hormone that caused consistent changes in the concentrations of this element.
   - The addition of Sr in vitro to equimolar concentrations with Ca to bovine serum forces the binding of this last element with the total serum proteins. An opposite phenomenon takes place when the experience is conducted with albumin bovine as a single protein.

b. Zinc and Copper:
   - The Zn bound to albumin varies more consistently than Zn bound to globulins after acute and chronic overload. This fact allows suggesting that the albumin could act as a buffer.
   - In the myocardial infarction, Cu in serum increases significantly after the 5th day after the heart attack, while the Zn in serum decreases from the 1st day; being the lowest values of Zn found on the 3rd day after the attack. The total Cu in serum showed an excellent correlation with the Cu bonded to the albumin and to the globulins (ceruloplasmin), while, the total Zn in serum only presents a positive correlation with the Zn bonded to the albumin.

c. Lead:
   - In the diagnosis and the control of the Pb poisoning, the study of the biomarkers of the hem biosynthesis pathway is very efficient, even more so than its concentration of Pb in blood and urine.
   - According to this, the Pb concentration in blood means the degree of Pb intake; the Pb concentration in urine, the degree of its elimination; the Pb concentration in urine after the administration of EDTA or some other chelating agent (Dimercaprol -BAL, British Anti-Lewisite- and penicilamine), the degree of its accumulation; the ALA-D activity, the degree of exposure and these ones are directly related to the levels of Pb in blood; the Zn-protoporphyrin IX concentration, the degree of intake and chronic poisoning evolution; and finally the coproporphyrins (coproporphyrin III) gives an indication of the severity of the poisoning.

- In recent years the concentration of Pb in blood has decreased significantly in the aragonese population.

d. Metals in cerebrospinal fluid (Ca, Mg, Zn, Cu, Fe and Mn):
   - The concentration of different metals jointly with the rates of cellular and protein concentration has been proved to be a useful tool for the understanding of the pathogenesis of some brain conditions.

4. References

Alda, JO. & Escanero, JF. Transport of calcium, magnesium and strontium by human serum proteins. *Rev. Esp. Fisiol.* 1985; 41: 145-150.

Aschner, M. & Gannon, M. Manganese uptake and efflux in cultured rat astrocytes. *J. Neurochem.* 1992; 58: 730-735.

Barbosa, F.; Tanus-Santos, JE.; Gerlach, RF. & Parsons, PJ. A Critical Review of Biomarkers Used for Monitoring Human Exposure to Lead: Advantages, Limitations, and Future Needs. *Environ. Health Perspect.* 2005; 113(12): 1669–1674.

Bergomi, M.; Vinceti, M.; Nacci, G.; Pietrini, V. & Bratter, P. Environmental exposure to trace elements and risk of amyotrophic lateral sclerosis: a population-based control study. *Environ. Res.* 2002; 89: 116-23.

Berlin, A. & Schaller, KH. European Standardized Method for the determination of delta-amino-levulinic acid dehydratase activity in blood. *Z. Klin. Chem. Klin. Biochem.* 1974; 12: 389-390.

Berlin, A.; Schaller, KH.; Grimes, H.; Langevin, M. & Trotter J. Environmental exposure to lead: analytical and epidemiological investigations using the European Standardized Method for blood delta-amino-levulinic acid dehydratase activity determination. *Int. Arch. Occup. Environ. Health* 1977; 39: 135-141.

Campbell, BC.; Brodie, MJ.; Thompson, GG.; Meredith, PA.; Moore, R. & Goldberg, A. Alterations in the activity of the enzymes of haem biosynthesis in lead poisoning and acute hepatic porphyria. *Clin. Sci. Mol. Med.* 1977; 53: 335-340.

Carpenter, MB. Neuroanatomía Humana de Strong y Edwin (5ª Edición, 2ª reimpresión). Buenos Aires: El Ateneo, 1985, 1-19.

Cartón, JA. Saturnismo: epidemiología y diagnóstico. *Med. Clin. (Barc)* 1985; 84: 492-499.

Cartón, JA. Saturnismo. *Med. Clin. (Barc)* 1988; 91: 538-540.

Castellano, MªC.; Soteras, F.; Córdova, A.; Elósegui, LMª. & Escanero, JF. Zinc distribution between protein serum ligands in rats: acute and chronic overload of zinc. *Med. Sci. Res.* 1988; 16: 1229-1230.

Córdova, A.; Soteras, V.; del Villar, V.; Elósegui, LMª. & Escanero, JF. Efecto de la tiroparatiroidectomía, la parathormona y la calcitonina sobre el estroncio plasmático en rata. *Rev. Esp. Fisiol.* 1990; 46(2): 139- 146.

Chausmer, AB.; Weiss, P. & Wallach, S. Effect of thyrocalcitonin on calcium exchange in rat tissues. *Endocrinology* 1965; 77: 1151-1154.

Comar, CL. In: "Strontium metabolism", Leniham, JMA; Loutit, JF; Martin, JH. Eds. New York: Academic Press, 1967; 17-31.

Comar, CL.; Wasserman, RH.; & Nold, MM. Strontium-Calcium Discrimination Factors in the Rat. *Proc. Soc. Exp. Biol., N.Y.*, 1956; 92(4): 859-863.

Escanero, JF. Inferencia del estroncio en el metabolismo del calcio. Tesis Doctoral. Departamento de Fisiología. Facultad de Medicina. Universidad de Zaragoza, 1974.

Escanero, JF.; Carre, M. & Miravet, L. Effets des différents métabolites de la vit. D_3 et de la concentration calcique sur l´absorption intestinale de strontium. *C. R. Soc. Biol.* 1976; 170: 47-53.

Escanero, JF. & Córdova, A. Effects of glucagon on serum calcium, magnesium and strontium levels in rats. *Miner. Electrolyte Metab.* 1991; 17: 190-193.

Elwell, WT. & Gidley, JAF. Atomic-Absorption Spectrophotometry (2nd edition). International Series of Monographs in Analytical Chemistry (vol. 6). Oxford : Pergamon Press, 1966.

Gálvez, M.; Elósegui, LMª.; Guerra, M.; Moreno, JA. & Escanero, JF. Zinc Exchange between erythrocytes and médium with and without albumin at different temperaturas. In: "Metal Ions in Biology and Medicine", Anastassoupoulos, J; Collery, P; Theophanides, T; Etienne, JC, eds. Paris: John Libbey Eurotext, 1992, 2: 89-90.

Gálvez, M.; Moreno, JA.; Elósegui, LMª. & Escanero, JF. Zinc uptake by human erythrocytes with and without serum albumin in the medium. *Biol. Trace Elem. Res.* 2001; 84: 45-56.

Gálvez, M.; Moreno, JA.; Elósegui, LMª. & Escanero, JF. Zinc uptake by human erythrocytes. 1. Effect of Na and K in medium at different and a medium at different temperatures. In: "Metal Ions in Biology and Medicine", Anastassoupoulos, J; Collery, P; Theophanides, T; Etienne, eds. Paris: John Libbey Eurotext, 1996a, 4: 218-221.

Gálvez, M.; Moreno, JA.; Elósegui, LMª. & Escanero, JF. Zinc uptake by human erythrocytes. 2. Effetc of the temperature on Zn-uptake sensitive to the stilbenes. In: "Metal Ions in Biology and Medicine", Anastassoupoulos, J; Collery, P; Thephanides, T; Etienne, JC, eds. Paris: John Libbey Eurotext,1996b, 4: 222-224.

Giroux, EL. & Henkin, RI. Competition for zinc among serum albumin and amino acids. *Biochim. Biophys. Acta* 1972; 273: 64-72.

Goldberg, A. Lead poisoning and haem biosynthesis. *Br. J. Haematol,* 1972, 23: 521-524.

Goldberg, A., Meredith, PA., Miller, S., Moore, MR. & Thomson, GG. Hepatic drug metabolism and haem biosynthesis in lead poisoned rats. *Br. J. Pharmacol.* 1978, 62: 529-536.

Gómez, E.; de Diego, C.; Orden, I.; Elósegui, LMª.; Borque, L. & Escanero, JF. Longitudinal study of serum cooper and zinc levels and their distribution in blood proteins alter acute myocardial infarction. *J. Trace Elements Med. Biol.* 2000; 14: 65-70.

González López, JM. Influencias del plomo en el metabolismo de las porfirinas. Tesis Doctoral. Resúmenes De Tesis Doctorales, Curso 1989-1990. Universidad de Zaragoza (España), Comisión de Doctorado, 1992, 65. I.S.B.N.: 84-7733-314-9.

González-Romarís, EMª.; Idoate-Cervantes, I.; González-López, JM. & Escanero-Marcén, JF. Concentration of calcium and magnesium and trace elements (zinc, cooper, iron and manganeso) in cerebrospinal fluid: A try of a pathophysiological classification. *J. Trace Elements Med. Biol.* 2011; 25 Supl.: S45-S49.

Granick, JL., Sassa, S. & Kappas, A. Some Biochemical and clinical aspects of lead intoxication. *Advan. Clin. Chem.* 1978, 20: 287-339.

Gurer-Orhan, H.; Sabır, HU. & Özgüneş, H. Correlation between clinical indicators of lead poisoning and oxidative stress parameters in controls and lead-exposed workers. *Toxicology,* 2004; 195(2–3): 147-154.

Guyton, AC. Anatomía y fisiología del sistema nervioso. Neurociencia básica, 2ª reimpresión. Buenos Aires: Editorial Médica Panamericana S.A., 1990, 75-84, 21-36 y 129-149. Edición original: Guyton, AC. Basic Neuroscience. Anatomy and Physiology. Philadelphia: W.B. Saunders, 1987.

Hammond, C. & Tritsch, D. Neurobiologie cellulaire. Paris: Doin Éditeurs, 1990, 439-462.

Hazell, AS. Manganese decreses glutamate uptake in cultured astrocytes. *Neurochem. Res.* 1997; 22: 1443-1447.

Hernberg, S. & Nikkanen, J. Effect of lead on delta-amino-levulinic acid dehydrase - A selective review - . *Pravoc Lék.* 1972; 24: 2-3.

Hessel, DW. A simple and rapid quantitative determination of lead in blood. *At. Absorpt. Newsletter* 1968; 7: 55.

Izquierdo-Álvarez, S.; Calvo-Ruata, MªL.; González-López, JM.; García de Jalón-Comet, A. & Escanero-Marcén, JF. The Need to Update Reference Values for Lead in Zaragoza, Spain. *Biol. Trace Elem. Res.* 2008; 123: 277-280.

Jacob, K.; Sommer, W.; Meyer, HD. & Vogt, W. Ion-pair high-performance liquid chromatographic separation of porphyrin isomers. *J. Chromatogr.* 1985; 349: 283-293.

Johnson, JW. & Ascher, P. The NMDA receptor and its channel modulation by magnesium and by glicine. In: Lodge, D, ed. Excitatory amino acids in health and disease. New York: John Wiley and Sons 1988: 143-64.

Kamel, F.; Umbach, DM.; Hu, H.; Munsat, TL.; Shefner, JM.; Taylor, JA. et al. Lead exposure as a risk factor For amyotrophiv lateral sclerosis. *Neurodegener. Dis.* 2006; 2: 195-201.

Kamel, F.; Umbach, DM.; Lehman, TA.; Park, LP.; Munsat, TL.; Shefner, JM. et al. Amyotrophic lateral sclerosis, lead, and genetic susceptibility: polymorphism in in test delta-aminolevulinic acid dehydratase and vitamin D receptor genes. *Environ. Health Perspect.* 2003; 111: 1335-9.

Kapaki, E.; Zournas, C.; Kanias, G.; Zambelis, T.; Kakami, A. & Papageorgiou, C. Essential trace elements Alterations in amyotrophic lateral sclerosis. *J. Neurol. Sci.* 1997; 147: 171-5.

Kuliseusky, J. & Puyols, J. Pallidal hyperintensity on magnetic resonance imaging in cirrotic patiens: Clinical correlation. *Hepatology* 1992; 16: 1382.

Levine, J.; Rapoport, A.; Mashiah, M. & Dolev, E. Serum and cerebrospinal levels of calcium and magnesium in acute versus remitted schizophrenic patiens. *Neuropsychobiology* 1996; 33(4): 169-72.

McGuire, V.; Logstreth, WT. Jr; Nelson, LM.; Koepsell, TD.; Checkoway, H.; Morgan, MS. et al. Occupational exposure and amyotrophic lateral esclerosis: a population-based control study. *Am. J. Epidemiol.* 1997; 145: 1076-88.

Mauzerall, D. & Granik, S. The occurrence and determination of δ-aminolevulinic acid and porphobilinogen in urine. *J. Biol. Chem.* 1956; 219: 435-446.

Meredith, PA.; Moore, MR. & Goldberg, A. Erythrocyte δ-aminolevulinic acid dehydratase activity and blood Protoporphyrin concentrations as indices of lead exposure and altered haem biosynthesis. *Clin. Sci.* 1979; 56: 61-69.

Meyer, HD. Porphyrins. In: Henschen, A; Hupe, KP; Lottspeich, F; Voelter, W, eds. High Performance Liquid Chromatography in Biochemistry. Weinheim (FRG): VCH Verlagsgessellschaft GmbH 1985; 445-479.

Meyer, HD.; Jacob, K. & Vogt, W. Rapid and Simple Direct Determination of Porphyrins in Urine by Ion-Pair Reversed-Phase High Performance Liquid Chromatography. *Journal of HRC & CC* 1980a; 85-86.

Meyer, HD.; Jacob, K. & Vogt, W. Ion-Pair-Reversed-Phase High-Performance Liquid Chromatographic Determination of Porphyrins from Red Blood Cells. *Chromatographia* 1982; 16: 190-191.

Meyer, HD.; Jacob, K.; Vogt, W. & Knedel, K. Diagnosis of porphyries by ion-pair high-performance liquid Chromatography. *J. Chomatogr.* 1980b; 199: 339-343.

Nolte, J. El Cerebro Humano. Introducción a la anatomía funcional. Primera Edición española. Madrid: Mosby/Doyma libros, 1994, pp. 33-75. Edición original: Nolte, J. The Human Brain. Third edition. Mosby - Year Book, Inc, MCMXCIII.

Norenberg, MD. Astroglial dysfunction in hepatic encephalopathy. *Metab. Brain Dis.* 1998; 13: 319-35.

Pearson, PJ. & Slavin, W. A rapid Zeeman graphite furnace atomic absorption spectrophotometric method for the determination of lead in blood. *Spectrochem. Acta* 1993; 48: 925-939.

Piomelli, S. Free erythrocyte porphyrins in the detection of undue absorption of Pb and Fe deficiency. *Clin. Chem.* 1977; 23(2): 264-269.

Sakai, T. Biomarkers of Lead Exposure. *Ind. Health* 2000; 38: 127–142.

Sakai, T., Yanagihara, S., Kunugi, Y. & Ushio, K. relationships between distribution of lead in erythrocytes in vivo and in vitro and inhibition of ALA-D. *Br. J. Ind. Med.* 1982, 39: 382-387.

Schaller, KH. & Berlin, A. Δ-Aminolaevulinate Dehydratase. In: Methods of Enzymatic Analysis (Volume IV, Third Edition). Bergmeyer, HU, Editor-in-Chief; Bergmeyer, J; Graβl, M; eds. Weinheim: Verlag Chemie GmbH, 1984; 363-368.

Schwartz, S.; Berg, MH.; Bossenmaier, I. & Dinsmore, H. Determination of porphyrins in biological materials. In: Methods of Biochemical Analysis, Glick, D., ed. New York: Interscience, 1960, Vol. 8, 221-293.

Skoog, DA. & West, DM. Análisis Instrumental (2ª Edición). México, D.F.: Nueva Editorial Interamericana, S.A. de C.V., 1984, 158-177 y 317-369.

Skoog, DA.; Holler, FJ. & Nieman, TA. Principios de Análisis Instrumental (5ª Edición). Madrid: McGraw-Hill/Interamericana de España, S.A.U., 2001, pp. 122-150 y 219-244.

Skoog, DA.; West, DM.; Holler, FJ. & Crouch, SR: Fundamentos de Química Analítica (8ª Edición, 2ª reimpresión). Madrid: Paraninfo, 2008, 719-723 y 870-880.

Taylor, A.; Branco, S.; Halls, D.; Patriarca, M.; & White, M. Atomic spectrometry update. Clinical and biological materials, foods and beverages. *J. Anal. Atom. Spectr.* 2002; 17: 414- 455.

Thomson, AM. A magnesium-sensitive post-synaptic potential in rat cerebral cortex resembles neuronal responses to N-methylaspartate. *J. Physiol.* 1986; 370: 531-49.

Tomokuni, K. & Ogata, M. Relationship between lead concentration in blood and biological response for porphyrin metabolism in Workers occupationally exposed to lead. *Arch. Toxicol.* 1976, 35: 239-246.

Trasobares, EM. Plomo y mercurio en sangre en una población laboral hospitalaria y su relación con factores de exposición. Tesis Doctoral. Madrid: Universidad Complutense, 2010. ISBN: 978-84-693-6339-3.

Vercruysse, A., ed. Techniques and Instrumentation in Analytical Chemistry, Volume 4. Evaluation of Analytical Methods in Biological Systems, Part B: Hazardous Metals in Human Toxicology. Amsterdam, The Netherlands: Elsevier Science Publishers, B.V. 1984.

Weissenborn, K. Pallidal lesion in patiens with liver cirrosis: Clinical and MRI evaluation. *Metab. Brain Dis.* 1995; 10: 219-231.

Willard, HH.; Merritt, LL.; Dean, JA. & Settle, FA. Jr. Métodos Instrumentales de Análisis. México, D.F.: Grupo Editorial Iberoamérica, 1991, 95-100 y 219-252.

2

Estimation of the Velocity of the Salivary Film at the Different Regions in the Mouth – Measurement of Potassium Chloride in the Agar Using Atomic Absorption Spectrophotometry

Shigeru Watanabe
Meikai University
Japan

1. Introduction

Saliva is secreted into the mouth at a rate of 0.3 to 0.4 ml per minute. Retained saliva in the mouth physiologically triggers swallowing to carry the saliva out of the mouth. Dawes (1983) have reported the volume of saliva in the mouth just before swallowing, the rate of swallowing, the volume swallowed per swallow, and the volume of saliva in the mouth just after swallowing. Clearance of materials from the mouth is facilitated by alternately-performed saliva secretion and swallowing, and thereby the oral environment is maintained relatively constant (Fig.1). The unstimulated salivary flow rate and saliva volume in a single swallowing have the most influence on the efficiency of clearance.

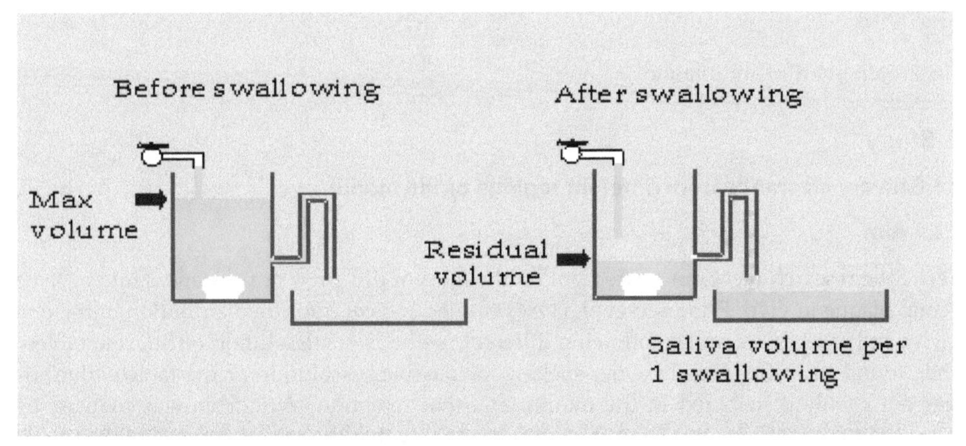

Fig. 1. Saliva volume in the mouth

Saliva is a crucial factor for protection of the oral environment. The rate of oral clearance of sugar and acid is inversely related to the onset and progression of dental caries, as shown particularly in persons with severe hyposalivation.

Saliva secreted into the mouth flows slowly as a thin film, over the tooth surfaces and mucosa and is cleared from the mouth by swallowing (Fig. 2). However, saliva does not flow equally throughout the mouth, and there are differences in the different areas. Measurement of the volume of saliva and velocity of the salivary film at different locations in the mouth are important for understanding the site-specificity of dental caries and periodontal disease.

Using agar as an artificial-plaque, we have conducted studies on the five following items by measuring the clearance of potassium chloride from the agar using an atomic absorption spectrophotometer.

1) Salivary clearance from different regions of the mouth. 2) Salivary clearance in children with complete primary dentitions. 3) Influence of the location of the parotid duct orifice on oral clearance. 4) Effect of salivary flow rate on fluoride retention in the mouth. 5) Estimation of the velocity of the salivary film at different locations in the mouth.

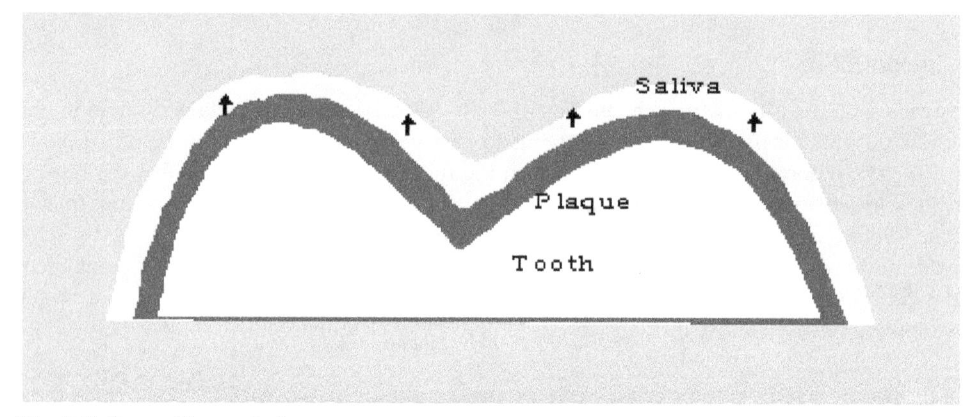

Fig. 2. Salivary film and plaque

2. Study

2.1 Salivary clearance from different regions of the mouth

2.1.1 Aim

Very little research has been carried out on the rates of diffusion of substances into or from dental plaque in vivo. Primosch et al. (1986) studied topical fluoride distribution in the oral cavity and rates of clearance following different methods of dissolution of fluoride tablets. They found that after the chewing, sucking, or passive dissolution of the tablets, fluoride was not evenly distributed in the mouth, and that retention of fluoride was reduced by increased salivary flow rate. Thus, it would seem likely that the rate of renewal of the film of saliva over plaque must influence diffusion rates into and from plaque.

The aim of this study was to determine the velocity of the salivary film by determining the rate of diffusion of potassium chloride from an artificial plaque at different sites in the mouth.

2.1.2 Materials and methods

- Determination of the rate of potassium chloride clearance:

A 1-mol/L solution of potassium chloride was mixed with sufficient agarose (Electrophoresis Purity Reagent; BioRad Laboratories, Richmond, CA) to give a 1.0% solution which was heated until the agarose dissolved. The acrylic chambers (Fig. 3) to hold the gel were rectangular (16 mm in length, 8 mm wide, and 1.5 mm thick) with a cylindrical central depression (6 mm diameter and 1.5 mm depth).

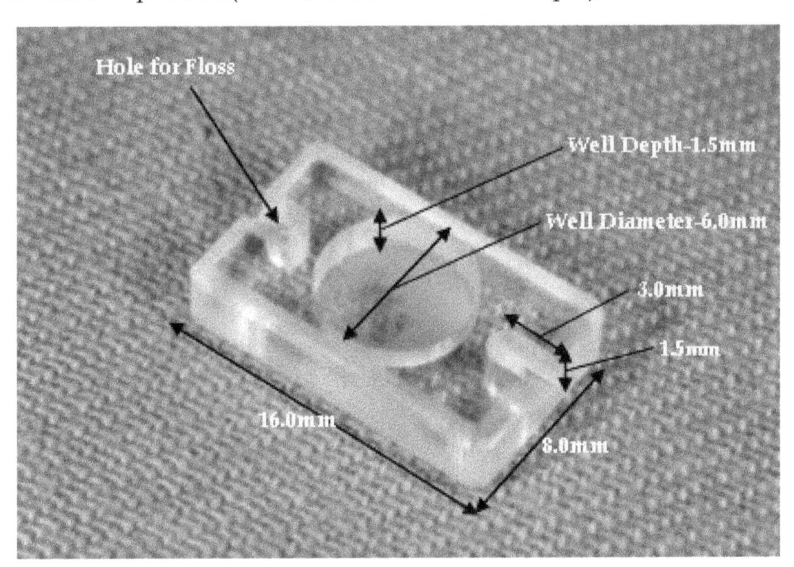

Fig. 3. Diagram of design of the diffusion chambers

The weight of the agarose held in the center well of each chamber was measured six times using an electronic balance (FX-3200; A&D, Tokyo, Japan), and chambers in which the mean weight of agarose was more than 2 SD from the mean were excluded.

Two chambers were initially covered with a layer of Parafilm (American Can, Greenwich, Conn., USA), were attached bilaterally by floss to the teeth, with the gel surface away from the teeth. The chambers were attached to the upper first molars for measurement of the posterior sites (UPB) and to both upper incisors for measurement of the anterior site (UAB) (Fig. 4).

After temperature and salivary flow equilibration, the Parafilm was removed at time 0. The first diffusion chamber was removed from the mouth after being exposed to saliva for a selected period of time and the gel was transferred to flasks containing 400 ml of (100 ppm) sodium chloride. Subsequently the second chamber was removed and the potassium chloride extracted by the same procedure. The fluid was agitated intermittently for 90 min, and the potassium concentration was assayed by atomic absorption spectrophotometry (Shimadzu AA-6105, Kyoto, Japan). The times were chosen so that between about 30 and 60% of the potassium chloride would have diffused from the agarose discs. The initial KCI concentration in the agarose discs, which had not been placed into the mouth, was also measured.

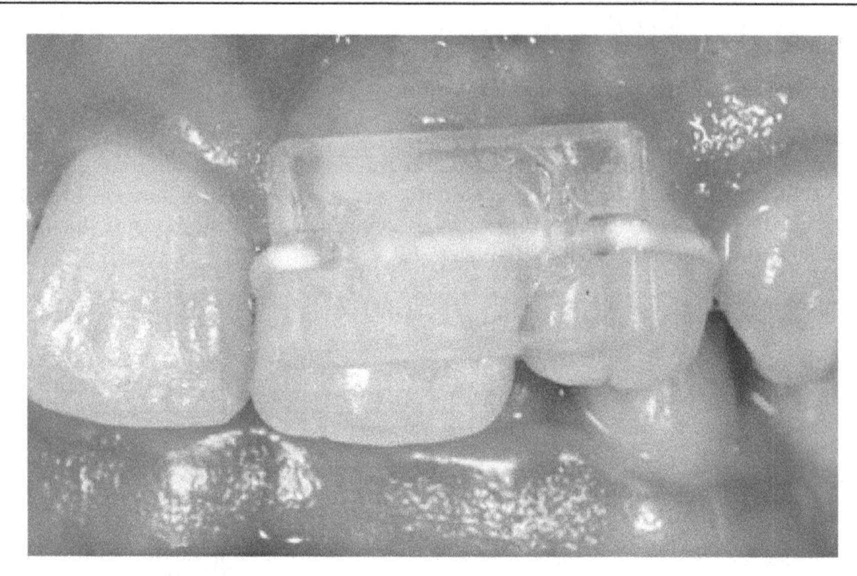

Fig. 4. Acrylic chamber attached to the upper central incised

- Calculation of the half-time (the time for half the KCl to diffuse from the gel) (Lecomte & Dawes, 1987) for clearance.

The rate of potassium chloride clearance from the gels into a large, stirred volume was determined. One involved suspending the filled chambers in one liter of 100 ppm NaCl, stirred by a magnetic stirrer, either at room temperature or at 37℃. The diffusion chambers were taken from the fluid at selected time intervals and the gels transferred quantitatively with a sewing needle to flasks containing 500 ml of 100 ppm sodium chloride. The fluid was agitated intermittently for 90 min, since preliminary studies showed that the remaining potassium chloride was extracted from the gel in this time interval. The potassium concentration was also measured in identically prepared agarose discs which had not been put into the 100 ppm NaCl, to give the initial concentration.

A least-squares straight line was fitted, by computer, to the potassium concentration plotted against the square root of time. This gives a very good approximation of the theoretical clearance curve until about 65% of the diffusant has been lost from the gel (see 2-1-5). From the results, the half-time was calculated.

2.1.3 Subjects and locations

The subjects were 6 adults with a mean age of 26 years. They had a complete dentition up to the second molar and no malocclusion.

Seven different sites in the mouth were chosen for measurements. These were the Lower anterior lingual (LALi) and buccal (LAB), lower posterior buccal (LPB) and lingual (LPLi), upper posterior lingual (UPLi) and buccal (UPB), and upper anterior buccal (UAB). The flow rate of unstimulated whole saliva was measured on each occasion for 5 min by being allowed to drip off the lower lip into a weighed container.

2.1.4 Result

The half-times in the mouth varied with locations and with salivary flow rate, as shown in Fig. 5 When the flow rate was unstimulated, the shortest halftimes occurred in the LALi site and the longest in the UAB site. In both groups, the difference was significant at $p<0.001$.

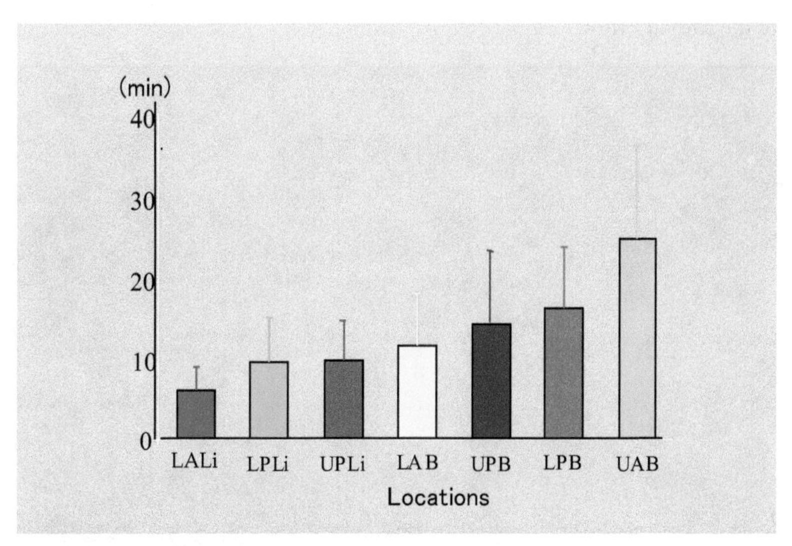

Fig. 5. Half-time when salivary flow rate is unstimulated

2.1.5 Discussion

In this study, we measured the concentration of residual potassium in agarose gel to determine the velocity of salivary flow for the 7 different sites. The reason why potassium chloride was used as the target substance (with an agarose gel used as artificial plaque) is that it is readily soluble in water, harmless, has a low molecular weight enabling it to diffuse easily, is present at a low concentration (around 20 mM) in saliva and can be measured relatively easily. Because the potassium concentration in the agarose gel used in this study was much higher than that in the saliva of the subjects, it was unlikely that the potassium concentration in the saliva affected that in the gel.

The relationship between time and the quantity of potassium diffused from the gel into the saliva was pre-determined in a pilot study. Clearance was evaluated by determining the half-time, the time at which the concentration at time 0 is reduced to half, from the relationship between 3 time points, including time 0, and potassium concentration, as well as by comparing the mean half-time between different sites. Since the correlation between time and the quantity of potassium eluted from the agarose gel was found to decrease in the early and late phases of the test (Lecomte P, Dawes C, 1987), the time to hold the holder in the mouth was determined so that the half-time would be almost at the mid-point of the test. For the measurement of potassium concentration, sodium chloride solution was used as the solvent to avoid errors in measurements due to ionization of potassium.

2.2 Salivary clearance in children with complete primary dentitions

2.2.1 Aim

Nursing bottle caries (Fig.6) is a specific form of rampant decay on the buccal surface of the upper anterior primary teeth. Some etiological factors, such as the types of microorganisms, tooth structure, and diet, have been reported, but there is little information about the influence of the salivary flow rate.

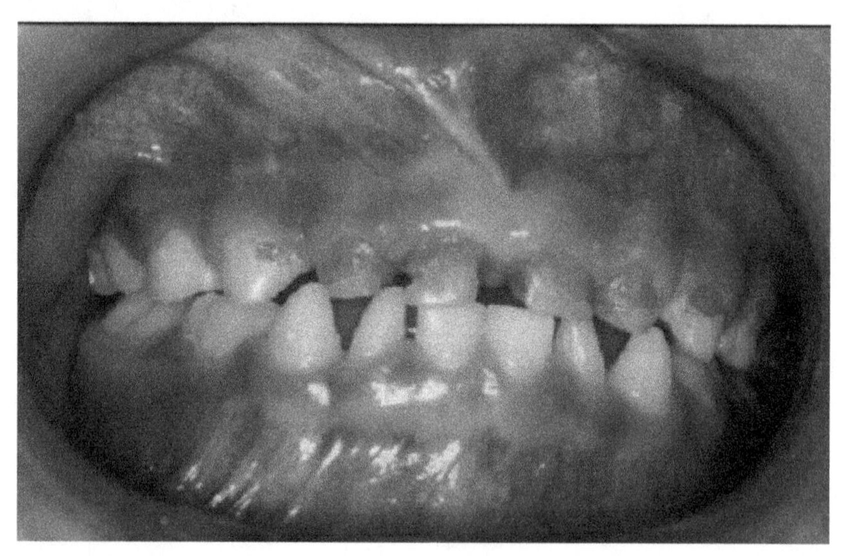

Fig. 6. Nursing bottle caries

Very little research has been carried out on the salivary flow rate or salivary clearance in children. Although the average thickness of the salivary film covering teeth and oral mucosa in children is essentially identical with values reported for adults, marked differences were found between children and adults for such parameters as unstimulated and stimulated whole-salivary flow rates, the volume of saliva in the mouth before and after swallowing, and the surface area of the mouth.

The aim of this study was to evaluate the rates of salivary clearance at different locations in the mouths of children and the effect of the spaces in the primary dentitions to determine whether prolonged clearance would occur in sites particularly susceptible to nursing bottle caries.

2.2.2 Materials and methods

The determination of the rate of potassium chloride clearance was done by the same methods as in study 1. A 1-mol/L solution of potassium chloride was mixed with sufficient agarose (Electrophoresis Purity Reagent; BioRad Laboratories, Richmond, CA) to give a 1.0% solution which was heated until the agarose dissolved. The acrylic chambers (Fig. 3) to hold the gel were rectangular (16 mm in length, 8 mm wide, and 1.5 mm thick) with a cylindrical central depression (6 mm diameter and 1.5 mm depth). The potassium concentration in agarose was analyzed by absorption spectrophotometry.

The subjects were 4 boys and 8 girls, 5 years of age, who were all in good health and with complete primary dentitions. Six subjects had primary spaces (located mesial to the maxillary and distal to the mandibular canines) and developmental spaces (present between the remaining teeth) (Fig. 7), and the other 6 subjects had no spaces in their arches (Fig. 8). The mean values of right and left primary spaces and total developmental spaces for 6 subjects who have spacing arch were 1.1 ± 0.4 mm, 1.2 ± 0.4, and 4.2 ± 1.9 mm for the upper dentition and 0.8 ± 0.3 mm, 0.7 ± 0.3, and 3.8 ± 1.4 mm for the lower dentition, respectively.

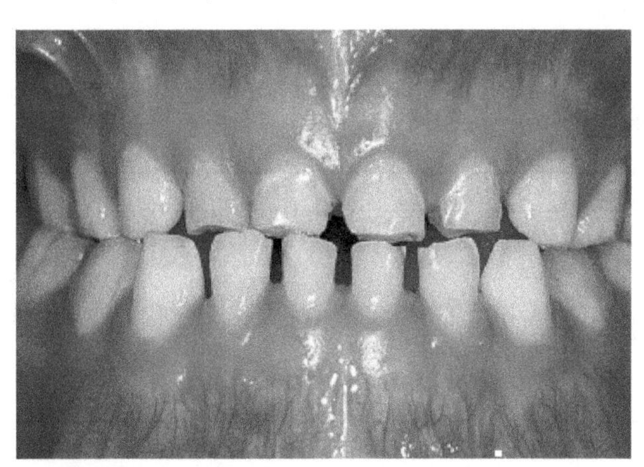

Fig. 7. Spacing arch at 5 years old.

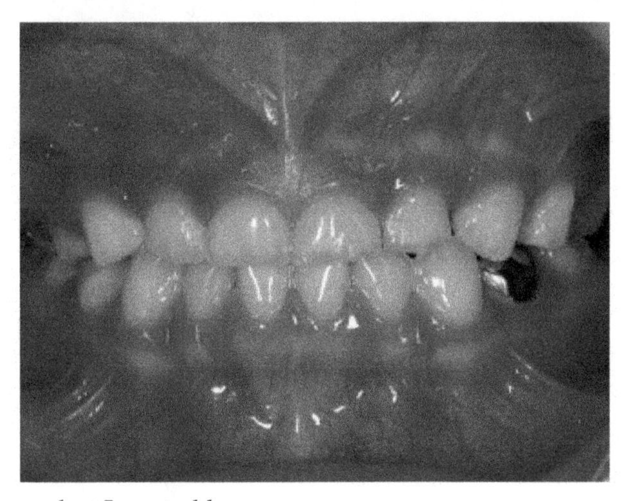

Fig. 8. No spacing arch at 5 years old.

Seven different sites in the mouth were chosen for measurements. These were the lower anterior lingual (LALi) and buccal (LAB), lower posterior buccal (LPB) and lingual (LPLi), upper posterior lingual (UPLi) and buccal (UPB), and upper anterior buccal (UAB). The flow rate of unstimulated whole saliva was also measured on each occasion for 5 min by being allowed to drip off the lower lip into a weighed container.

The rate of potassium chloride clearance from the gels into a large volume (1 liter) of 100 ppm NaCl fluid at 37℃ stirred by a magnetic stirrer was determined for the estimation of the half-time.

2.2.3 Result

The mean half-times for in vitro clearance into the large volume of stirred 100 ppm NaCl at 37°C was 3.9 ± 0.5 min.

The half-times in the mouth varied with location as shown in Table 1. The half-times of all sites for the spacing arch and of the LALi site for the no-spacing arch were reduced ($p<0.05$) as compared with the values when the flow rate was unstimulated.

( )SD

	Location						
	LALi	UPLi	LPLi	UPB	LAB	LPB	UAB
Spacing arch	5.9***	9.6*	9.4*	14.2	11.5	16.1	24.6
	(±2.8)	(±4.8)	(±5.4)	(±8.8)	(±6.4)	(±7.4)	(±11.4)
No-spacing	5.3 ***	9.1 **	10.1 *	13.8	13.8	17.4	25. 9
arch	(±2.1)	(±4.4)	(±5.3)	(±6.8)	(±7.0)	(±8.1)	(±9.4)

Statistical analyses werw carried out between UAB site and the other sites in each group: *$p<0.05$, **$p<0.01$, ***$p<0.001$.
Mean unstimulated salivary flow rates were 0.47±0.2 ml/min.

Table 1. Half-times (mean ± SD) and salivary flow rates when salivary flow was unstimulated

When the saliva flow rate was stimulated (Table 2), the shortest halftimes occurred in the LALi site and the longest in the UAB site. The clearance from the LAB site in the spacing arch showed almost the same value as those from the LALi sites in both groups.

Halftimes(mean±SD)and salivary flow rates when salivary flow was stimulated

	Location		
	LALi	LAB	UAB
Spacing arch(n=6)			
Halftime, min	4.0±0.3**	4.5±0.8**	11.3±4.5
Flow rate, ml/min	3.4±1.5	4.1±2.1	3.4±1.7
No-Spacing arch(n=6)			
Halftime, min	4.0±0.2**	9.4±2.8*	15.4±4.2
Flow rate, ml/min	3.7±1.2	3.9±1.7	4.8±1.7

Halftime in large volume at 37°C=3.9 ± 0.5min.Statistical analyses were carried out detween UAB site and the other sites in each group :*$p<0.05$ **$p<0.01$

Table 2. Halftimes (mean ± SD) and salivary flow rates when salivary flow was stimulated

2.2.4 Discussion

The present study showed that the rate of clearance of substances from agarose gels into saliva varies markedly in different regions of the primary dentition. The location closest to the submandibular and sublingual ducts (LALi) showed the lowest half-time, whereas the UAB site had a clearance half-time 6.5 times longer than that for clearance into a large volume in vitro. As the opening of the parotid duct is situated on the rearward of the upper second primary molar in the children's mouth, there was a relatively long half-time in the UPB site. Since the mean salivary film thickness in 5-year-old children has been estimated to have almost the same value (0.06-0.09 mm) (Watanabe and Dawes, 1990) as in adults (Collins and Dawes, 1987), these results suggest that the velocity of the salivary film varies in different regions.

The relative order of the half-times at the different sites in the no-spacing arches was identical with that found in a study on adult subjects when saliva flow was unstimulated. Although in the spacing arch, the LAB site had a shorter clearance half-time than the UPB site. These results may be due to the fact that in the spacing arches, the tongue pushes out a portion of saliva from the lingual to the buccal side during swallowing, and this is in accordance with clinical findings that these sites are not susceptible to caries. The ideal arch in the primary dentition has spacing between the teeth (Pinkham et al, 1988), but Foster and Hamilton (1969) reported that only 33% had spacing between all the incisors in the upper and lower arches and that only 12% had spacing between all teeth in both arches in 100 British children aged 30-36 months.

Although it is known that nursing bottle caries depends on the feeding pattern in infancy, the present results suggest that the upper anterior buccal site in a no-spacing arch will be the most cariogenic site in a child's mouth because it has the lowest rate of salivary clearance.

2.3 Influence of the location of the parotid duct orifice on oral clearance

2.3.1 Aim

The rate of oral clearance was shown to vary markedly at different locations in the mouth. Oral clearance is slower for teeth in the maxilla than for those in the mandible and slower for the buccal surfaces of the teeth than for the lingual. Oral clearance on the labial surface of the upper anterior region is the slowest, while that for the lingual surface of the lower anterior region is the fastest. The lingual surface of the lower anterior region is near the openings of the ducts of the submandibular and sublingual glands, which probably accounts for the fastest rate of oral clearance being there. The effect of unstimulated parotid saliva on clearance around the maxillary first molar is not very striking, perhaps because the volume ratio of parotid saliva to total saliva is only about 15% at rest for each side. However, with stimulation, the proportion of parotid saliva increases, increasing clearance over the maxillary first molar, which is closest to the parotid duct. Few studies have examined positional relationships between the parotid duct orifice and the maxillary molars or individual differences in this positional relationship (Suzuki et al. 2009). The present study sought to ascertain the location of the parotid duct orifice in relation to the maxillary molars and whether oral clearance at locations 1 cm mesial and distal to the duct opening would be as rapid as that directly opposite the opening of the duct.

2.3.2 Materials and methods

2.3.2.1 Location of the parotid duct orifice

- Subjects

These were 35 consenting adults (20 men, 15 women) with a mean age of 27.1 years (range, 23-35 years). They had a complete dentition up to the second molar and no malocclusion. In each subject, plaster models were made after taking impressions of the upper and lower dentitions.

- Impressions of the right and left parotid duct orifice

Before taking an impression of the parotid duct orifice, a 2-mm hole was made at the centre of an adhesive therapeutic agent for aphthous stomatitis (Aftach; Teijin, Tokyo) and the agent was placed on the mucosa so that the hole matched the parotid duct orifice. Next, using a vinyl siloxane impression material (Stat BR; Car Japan, Tokyo), an impression of the buccal tooth surfaces and mucosa around the Aftach was taken with the teeth in centric occlusion to localize the duct opening in relation to the teeth.

- Reference plane setting

To take standard photos, a horizontal reference plane was set for each maxillary plaster model. This was a triangular plane defined by the occlusal plane at the maxillary midline and the distobuccal cusp of the left and right maxillary first molars.

- Taking standard photos

The standard plane was set horizontally and the plaster model was matched with the impression of the parotid duct orifice. In order to take standard photos from the same angle, the line connecting the disto- and mesio-buccal interdental papillae of the maxillary first molar was set orthogonal to the imaging direction.

- Location of the parotid duct orifice

After defining the reference plane on standard photos as the X axis and the line perpendicular to the X axis passing through the distal plane of the first molar as the Y axis, the location of the parotid duct orifice was measured in relation to the reference point.

In one subject the location of one parotid duct was determined six times in order to assess the reliability of the method.

2.3.2.2 Oral clearance on the buccal surface of the upper molar region

- Subjects

Subjects comprised 12 (8 men, 4 women mean age 28.3 years) of the original 35 subjects whose parotid duct orifice fell within 1 SD of the mean values for the X and Y coordinates obtained in Study 2-3-2-1.

- The rate of secretion by the parotid gland

The subjects had not eaten for at least one hour prior to the study and the studies were done in either the mid-morning or mid-afternoon. In the 12 subjects for whom oral clearance was measured, Lashley cups were attached over the left and right parotid duct orifices and with the agar holders in position, parotid saliva was collected on 5 separate occasions for a 5-min period without stimulant.

- Diffusing substance and agar holder

Oral clearance was assessed using the same methods of the study 2-1. 1% agar containing 1 mol/l potassium chloride was placed into cylinders (diameter, 4 mm; depth, 1 mm) held by an acrylic holder (width, 30 mm; height, 10 mm; thickness, 2 mm). The open surfaces of the cylinders were initially covered with microscope slides to allow the agar to set. In each agar holder, 3 cylinders were placed horizontally at 6-mm intervals (Fig. 9).

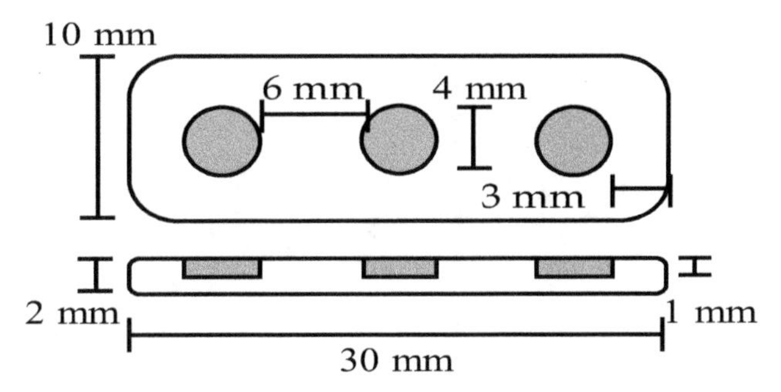

Fig. 9. Agar holder

The cylinders were attached to the teeth using Hydroplastic (TAK, Tokyo) so that the central cylinder would be on the buccal surface of the first molar, at the coordinates of the mean X and Y values obtained in Study 2-4-2-1. After salivary secretion stabilized, which took about 1 minute when parotid flow was measured with a Lashley cannula, the Parafilm was removed to initiate the experiment. On separate occasions, the holder was retained for 5, 10, 20 or 40 min without stimulant. At each time point, the concentration of residual potassium in the agar was measured for calculation of the half- time (the time for half of the potassium chloride to diffuse from the gel), as described by the study 1. Concentrations of potassium were measured by removing the agar cylinders from the holder, soaking each in 300 ml of 100 ppm sodium chloride solution for 90 min, and measuring the levels of eluted potassium by atomic absorption spectroscopy using an ANA-182 spectroscope (Tokyo Koden, Tokyo). The experiment was performed 3 times on both sides of each subject, and mean values were calculated. During the experiment, subjects were asked to refrain from touching the agar holder with their tongue or talking.

2.3.3 Result

Along the X axis, the location of the left and right parotid duct orifices varied within a range of −7.5 to +6.1 mm (Mean ± S.D.) from the reference point. Mean location (-0.36 ± 3.76) was just mesial to the reference point. Along the Y axis, the orifice was always located on the positive side of the reference point, ranging from +3.8 to +10.4 mm (mean value: 7.21 ± 2.15) (Fig. 10). This suggests that the parotid duct orifice is located above the reference plane near the contact surface between the maxillary first and second molars. Also, ranges of 13 mm in the mesiodistal direction and 6 mm in the perpendicular direction were noted, showing that there was a high degree of inter-individual variation. No significant left-right differences were identified. The intra-individual right-left differences were significantly less ($P < 0.001$) than the overall variability among subjects.

The unstimulated parotid saliva flow rates for left and right sides were 0.02 ± 0.02 and 0.02 ± 0.02 ml/min, respectively and no significant difference was found between results for the two sides. No significant differences in half-time could be detected between comparable left and right regions. Fig.11 shows the half-times for the right and left sides without stimulant. The half-time of the central cylinder was the shortest, followed by the mesial and then the distal cylinders, in that order, for both left and right sides. The half-time values among the 3 cylinders were all significantly different.

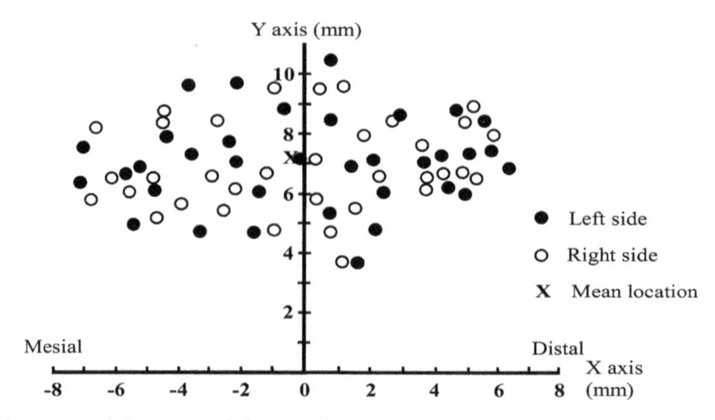

Fig. 10. The location of the parotid duct orifice. The symbols indicate the individual results for the 35 subjects. The x indicates the mean position of the duct orifice.

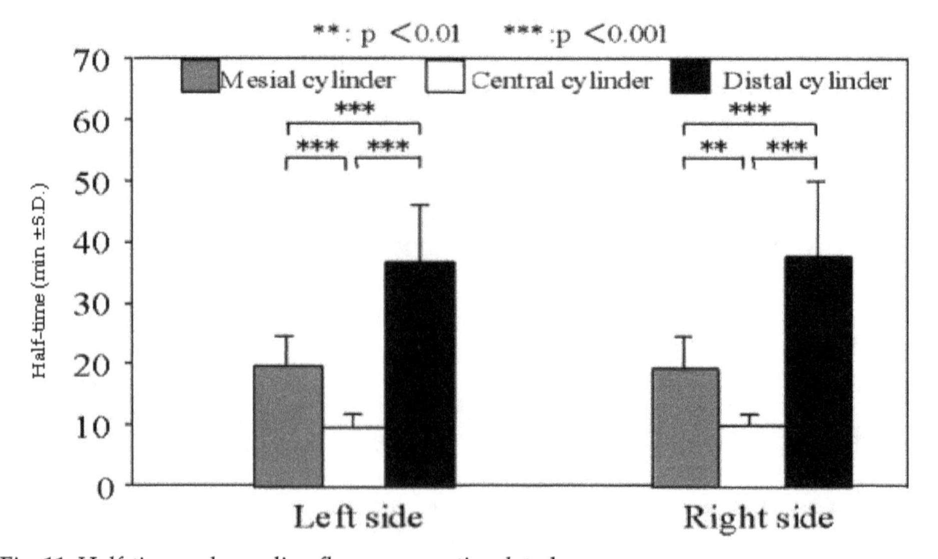

Fig. 11. Half-times when saliva flow was unstimulated

2.3.4 Discussion

The main finding from the second study was that clearance from a site directly opposite the opening of the parotid duct was significantly faster than from sites only one cm either mesially or distally. When salivary flow was unstimulated or stimulated, the clearance half-times mesial or distal to the duct opening were two or more times longer than those opposite the duct opening.

The present results are in conformity with those of Weatherell et al. (1986) who found that when a fluoride tablet was placed in the buccal vestibule, the fluoride concentration peaked in the fluid adjacent to the tablet but was much lower both mesially and distally. The previous reports and our results suggest that when parotid saliva exits the parotid duct, it primarily flows downwards and then, from the results of Weatherell et al. (1986), probably lingually over the occlusal surface of the teeth, rather than flowing mesially or distally in the buccal sulcus. If it flowed primarily in either of these two directions, one would have expected very little difference between the clearance rates from the mesial or distal agar cylinder and that from the cylinder positioned over the parotid duct opening. Sass and Dawes (1977) also reported that very little parotid saliva appeared to flow mesially when flow was either unstimulated or stimulated by the use of chewing gum.

In conclusion, the degree of individual variation in the location of the parotid duct orifice is great and its exact location will markedly affect oral clearance at different positions on the buccal surfaces of the upper molars.

2.4 Effect of salivary flow rate on fluoride retention in the mouth

2.4.1 Aim

Salivary clearance rates in different parts of the mouth are known to vary. The clearance half-times on the buccal surfaces of the upper anterior teeth were the longest of any site in the mouth. These show that the saliva secreted into the oral cavity is not perfectly mixed. Weatherell et al (1986) reports the difference by the fluoride distribution in the mouth after fluoride rinsing. Duckworth and Morgan (1991) and Heath et al. (2001) have also reported oral fluoride retention after use of fluoride rinse. These researches demonstrate the mechanism of the salivary clearance reported by Dawes (1983). According to Lear et al (1965), the salivary flow rate in the sleep is almost similar to the zero, but there are few reports the clearance of the fluoride in the sleep.

The purpose of this research was to measure the site-specificity of fluoride clearance when the subjects were awake and when they had been sleeping.

2.4.2 Materials and methods

40 mg of NaF and 5 ml distilled water were mixed with 0.15 g agarose which was heated until the agarose dissolved. Aliquots were pipetted into holders (diameter 4 mm, depth 1 mm) and these were bonded onto mouthguards produced from plaster casts of each subject (Fig.12).

The bonding sites were on the labial of maxillary incisors (UAB), the buccal of left maxillary molars (UPB) and the lingual of lower incisors (LAL). When the subjects were awake, the upper and lower mouthguards were fixed in the mouth and exposed to saliva for 15, 45 minutes. The agarose was taken out of the holder and put into 2 ml of distilled water mixed with 0.1 ml of the total ion strength adjustment buffer (TISABⅢ, Thermo Orion, IL, USA) for 90 minutes and the fluoride concentration was measured by atomic absorption spectrophotometry (Shimadzu AA-6105, Kyoto, Japan) as described in study 1. The fluoride concentration of the agarose held in the holder of each mouthguard was measured six times, and holders in which the mean concentration of agarose was more than 2 SD from the mean were excluded. To examine the retention of fluoride in the mouth during sleep, the mouthguards were placed before going to bed (0:00 a.m.) and removed at 6:30 a.m. and the fluoride concentration measured by a fluoride electrode (Thermo Fisher Scientific , MA, USA). The subjects, 6 adults who were all in good health and whose salivary flow rates exceed 0.3 ml/min were selected. Before the experiment, the subjects were explained the purpose and got their cooperation. In order to determine the effects of site specificity of salivary clearance, the data were analyzed by analysis of variance in randomized blocks and by Duncan`s New Multiple Range Test.

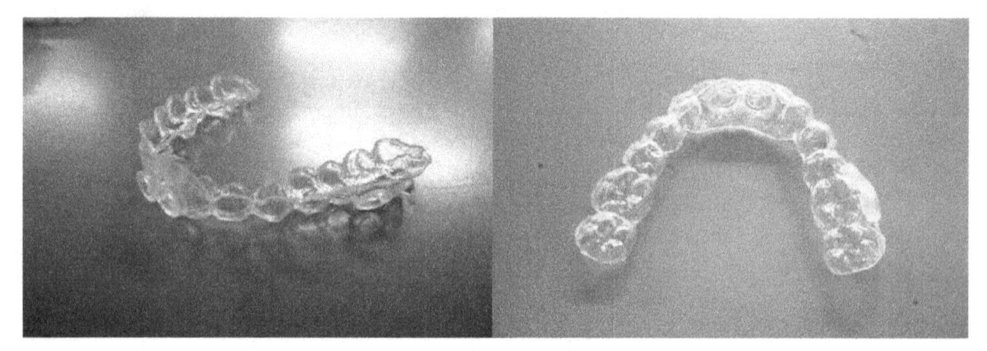

Fig. 12. Mouthguard with agarose holders. (Left: Upper, Right: Lower)

2.4.3 Results

Fig.13 showed the comparison of the mean half-times of each place, expressed as a standard in the value at LAL. The half-times were lowest in LAL and were highest in UAB. There were significant differences between the LAL and UAB ($p<0.01$), and between the LAL and UPB ($p<0.05$).

Table 3 showed the comparison of the mean volume of fluoride retention at 6:30 am. when the subjects had been sleeping. The fluoride concentrations were expressed as a percentage of that of the initial agarose which did not expose to saliva in the mouth. The values in LAL were also lowest, and UAB were highest. There were significant differences between the LAL and UAB ($p<0.05$) and between the LAL and UPB ($p<0.05$).

Most studies on fluoride clearance in the mouth have been carried out when the subjects were awake, and there is little information when they were sleeping. Ekstrand et al. (1986) and Featherstone et al.(1986) have suggested that fluoride, even at low concentrations, is necessary in the oral fluids to obtain maximum caries inhibition and have concluded that continuous or frequent elevation of the fluoride concentration in the oral fluids would be advantageous.

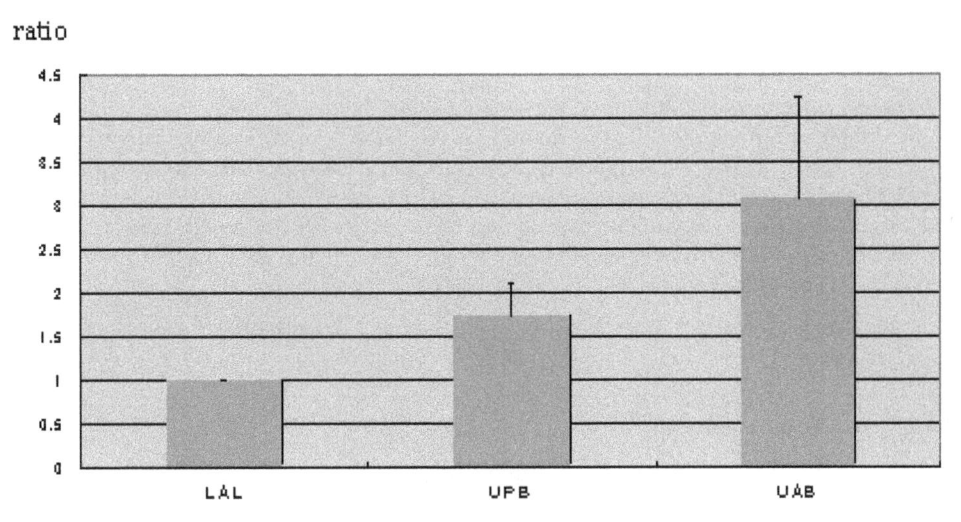

Fig. 13. The comparison of the mean half-times of each place, expressed in relation to the value for LAL. ($p<0.01$: LAL vs. UAB, $p<0.05$: LAL vs. UPB)

	LAL	UPB	UAB
	Mean S.D.	Mean S.D.	Mean S.D.
Subject A	1.2 ±0.3	1.5 ±0.2	3.4 ±2.2
Subject B	1.6 ±0.6	2.6 ±1.7	4.0 ±1.7
Subject C	0.9 ±0.2	0.6 ±0.2	2.6 ±1.8
Subject D	1.8 ±0.6	1.8 ±0.7	3.0 ±2.4
Subject E	1.7 ±0.8	4.1 ±1.2	3.3 ±2.6
Subject F	2.1 ±0.8	2.3 ±1.0	3.5 ±1.4
Mean	1.6 ±0.6*	2.2 ±1.2*	3.3 ±1.8

(*$p<0.5$: Significantly different from the mean volume of UAB)

Table 3. The mean volume (%) of fluoride retention at 6:30 a.m. when the subject had been sleeping

In this study it was shown that the fluoride concentration in the saliva was kept at high level for a long time during sleeping. In order to prevent dental caries at the buccal surfaces of the upper anterior teeth, it seems to be good to use a fluoride rinse before going to bed.

2.5 Estimation of the veloci1ty of the salivary film at different locations in the mouth

2.5.1 Aim

Although a great deal of information is available about the overall flow rate of whole saliva in man, there is no quantitative information on the velocity of flow of the salivary film in different regions of the mouth. Once secreted into the oral cavity, saliva forms a thin film, approximately 0.1 mm thick, which moves around inside the mouth until it is eventually swallowed. The higher the saliva secretion rate, the more frequently swallowing occurs, and the cleaner the mouth will be remain, However, this salivary film does not distribute evenly or reach all parts of the mouth.

The aim of this study was to estimate of the velocity of the salivary film at different locations in the mouth.

2.5.2 Materials and methods

- The equipment used in the salivary film velocity studies.

An extraoral device was used to adjust the flow rate of a 0.1-mm-thick film of artificial saliva over an agarose disk to determine the clearance half-time in the same manner as that performed intraorally (Dawes et al, 1989). Then, from the relationship between the intraoral and extraoral half-times, the salivary film velocities of the UAB and UPB sites were estimated. The half-time at UAB and UPB were evaluated by the method of study 2.1.

Fig. 14 shows the equipment used. The diameter of the well in the lower part of the device was 6 mm, the same as the width of the 0.1-mm-deep slot in the upper part. Thus, the fluid was directed over the surface of the gel. The well was 4 mm from the end of the device.

The well in the lower part of the device was filled with 1 mol/L KCl in 1 % agarose, as described for study 2.1 (Fig.3) and the upper and lower parts of the device were held together with three spring clamps.

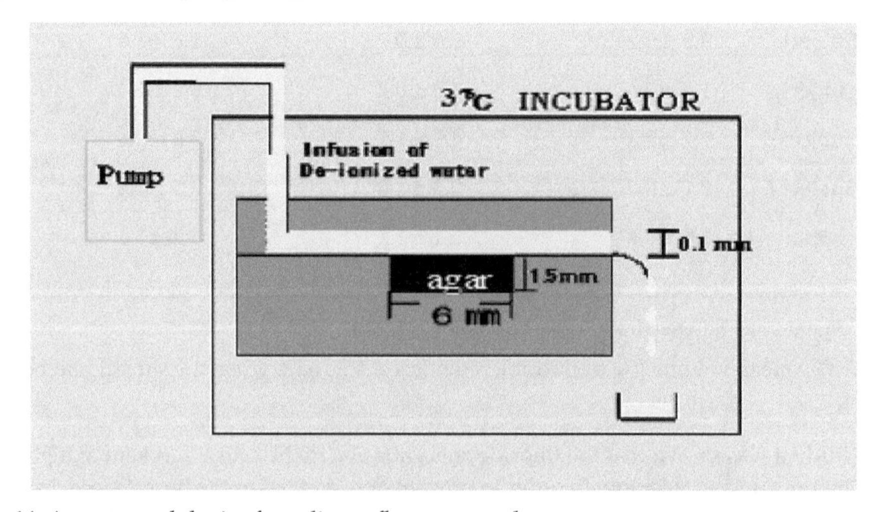

Fig. 14. An extraoral device for salivary flow rate study.

The device was maintained at 37 °C, and de-ionized water at the same temperature was infused with an infusion pump (Model 2000 IW, Harvard Apparatus Co., USA), the flow rate of which was adjustable over a wide range. The pump activated a 5-mL syringe connected to the device via polyethylene tubing.

For an experiment to be initiated, the flow rate of the pump was set to 1.07 ml/min. As soon as the water filled the tubing and completely covered the gel, a stop watch was started, and the flow rate of the pump was set to the desired value. After a pre-determined time, the pump was stopped, the two halves of the acrylic device were separated, and the agarose gel was removed with a needle and transferred to an appropriate volume of 100-ppm NaCl. The potassium concentration was determined by atomic absorption spectrophotometry.

For each flow rate, the experiment was repeated using five different gels for different durations to enable up to 70% of the KCl to be cleared from the gel. For each flow rate, a control gel that had not been exposed to water was used to determine the initial potassium concentration. The experiment was repeated three times for each flow rate. A least-squares straight line was fitted by computer to the potassium concentration plotted against the square root of time, and the half-time was calculated.

2.5.3 Results

There was a significant difference in the mean clearance half-time for the UAB between unstimulated (51.2 ± 19.3 min) and stimulated (40.1 ± 15.1 min) ($P < 0.05$) salivary flow rates. A significant difference was also found in the mean half-time for the UPB between unstimulated (19.2 ± 6.9 min) and stimulated (12.1 ± 5.2 min) ($P < 0.01$) salivary flow rates.

The results on the effect of velocity on the clearance half-time are shown in Table 4. With the flow rates set, the film velocity varied from 0.67 to 100 mm/min. The clearance half-times were inversely related to the velocity of fluid flow, and varied from 2.2 min to 58.3 min.

Fluid Flow Rate (ml/min)	Velocity of Fluid flow (mm/min)	Half-time (min) 6 mm in diam
0.06	100	2.2 ± 1.9
0.005	8.33	8.3 ± 1.2
0.003	5.00	15.1 ± 1.9
0.001	1.66	21.5 ± 2.8
0.0005	0.83	39.4 ± 4.2
0.0004	0.67	58.3 ± 6.1

Table 4. Effect of velocity of fluid flow on the mean half-time ± S.D. for clearance of KCL from an agarose gel, 1.5 mm in depth

Table 5 shows the in vivo clearance half-times for the UAB and the UPB as well as the estimated velocities of flow of the salivary film, as determined from the data in Table 1.

When salivary flow was unstimulated, the velocity of the salivary film of the UAB was estimated as 0.8 mm/min, whereas for the UPB it was estimated as 40.1 mm/min. When salivary flow was stimulated, the velocity of flow for the UAB was estimated as 2.3 mm/min, whereas for the UPB it was estimated as 12.1 mm/min.

Site	Salivary Flow	Clearance Half-time* (min)	Estimated Velocity of Salivary Film** (mm/min)
UAB	U	51.2 ± 19.3	0.8
	S	40.1 ± 15.1	2.3
UPB	U	19.2 ± 6.9	40.1
	S	12.1 ± 5.2	12.1

U = unstimulated (mean S.D = 0.42 ± 0.21 ml/min).
S = stimulated (mean S.D. = 4.5 ± 2.3 ml/min).
* = in vivo data
** = from the data for the half-time in Table 1 of this study.

Table 5. Estimated velocity of the salivary film at UAB and UPB sites in the mouth

2.5.4 Discussion

Extraoral device for estimating salivary velocity at both sites:

Lagerlöf and Dawes (1984) measured oral salivary volume immediately before and after the onset of swallowing and reported the mean volumes to be 1.07 and 0.77 ml, respectively. Collins and Dawes (1987) and Watanabe and Dawes (1990) measured the surface area of the mouth, and based on the oral salivary volumes reported by Lagerlöf and Dawes (1984)., they estimated the mean thickness of the salivary film in the mouth to be 0.1 mm. The extraoral device on which a 0.1-mm-thick salivary film flows on an agarose gel was designed on the basis of these reports to reproduce the situation in the mouth extraorally. Artificial saliva was allowed to flow onto an agarose gel in the same holder as that used in the mouth at different flow rates to determine the relationship between flow rate and clearance half-time, based on which salivary velocity at the two sites in the mouth was calculated. The velocity estimated with this device appears to be more useful for comparing salivary velocity between the different sites in the mouth than in determining actual salivary velocity. The mean half-time at stimulated salivary flow was 12.1 ± 5.2 min in the present study, which is substantially different from that obtained at unstimulated salivary flow. This may be attributable to the substantial difference in the secretion rate of saliva between when saliva is unstimulated and stimulated condition.

3. Conclusion

In clinical dentistry, it is generally accepted that the mandibular front teeth has low caries sensitivity, and the maxillary front teeth has high (e.g. nursing bottle caries) and the caries incidence of the buccal side of teeth in old person is higher than the lingual side. The results of this study have confirmed these clinical situations.

The author has also concluded that the velocity of salivary film is different by each site in the oral cavity and the slow velocity of the salivary film over the different surfaces of the teeth will retard clearance from diffusants of plaque such as acid. This suggests that the pH of each site in the oral cavity is slso different (Watanabe 2008),

4. Acknowledgment

The author thanks to emeritus proffesor C. Dawes (University of Manitoba) for helpful suggestions and comments.

5. References

Collins, L.M.C., & Dawes, C. (1987). The surface area of the adult human mouth and thickness of the salivary film covering the teeth and oral mucosa. *Journal of Dental Research,* Vol.66, No.8, pp. 1300 -1302, ISSN

Dawes, C. (1983). A mathematical model of salivary clearance of sugar from the oral cavity. *Caries Research,* Vol.17, No.4, pp. 321-334, ISSN 0008-6568

Dawes, C. & Watanabe, S., Biglow-Lecomte, P., & Dibdin, GH. (1989). Estimation of the velocity of the salivary film at some different locations in the mouth. *Joural of Dental Research,* Vol.68, No.11, pp.1479-1482, ISSN

Duckworth, RM., & Morgan, SN. (1991). Oral fluoride retention after use of fluoride dentifrices. *Caries Research,* Vol.25, No.2 , pp. 123-129, ISSN 0008-6568

Ekstrand, J. Lagerlöf, F. & Oliveby, A.(1986). Some aspects of the kinetics of fluoride in saliva; In: *Factors relating to demineralization and remineralisation of the teeth.* Leach, SA., & Edgar, WM. (Ed), ISBN 0-947946-73-X , pp 91-98, IRL Press, Oxford, UK

Foster, TD., & Hamilton, NC. (1969). Occlusion in the primary dentition. Study of children at 2.5 to 3 years of age. *British Dent Journal, Vol.*126, No.2, pp. 76-79, ISSN

Featherstone, JDB., Oreilly, MM., Shariati, M., & Brubler, S.(1986). Enhancement of remineralisation in vitro and in vivo, In *Factors relating to demineralization and remineralisation of the teeth.* Leach, SA., & Edgar, WM. (Ed), ISBN 0-947946-73-X , pp 23-34, IRL Press, Oxford, UK

Heath, K., Singh, V., Logan, R., & McIntyre. J.(2001). Analysis of fluoride levels retained intraorally or ingested following routine clinical applications of topical fluoride products. *Australian Dental Journal,* Vol.46, No. 1, pp.24-31, ISSN 00450421

Lagerlöf, F., & Dawes, C.(1984). The volume of saliva in the mouth before and after swallowing, *Journal of Dental Research.,* Vol.63, No.5 , pp. 618-621, ISSN 0022-0345

Lecomte, P., & Dawes, C.(1987).The influence of salivary flow rate on diffusion of potassium chloride from artificial plaque at different sites in the mouth. *Journal of Dental Research,* Vol.66, No.11, pp.1614-1618, ISSN

Pinkham, JR., Casamassimo, PS., Fields, HW., McTigur., DJ., & Nowak, AJ.(1988) *Infancy through adolescence.* Pediatric Dentistry, ISBN 0-7216-2106-6, Philadelphia, Saunders, USA

Primosh, RE., Weatherell, JA., & Strong, M.(1986). Distribution and retention of salivary fluoride from a sodium fluoride tablet following various intraoral dissolution methods. *Journal of Dental Research,* Vol.65, No.7, pp. 1001-1005, ISSN 0022-0345

Sass, R., & Dawes, C.(1997).The intra-oral distribution of unstimulated and chewing-gum-stimulated parotid saliva. *Archives Oral Biology*, Vol.42, No.7, pp. 469-474, ISSN 0003-9969

Suzuki, A., Watanabe, S., Ono, Y., Ohashi, H., Pai, C., Xing, X., & Wang, X. (2009). Influence of the location of the parotid duct orifice on oral clearance. *Archives of Oral Biology*, Vol.54, No.3, pp. 274-278, ISSN 0003-9969

Watanabe, S. & Dawes, C.(1992) Salivary flow rate and salivary film thickness in five-year-oldchildren. *Journal of Dental Research.*, Vol.69, No.5, pp. 1150-1153 0003-9969

Watanabe, S. (1992). Salivary clearance from different regions of the mouth in children. *Caries Research*, Vol.26, No.6, pp. 423-427 ISSN 0008-6568

Watanabe, S., Ogihara, T., Takahashi, S. 、Watanabe, K., Xuan, K., & Suzuki, S.(2010). Salivary clearance and pH in the different regions. IADR *General Session. http://iadr.confex.com/iadr/2010barce/webprogram/Paper132884.html*, (accessed 2012-05-08)

Weatherell, JA., Strong, M., Robinson, C., & Ralph, JP. (1986). Fluoride distribution in the mouth after fluoride rinsing. *Caries Research*, Vol.20, No.2, pp. 111-119, ISSN 0008-6568

3

Analysis of Environmental Pollutants by Atomic Absorption Spectrophotometry

Cynthia Ibeto[1], Chukwuma Okoye[2], Akuzuo Ofoefule[1] and Eunice Uzodinma[1]
[1]*Biomass Unit, National Center for Energy Research & Development, University of Nigeria, Nsukka, Enugu State,*
[2]*Department of Pure and Industrial Chemistry, Faculty of Physical Sciences, University of Nigeria, Nsukka, Enugu State, Nigeria*

1. Introduction

Environmental pollution as a result of man's increasing activities such as burning of fossil fuels and automobile exhaust emission has increased considerably in the past century due mainly to significant increases in economic activities and industrialization. Burning of fossil fuels and petroleum industry activities have been identified as primary sources of atmospheric metallic burden leading to environmental pollution. Several studies have shown that heavy metals such as lead, cadmium, nickel, manganese and chromium amongst others are responsible for certain diseases (Hughes, 1996). In general, heavy metals are systemic toxins with specific neurotoxic, nephrotoxic, fetotoxic and teratogenic effects. Heavy metals can directly influence behavior by impairing mental and neurological function, influencing neurotransmitter production and utilization, and altering numerous metabolic body processes. Systems in which toxic metal elements can induce impairment and dysfunction include the blood and cardiovascular, eliminative pathways (colon, liver, kidneys, skin), endocrine (hormonal), energy production pathways, enzymatic, gastrointestinal, immune, nervous (central and peripheral), reproductive and urinary that have lethal effects on man and animals. These diseases include abdominal pain, chronic bronchitis, kidney disease, pulmonary edema (accumulation of fluid in the lungs), cancer of the lung and nasal sinus ulcers, convulsions, liver damage and even death (Hughes, 1996).

Heavy metals get into the environment: water, soil, air and land through activities like intense agriculture, power generation, industrial discharges, seepage of municipal landfills, septic tank effluents e.t.c. Many authors have reported high levels of heavy metal ions in the soil, rivers and groundwater in different areas of Nigeria (Ibeto & Okoye, 2010a). To save the environment from further deterioration and also maintain sound public health, a strategy can be effectively utilized which is the use of organic materials such as municipal solid waste, agricultural waste and industrial waste to produce biogas. Biogas is a suitable alternative fuel which burns with similar properties to natural gas. Unlike natural gas, it is clean and has no undesirable effects on the environment. It is a mixture of gases consisting of around 60 to 70% of methane produced by the process of anaerobic digestion in a digester. The effluent of this process is a residue rich in the essential inorganic elements needed for healthy plant growth known as bio fertilizer, which when applied to the soil enriches it with no detrimental effects on the environment. Many authors have also reported the utilization of various wastes found in the environment, ranging from animal wastes, plant wastes to leaf litters and food wastes (Ofoefule et al., 2010; Uzodinma et al., 2011). It is also recommended that other alternative fuels such as bioethanol, which are becoming increasingly important not only because of the diminishing petroleum reserves, but also because of the environmental consequences of exhaust gases from petroleum fueled engines be made available for use in Nigeria. Good quality biodiesel fuel which is derived from triglycerides has attracted considerable attention during the past decade as a renewable, biodegradable and non-toxic fuel producing less particulate matter, hydrocarbons, aromatics, carbon-monoxide and soot emissions when burnt in the engines. Its production, marketing and use should therefore be highly encouraged as is the case in Europe, America and some other parts of the world.

Several spectroscopic methods have been used to monitor the levels of heavy metals in man, fossil fuels and environment. They include; flame atomic absorption spectrometry (AAS), atomic emission spectroscopy (AES), graphite furnace atomic absorption spectrometry (GFAAS), inductively coupled plasma-atomic emission spectroscopy (ICP/AES), inductively coupled plasma mass spectrometry (ICP/MS), x-ray fluorescence spectroscopy (XRFS), isotope dilution mass spectrometry (IDMS), electrothermal atomic absorption spectrometry (ETAAS) e.t.c. Also other spectroscopic methods have been used for analysis of the quality composition of the alternative fuels such as biodiesel. These include Nuclear magnetic resonance spectroscopy (NMR), Near infrared spectroscopy (NIR), inductively coupled plasma optical emission spectrometry (ICP-OES) e.t.c.

2. Sources of heavy metal pollution of the environment

2.1 Lead

Lead is a common industrial metal that has become widespread in air, water, soil and food. It is a naturally occurring metal that has been used in many industrial activities and therefore many occupations may involve exposure to it such as auto-mechanic, painting, printing, welding e.t.c putting the workers at risk of potential high exposure. In the atmosphere, lead exists primarily in the form of $PbSO_4$ and $PbCO_3$. Lead in paints and automobile exhausts are still recognized for its toxicity (Hughes, 1996). Episodes of poisoning from occasional causes such as imperfectly glazed ceramics (Matte et al., 1994), the use of medicines which may contain as much as 60% lead available from Asian healers and cosmetic preparations, may affect any age group and cases may present as acute emergencies (Bayly et al., 1995). The main source of adult exposure is food, air inhalation accounts for 30% and water of 10% (John et al., 1991).

Some individuals and families may be exposed to additional lead in their homes. This is particularly true of older homes that contain lead based paint. In an attempt to reduce the amount of exposure due to deteriorating leaded paint, the paint is commonly removed from homes by burning, scraping or sanding. These activities have been found to result to at least temporarily, in higher levels of exposure for families residing in those homes. Special population at risk of high exposure to tetra ethyl lead produced by reacting chloroethane with a sodium-lead alloy, includes workers at hazardous sites and those involved in the manufacturing and dispensing of tetraethyl lead (Gerbeding, 2005a). The production process is illustrated with the equation below.

$$4\ NaPb + 4\ CH_3CH_2Cl \rightarrow (CH_3CH_2)_4Pb + 4\ NaCl + 3\ Pb$$

Individuals living near sites where lead was produced or sites where lead was disposed and also hazardous waste sites where lead has been detected in some environmental media also may be at risk for exposure (Hazdat, 2005).

2.2 Cadmium

The principal form of cadmium in air is cadmium oxide, although some cadmium salts, such as cadmium chloride, can enter the air, especially during incineration. Environmental discharge of cadmium due to the use of petroleum products, combustion of fossil fuels (petroleum and coal) and municipal refuge contribute to airborne cadmium pollution (De Rosa et al., 2003) and possibly introduce high concentrations of this potential reproductive toxicant into the environment. This may be particularly true for Nigeria where refuse are burnt without control. In addition, humans may be unwittingly exposed to cadmium via contaminated food or paper (Wu et al., 1995) cosmetics and herbal folk remedies (Lockitch, 1993). All these factors put Nigerian population at high risk of cadmium toxicity (Okoye, 1994).

The greatest potential for above average exposure of the general population to cadmium is from smoking which may double the exposure of a typical individual. Smokers with additional exposure are at highest risk (Elinder, 1985). Soil distribution of urban waste and sludges is also responsible for significant increase in cadmium content of most food crops (WHO, 1996). Persons who have cadmium-containing plumbing, consume contaminated drinking water or ingest grains or vegetables grown in soils treated with municipal sludge or phosphate fertilizer may have increased cadmium exposure (Elinder, 1985). Persons who consume large quantities of sun flower kernels can be exposed to higher levels of cadmium. Reeves & Vanderpool (1997) identified specific groups of men who were likely to consume sunflower kernels. The groups included: basket ball and soft ball players, delivery and long distance divers and line workers in sunflower kernel processing plants.

2.3 Nickel

A person may be exposed to nickel by breathing air, drinking water, or smoking tobacco containing nickel. Skin contact with soil, bath or shower water, or metals containing nickel, as well as metals plated with nickel can also result in exposure. Coins contain nickel. Some jewellery are plated with nickel or made from nickel alloys (Gerbeding, 2005b). Exposure of an unborn child to nickel is through the transfer of nickel from the mother's blood to fetal blood. Likewise, nursing infants are exposed to nickel through the mother to breast milk. However, the concentration of nickel in breast milk is either similar or less than the concentration of nickel in infant formulas and cow's milk. Children may also be exposed to nickel by eating soil. Normally, the exact form of nickel one is exposed to is not known. It could be in form of nickel sulphate, nickel oxide, nickel silicate, iron-nickel oxides, nickel subsulfide or metallic nickel (Gerbeding, 2005b).

Patients may be exposed to nickel in artificial body parts made from nickel-containing alloys which are used in patients in joint prostheses, sutures, clips, and screws for fractured bones. Corrosion of these implants may lead to elevated nickel levels in the surrounding tissue and to the release of nickel into extracellular fluid. Serum albumin solutions used for intravenous infusion fluids have been reported to contain as much as 222 µg nickel/L, but are very rarely encountered. Dialysis fluid has been reported to contain as much as 0.82 µg nickel/L. Studies of nickel in serum pre- and post-dialysis show between 0 and 33% increases in nickel concentrations in patients (IARC, 1990).

2.4 Manganese

Populations living in the vicinity of ferromanganese or iron and steel manufacturing facilities, coal-fired power plants, or hazardous waste sites are exposed to elevated manganese particulate matter in air or water, although this exposure is likely to be much lower than in the workplace (Koplan, 2000a). Manganese is eliminated from the body primarily through the bile. Interruption of the manufacture or flow of bile can impair the body's ability to clear manganese. Several studies have shown that adults and children as well as experimental animals with cholestatic liver disorders have increased manganese levels in their blood and brain and are at risk from potentially increased exposure to manganese due to their decreased homeostatic control of the compound (Devenyi et al., 1994). In addition to oral diets, people on partial and total parenteral nutrition may be exposed to increased amounts of manganese. Forbes & Forbes (1997) found that of 32 patients receiving home parenteral nutrition due to digestive problems, 31 had elevated serum manganese levels (0.5–2.4 mg/L compared to normal range of 0.275–0.825 mg/L).

In comparison to other groups within the general population, persons living close to high density traffic areas, automotive workers, and taxi drivers may be exposed to higher concentrations of manganese arising from the combustion of methylcyclopentadienyl manganese tricarbonyl (MMT). MMT is actually a fuel additive developed in the 1950s to increase the octane level of gasoline and thus improve the antiknock properties of the fuel. Farmers, people employed as pesticide sprayers, home gardeners, and those involved in the manufacture and distribution of maneb and mancozeb may also be exposed to higher concentrations of these pesticides than the general public. People who ingest fruits and vegetables that have been treated with these pesticides and that contain higher-than-usual residues of the compounds (due to incomplete washing or over-application) may be exposed to increased concentrations of the pesticides. It is possible that medical workers may be exposed to higher concentrations of mangafodipir than the general population, although exposure routes other than intravenous are not expected to pose a significant risk (Koplan, 2000a). Manganese in the environment is in the form of their oxides or carbonates e.g MnO_2, $MnCO_3$ e.t.c.

2.5 Chromium

Blue prints, primer paints, household chemicals and cleaners, cements, diesel engines utilizing anti-corrosive agents, upholstery dyes, leather tanning processes, welding fumes, battery, rubber, dye, candles, printers and matches are occupational and environmental sources of chromium (Koplan, 2000b). In addition to individuals who are occupationally exposed to chromium, there are several groups within the general population that have potentially high exposures (higher than background levels) to chromium. These populations include individuals living in proximity to sites where chromium was produced or sites where chromium was disposed. Persons using chromium picolinate as a dietary supplement will also be exposed to higher levels of chromium than those not ingesting this product (Anderson, 1998). People may also be exposed to higher levels of chromium if they use tobacco products, since tobacco contains chromium. Workers in industries that use chromium are one segment of the population that is especially at high risk to chromium exposure. Occupational exposure from chromate production, stainless steel welding, chromium plating, and ferrochrome and chrome pigment production is especially significant since the exposure from these industries is to chromium (VI) (EPA 1984a).

Persons using contaminated water for showering and bathing activities may also be exposed via inhalation to potentially high levels of chromium(VI) in airborne aerosols. Elevated levels of chromium in blood, serum, urine, and other tissues and organs have also been observed in patients with cobalt-chromium knee and hip arthroplasts (Koplan, 2000b). Chromium in the environment can exist in many forms e.g chromium trioxide, potassium dichromate, sodium dichromate, potassium chromate, sodium chromate or ammonium dichromate e.t.c.

3. Environmental pollution

3.1 Fossil fuels combustion

The major sources of heavy metal pollution in urban areas of Africa are anthropogenic while contamination from natural sources predominates in rural areas. Anthropogenic sources of pollution include those associated with fossil fuel i.e. the non-renewable energy resources of coal, petroleum or natural gas (or any fuel derived from them) combustion, mining and metal processing (Nriagu, 1996). Fossil fuel consumption in Nigeria has risen ten-fold in the last two decades and consumption by urban households accounts for a large percentage, a trend which is expected to continue in the future. In a survey on urban household energy use patterns in Nigeria with respect to fuel preferences, sources and reliability of energy supply, it was found that kerosene, fuel wood, charcoal and electricity are the major fuels for urban use in Nigeria. Dependence on biomass fuels is rapidly giving way to the use of fossil fuels. Pollution problems associated with incidents of oil spills around automobile repair workshop resulting in metal contamination have been the subjects of many reports (Onianwa et al., 2001). Lead, cadmium, nickel, manganese and chromium are associated with automobile related pollution. They are often used as minor additives to gasoline and various auto-lubrication and are released during combustion and spillage (Lytle et al., 1995).

It is estimated that 8.5 million kg of nickel are emitted into the atmosphere from natural sources such as windblown dust and vegetation each year. Five times that quantity is estimated to come from anthropogenic sources (Nriagu & Pacyna, 1988) and the burning of residual and fuel oil is responsible for 62% of anthropogenic emissions. Chromium is released into the atmosphere mainly by anthropogenic stationary point sources including industrial, commercial and residential fuel combustion via the combustion of natural gas, oil, and coal. It has been estimated that emissions from the metal industry ranged from 35% to 86% of the total chromium and emissions from fuel combustion ranged from 11% to 65% of the total chromium. The main sources of manganese release to the air are industrial emissions, combustion of fossil fuels and re-entrainment of manganese-containing soils (EPA, 1987). High concentration of cadmium is released by human activities such as mining smelting operations and fossil fuel combustion. Coal, wood and oil combustion can all contribute cadmium to the atmosphere. It has been suggested that coal and oil used in classical thermal power plants are responsible for 50% of the total cadmium emitted to the atmosphere (Thornton, 1992). Anthropogenic sources of lead also include the mining and smelting of ore, manufacture of lead-containing products, combustion of coal and oil most notably leaded gasoline that may still be used in some countries including Nigeria. It is important to note that land is the ultimate repository for lead, and lead released to air and water ultimately is deposited in soil or sediment. For example, lead released to the air from leaded gasoline or in stack gas from smelters and power plants will settle on soil, sediment, foliage or other surfaces (Gerbeding, 2005a).

Atmospheric lead emissions in Nigeria have been estimated to be 2800 metric tonnes per year with most (90%) derived from automobile tail pipe (Nriagu et al., 1997). Lead in the form of tetra-ethyl lead $Pb(C_2H_5)_4$ is the most common additive to petrol to raise its octane number. Upon combustion in the petrol engine, the organic lead is oxidized to lead oxide according to the following reaction:

$$2Pb\,(C_2H_5)_4 + 27O_2 \longrightarrow 2PbO + 16CO_2 + 20H_2O$$

The lead oxide (PbO) formed, reacts with the halogen carriers (the co-additives) to form particles of lead halides- $PbCl_2$, PbBrCl, $PbBr_2$- which escape into the air through the vehicle exhaust pipes. By this, about 80% of lead in petrol escapes through the exhaust pipe as particles while 15-30% of this amount is air borne. Human beings, animal and vegetation are the ultimate recipients of the particulate (Ademoroti, 1996).

The lead level in Nigeria's super grade petrol is in the range 210-520 mg/L (Ademoroti, 1986). Automobile exhausts are also believed to account for more than 80% of the air pollution in some urban centres in Nigeria. The highest level of lead occurs in super grade gasoline with a concentration range of 600 to 800 mg/L (with a mean of 70μg/mL) and aviation gas with a concentration of 915μg/mL (Shy, 1990), which is much higher than permissible levels in some other countries. The comparable maximum levels in United States and Britain (UK) are 200μg/mL and 500μg/mL, respectively (Osibanjo & Ajaiyi, 1989). Automobiles in Nigeria may still be using leaded gasoline. Many cars are poorly maintained and characteristically emit blue plumes of bad odour and unburnt hydrocarbons (Baumbach et al., 1995), implying that a higher percentage of the lead in gasoline is emitted to the atmosphere.

Gasoline sold in most African countries contains 0.5–0.8g/L lead. In urban and rural areas and near mining centers, average lead concentrations are up to 0.5–3.0μg/m^3 in the atmosphere and >1000μg/g in dust and soils (Nriagu et al., 1996). In Nigeria, the level of lead in petrol is estimated at 0.7g/L. The national consumption of petrol in the country is estimated at 20 million litres per day with about 150 people per car. It is therefore predicted that at least 15 tonnes of lead is emitted into the environment through combustion of fossil fuel (Agbo, 1997). The annual motor gasoline consumption in 2000 was 56 litres per person. In 2005, Nigerian National Petroleum Corporation (NNPC) recorded domestic consumption of Premium Motor Spirit (Petrol) as 9,572,014,330 litres, while 2,361,480,530,000 litres of Automotive Gas Oil (Diesel) were equally recorded. Therefore, an average car in Nigeria uses over 1800 liters of petrol per year. The number of cars in Nigeria is assumed to be 3.27 million and fuel consumption for an average Nigerian car is 9.0 km L^{-1} (Ajao & Anurigwo, 2002). In 2006, 9.13 billion Litres of gasoline were consumed in Nigeria. Presently, Nigeria's daily fuel consumption stands at 30 million litres per day (Energy ministry, 2008). As reported by Agbo, 1997, the level of lead in petrol used in Nigeria is estimated at 0.7g/L. This probably is still the case as the use of leaded petrol may still be obtainable in Nigeria even though several countries have banned it's use (United Nations, 2006). It is therefore predicted that if it is so, at least 21,000kg of lead is emitted into the environment through combustion of petrol. Infact, Bayford & Co Ltd brought leaded petrol back to the UK market in 2000, primarily to service the needs of the classic car owners. As the only government approved supplier of leaded fuel in the UK, following the decision of the major oil companies to withdraw the product, Bayford remains committed to do all it can to make leaded fuel as widely available as possible and presently supplies over 36 petrol stations in the UK and still advertising for more supplies (Jonathan, 2008).

3.2 Indiscriminate disposal of waste

Anthropogenic sources of environmental pollution include those associated with industrial effluents, solid waste disposal and fertilizers (Nriagu, 1996). Heavy metals may enter soil and aquatic environments via sewage sludge application, mine waste, industrial waste disposal, atmospheric deposition and application of fertilizers and pesticides (Adaikpoh, et al., 2005). In Nigeria, recent reports indicate that the major contaminants found in drinking water especially from wells are heavy metals. These heavy metals find their way into the soil and groundwater through activities like intense agriculture, power generation, industrial discharges, seepage of municipal landfills, septic tank effluents, to mention a few. Infact, many authors have reported high levels of heavy metal ions in the soil, rivers and groundwater in different areas of Nigeria (Okuo et al., 2007). Indiscriminate disposal of toxic wastes therefore poses a great threat to human health.

4. Highlighted spectroscopic methods for heavy metals determination

4.1 Lead

Several analytical methods are available to analyze the level of lead in biological samples like blood. The most common methods employed are flame atomic absorption spectrometry (AAS). GFAAS and Anode stripping voltametry (ASV) are the methods of choice for the analysis of lead. In order to produce reliable results, background correction, such as Zeeman background correction that minimizes the impact of the absorbance of molecular species, must be applied. Limits of detection for lead using AAS are on the order of µg/mL (ppm) for flame AAS measurements, while flameless AAS measurements can detect blood lead levels at about 1ng/mL (Flegal & Smith, 1995). Inductively coupled plasma mass spectrometry (ICP-MS) is also a very powerful tool for trace analysis of lead and other heavy metals. ICP/MS not only can detect very low concentrations of lead but can also identify and quantify the lead isotopes present. Other specialized methods for lead analysis are X-ray fluorescence spectroscopy (XRFS), neutron activation analysis (NAA), differential pulse anode stripping voltametry, and isotope dilution mass spectrometry (IDMS). The most reliable method for the determination of lead at low concentrations is IDMS but due to the technical expertise required and high cost of the equipment, this method is not commonly used (Gerbeding, 2005a).

The primary methods of analyzing for lead in environmental samples are AAS, GFAAS, ASV, ICP/AES and XRFS. Less commonly employed techniques include ICP/MS, gas chromatography/photoionization detector (GC/PID), isotope dilution mass spectrometry (IDMS), electron probe X-ray microanalysis (EPXMA) and laser microprobe mass analysis (LAMMA). Chromatography (GC, HPLC) in conjunction with ICP/MS can also permit the separation and quantification of organometallic and inorganic forms of lead. Various methods have been used to analyze for particulate lead in air. The primary methods, AAS, GFAAS, and ICP/AES are sensitive to levels in the low $\mu g/m^3$ range (0.1–20 $\mu g/m^3$). Chelation/extraction can also be used to recover lead from aqueous matrices. GC/AAS has been used to determine organic lead, present as various alkyl lead species, in water. XRFS has been shown to permit speciation of inorganic and organic forms of lead in soil for source elucidation (Gerbeding, 2005a).

4.2 Cadmium

The most common analytical procedures for measuring cadmium concentrations in biological samples use the methods of atomic absorption spectroscopy (AAS) and atomic emission spectroscopy (AES). Methods of AAS commonly used for cadmium measurement are flame atomic absorption spectroscopy (FAAS) and graphite furnace (or electrothermal) atomic absorption spectroscopy (GFAAS or ETAAS). A method for the direct determination of cadmium in solid biological matrices by slurry sampling ETAAS has been described (Taylor et al., 2000).

Analysis for cadmium in environmental samples is usually accomplished by AAS or AES techniques, with samples prepared by digestion with nitric acid. Since cadmium in air is usually associated with particulate matter, standard methods involve collection of air samples on glass fiber or membrane filters, acid extraction of the filters, and analysis by AAS. Electrothermal inductively coupled plasma mass spectrometry (ETV-ICP-MS) has also been used to analyze size classified atmospheric particles for cadmium. The accuracy of the analysis of cadmium in acid digested atmospheric samples, measured by ACSV, was evaluated and compared with graphite furnace atomic absorption spectrometry (GFAAS) and inductively coupled plasma mass spectrometry (ICP-MS) (Koplan, 1999). Sediment and soil samples have been analyzed for cadmium using the methods of laser-excited atomic fluorescence spectroscopy in a graphite furnace (LEAFS), GFAAS and ETAAS preparation of the samples is generally accomplished by treatment with HCl and HNO_3.

Electrothermal vaporization isotope dilution inductively coupled plasma mass spectrometry (ETV-ID-ICP-MS) has been utilized for the analysis of cadmium in fish samples. Radiochemical neutron activation analysis (RNAA), differential pulse anodic stripping voltametry (ASV) and the calorimetric dithizone method may also be employed. The AAS techniques appear to be most sensitive, with cadmium recoveries ranging from 94 to 109% (Koplan, 1999).

4.3 Nickel

Analytical methods used in the determination of nickel in biological materials are the same as those used for environmental samples. Nickel is normally present at very low levels in biological samples. Atomic absorption spectrometry (AAS) and inductively coupled plasma-atomic emission spectroscopy (ICP-AES), with or without preconcentration or separation steps, are the most common methods. These methods have been adopted in standard procedures by EPA and the International Union of Pure and Applied Chemistry. Direct aspiration into a flame and atomization in an electrically heated graphite furnace or carbon rod are the two variants of atomic absorption. The latter is sometimes referred to as electrothermal AAS (ETAAS). Typical detection limits for ETAAS are <0.4 µg/L, while the limit for flame AAS and ICP-AES is 3.0 µg/L (Todorovska et al., 2002). Good precision was obtained with flame AAS after preconcentration and separation, electrothermal AAS, and ICP-AES. Inductively coupled plasma mass spectrometry (ICP-MS) techniques have been used to quantify nickel in urine with detection sensitivities down to approximately 1 µg/L. Voltammetric techniques are becoming increasingly important for nickel determinations since such techniques have extraordinary sensitivity as well as good precision and accuracy. Direct measurement of nickel in urine in the presence of other trace metals (e.g., cadmium, cobalt, and lead) was demonstrated using adsorption differential pulse cathodic stripping voltammetry at a detection limit of 0.027 µg/L (Gerbeding, 2005b).

The most common methods used to detect nickel in environmental samples are AAS, either flame or graphite furnace, ICP-AES, or ICP-MS. Nickel can also be analyzed in ambient and marine water using stabilized temperature graphite furnace atomic absorption (STGFAA) detection techniques as described in EPA methods 1639 and 200.12 respectively, which give limits of detection for nickel concentrations ranging between 0.65 and 1.8 µg/L and recoveries of >92%. Two other EPA standard test methods, 200.10 and 200.13, also use preconcentration techniques in conjunction with ICP-MS or graphite furnace AAS detection techniques, respectively, for analysis of nickel in marine water. One method uses activated charcoal to preconcentrate nickel in natural waters, followed by elution with 20% nitric acid and analysis by inductively coupled plasma-optical emission spectrometry (ICP-OES). This method achieved a detection limit of 82 ng/L (Gerbeding, 2005b).

4.4 Manganese

Flame atomic absorption analysis is the most straightforward and widely used method for determining manganese. In this method, a solution containing manganese is introduced into a flame, and the concentration of manganese is determined from the intensity of the colour at 279.5 nm. Furnace atomic absorption analysis is often used for very low analyte levels and inductively coupled plasma atomic emission analysis is frequently employed for multianalyte analyses that include manganese. Simple methods for the direct determination of Mn in whole blood by ETAAS have been described. Methods for measuring manganese therefore include spectrophotometry, mass spectrometry, neutron activation analysis and X-ray fluorimetry (Koplan, 2000a).

Atomic absorption spectrometry has been the most widely used analytical technique to determine manganese levels in a broad range of foods, as well as other environmental and biological samples. Tinggi et al., (1997) carried out a wet digestion technique using a 12:2 (v/v) nitric:sulfuric acid mixture for their determination, and for food samples with low levels of manganese, they found that the more sensitive graphite furnace atomic absorption analysis was required. Because manganese is often found at very low levels in many foods, its measurement requires methods with similarly low detection limits; these researchers identified detection limits of 0.15 mg/kg (ppm) and 1.10 µg/kg (ppb) for flame and graphite furnace atomic absorption spectrometry respectively (Tinggi et al. 1997).

A number of analytical methods for quantifying MMT in gasoline have been described including simple determination of total elemental manganese by atomic absorption and gas chromatography followed by flame-ionization detection (FID). In a certain method, in which MMT is detected in gasoline by gas chromatography coupled with flame photometric detection (FPD); the chemiluminescence of manganese is measured to determine MMT levels in a method that uses simple, inexpensive, and commercially available instrumentation (Koplan, 2000a).

4.5 Chromium

Prior to 1978, numerous erroneous results were reported for the chromium level in urine using electrothermal atomic absorption spectrometry (ETAAS) because of the inability of conventional atomic absorption spectrometry systems to correct for the high nonspecific background absorption. The use of GC-MS and ETAAS to determine ^{53}Cr and total Cr in biological fluids in order to investigate the distribution of Cr in lactating women following oral administration of a stable ^{53}Cr tracer have been reported. The authors detected ^{53}Cr in blood within 2 h of administration. They noted, however, that blood Cr changes in response to oral administration were variable and they considered that blood Cr was not tightly regulated. Similarly, the reported serum and plasma chromium concentrations of normal subjects have varied more than 5,000-fold since the early 1950s (Taylor et al., 2000).

The four most frequently used methods for determining low levels of chromium in biological samples are neutron activation analysis (NAA), mass spectrometry (MS), graphite spark atomic emission spectrometry (AES), and graphite furnace atomic absorption spectrometry (GFAAS). Of these four methods, only the GFAAS is readily available in conventional laboratories, and this method is capable of determining chromium levels in biological samples when an appropriate background correction method is used. The three commonly used methods that have the best sensitivity for chromium detection in air are GFAAS, instrumental neutron activation analysis (INAA), and graphite spark atomic emission spectrometry. Measurements of low levels of chromium concentrations in water have been made by specialized methods, such as inductively coupled plasma mass spectrometry (ICP-MS), capillary column gas chromatography of chelated chromium with electron capture detection (ECD), and electrothermal vaporization inductively coupled plasma mass spectrometry (Koplan, 2000b).

4.6 Biofuels

There are different spectrophotometric techniques for analysis of contaminants in biofuels. Simultaneous detection of the absorption spectrum and refractive index ratio with a spectrophotometer for monitoring contaminants in bioethanol has been carried out by Kontturi et al., 2011. Inductively Coupled Plasma Atomic Emission Spectrometry and optical emission spectral analysis with inductively coupled plasma (ICP-OES) have also been used to analyze biodiesel samples for trace metals (ASTM, 2007; ECS, 2006). An ICP-MS instrument fitted with an octopole reaction system (ORS) was used to directly measure the inorganic contents of several biofuel materials. Following sample preparation by simple dilution in kerosene, the biofuels were analysed directly. The ORS effectively removed matrix- and plasma-based spectral interferences to enable measurement of all important analytes, including sulfur, at levels below those possible by ICP-OES. A range of commonly produced biofuels was analysed, and spike recovery and long-term stability data was acquired. Also, suitably configured ICP-MS has been shown to be a fast and very sensitive technique for the elemental analysis of biofuels (Woods & Fryer, 2007).

A flow system designed with solenoid micro-pumps is proposed for fast and greener spectrophotometric determination of free glycerol in biodiesel. Glycerol was extracted from samples without using organic solvents. The determination involves glycerol oxidation by periodate, yielding formaldehyde followed by formation of the colored (3,5-diacetil-1,4-dihidrolutidine) product upon reaction with acetylacetone. The coefficient of variation, sampling rate and detection limit were estimated as 1.5% (20.0 mg L^{-1} glycerol, $n = 10$), 34 h^{-1}, and 1.0 mg L^{-1} (99.7% confidence level), respectively. A linear response was observed from 5 to 50 mg L^{-1}, with reagent consumption estimated as 345 µg of KIO_4 and 15 mg of acetylacetone per determination. The procedure was successfully applied to the analysis of biodiesel samples and the results agreed with the batch reference method at the 95% confidence level (Sidnei & Fábio, 2010).

5. Review of heavy metals in the environment using atomic absorption spectrophotometry

The negative effect on air quality will be unavoidable, if solid wastes are incinerated under uncontrolled conditions or left to biologically decompose in open areas, because waste gas will be given off to the atmosphere. Besides, heavy metals and hazardous organic pathogens are disseminated with organic wastes. Effluents from point sources change the characteristics of the receiving environment and its suitability for marinating its living communities and their ecological structure. Some metals when discharged into natural waters at increased concentration in sewage, industrial effluent or from mining and refining operations can have severe toxicological effects on aquatic environment and humans. Nigeria has a population of over 120 million. Degradation of water quality is most severe in the four states that contain 80 percent of the nations industries; Lagos, Rivers, Kano and Kaduna States, with the highest level of emission of 8000 tones of hazardous waste per year from Lagos State (Alamu, 2005).

5.1 Heavy metals in soils

In a study of soil samples of refuse dumps in Awka (Anambra State, Nigeria) the lead level (2467mg/kg) exceeded the limits set by the US Environmental Protection Agency. This study suggests that the refuse dumps in Awka may increase the level of environmental heavy metals in Nigeria (Nduka et al., 2006). Concentrations of cadmium, chromium, manganese, nickel and lead were determined in surface sediments of the Lagos Lagoon, Nigeria. The results revealed largely anthropogenic heavy metal enrichment and implicated urban and industrial waste and runoff water transporting metals from land - derived wastes as the sources of the enrichment. Okoye (1991) also reported that urban and industrial wastes discharged into the Lagos lagoon have had a significant impact on the ecosystem following the relative enrichment in the Lagoon fish with lead.

Several attempts have been made to assess the impact of the use of fossil fuels on the environment. Results obtained from the study on heavy metals (chromium, lead, cadmium, and nickel) concentrations and oil pollution in Warri area revealed that the concentrations of the heavy metals considered were higher in the oil-spilled sites relative to the control sites. Similarly, when compared with the European Community standards, the concentration is said to be quite significant. The results indicate the contribution of the oil industry to heavy metals contamination in the Niger-Delta area of Nigeria and that the operations of the oil industry in this study area have not been sufficiently accompanied by adequate environmental protection. To safeguard agricultural land in the area and hence human health, there is an urgent need for government to address the incidence of oil spills in this area (Essoka et al., 2006).

Concentrations of lead, cadmium, nickel, chromium and manganese were determined to assess the impact of automobiles on heavy metal contamination of roadside soil. The lead levels in polluted sites varied from 70 to 280.5μgg^{-1} and it rapidly decreased with depth. Similarly, mean concentrations of cadmium, nickel, chromium, and manganese were significantly higher at polluted sites and followed a decreasing trend with increase in depth. Correlation coefficients between heavy metals and traffic density were positively significant except for nickel. Profile samples showed that lead, cadmium, manganese were largely concentrated in the top 5cm confirming airborne contamination (Ramakrishnaiah & Somashekar, 2002).

In a study of the effect of traffic density on heavy metal content of soil and vegetation along roadsides in Osun State Nigeria, the concentration of the heavy metals decreased with increasing soil depth and horizontal distance from the road. Metal contamination correlated positively with traffic volume. Concentrations of lead, cadmium and nickel along the low traffic density were lower than the high traffic density (Amusan et al., 2003). Reclamation of auto repair workshop areas for residential and agricultural purposes makes high the risk assessment of heavy metal contamination (Ayodele et al., 2007). The levels of lead, cadmium and nickel were determined in the roadside topsoil in Osogbo, Nigeria, with the view to determining the effect of traffic density and vehicular contribution to the soil heavy metal burden. The levels of the metals at the high density roads were significantly higher than the corresponding levels at the medium and low traffic density roads. The average levels of lead, cadmium, and nickel in all road locations at a distance of 5m from the roads were 68.74±34.82, 0.60±0.31 and 8.38±2.40mg/kg respectively. Lead and cadmium were of average levels of 92.07±21.25 and 0.76±0.35 mg/kg respectively at a distance of 5m from the road at high traffic density roads, while the levels of nickel averaged 9.65±2.61mg/kg respectively. There was a rapid decrease in the level of the metals with distance, with the metal levels at a distance of 50m from the road almost reaching the natural background levels of the metals at the control sites (Fakayode & Olu-Owolabi, 2003a). The levels of the metals were also determined at the four major motor parks and at the seven mechanic workshop settlements. The levels of the metals at the motor parks and mechanic workshops were far above the levels at the control sites. The levels of lead, cadmium and nickel at the motor parks were 519±73.0, 3.6±0.8, and 7.3±4.6 mg/kg respectively, with the levels of lead, cadmium and nickel at the mechanic workshops averaging 729.57±110.93, 4.59±1.01 and 30.21±9.40mg/kg respectively (Fakayode & Olu-Owolabi, 2003a).

5.2 Heavy metals in food

Heavy metals have been analyzed and found to be in considerable quantities in Food in Nigeria. In the assessment of heavy metal levels in fish species of Lagos Lagoon, lead levels in the fishes were beyond W.H.O. acceptable limit of 1 ppm with a concentration range of 10.81-152.42 ppm (Akan and Abiola, 2008). Also, 86% and 84% of the 50 beverages (canned and non-canned respectively) obtained in Nigeria failed to meet the US EPA criteria for acceptable lead and cadmium levels in consumer products. 79.3% of the non-canned beverages showed lead levels that exceeded the US EPA's maximum contaminant level (MCL) of 0.015 mg/dm^3, 100% of the canned beverages had lead levels that were greater than the MCL. The range of the lead in the canned beverages was 0.002–0.0073 and 0.001–0.092 mg/dm^3 for the non-canned beverages. The cadmium levels ranged from 0.003–0.081 mg/dm^3 for the canned and 0.006–0.071 mg/dm^3 for non-canned beverages. About 85.71% of the canned beverages had cadmium levels that exceeded the maximum contaminant level (MCL) of 0.005 mg/dm^3 set by US EPA while 82.7% non-canned beverages had cadmium levels exceeding the MCL (Maduabuchi et al., 2006). In addition, Fakayode and Olu-Owolabi (2003b), reported that concentrations of lead and cadmium 0.59 mg/kg and 0.07 mg/kg respectively in chicken eggs in Ibadan were comparatively greater than levels found in other countries e.g lead concentrations of 0.048 ppm and 0.489 ppm obtained in China and India respectively and cadmium concentrations of 0.01 ppm and 0.004 ppm obtained in Canada and finland respectively.

Some reported works have also shown that planted crops and vegetations along major roads where there was high traffic volume contained high levels of lead content due to automobile exhaust. For instance, cadmium levels (0.12±0.03 - 0.28±0.03ppm) and nickel levels (3.02±0.14 - 6.50±0.25ppm) of staple foods (yam, cassava, cocoyam and maize) from oil-producing areas of Rivers and Bayelsa States of Nigeria were higher than those of non-oil producing areas (Abakaliki). Because of this high trace metal level, the staple foods from oil-producing areas examined are likely to be the major source of exogenous contamination of these metals in the populace (Akaninwor et al., 2005).

The concentration of cadmium has been found to be higher in some Nigerian foods as compared to those of some other countries as shown in Table 1.

5.3 Heavy metals in water

Groundwater and soil samples from 16 locations near petrol stations (PS) and mechanic workshops (MW) around Calabar, Nigeria, were analyzed for heavy metals and hydrocarbons to determine their concentrations and assess the impact of the PS and MW on groundwater in the area. Results show that mean concentrations of cadmium, chromium, manganese, nickel, and lead in groundwater are higher than the maximum admissible concentration (Nganje et al., 2007).

Results from the evaluation of ground water quality characteristics near two waste sites in Ibadan and Lagos revealed that some of the ground-water quality constituents determined exceeded the World Health Organization (WHO) standards for drinking water irrespective of source of pollution. Some of the ground-water samples were poor in quality in terms of cadmium, chromium, lead and nickel recorded (Ikem et al., 2002). The levels of heavy metals (cadmium, chromium, nickel, and lead) were analysed in the River Ijana (Ekpa-

Warri, Nigeria). Generally, excessive levels of the parameters of pollution above W.H.O. standards recommended for surface waters were observed (Emoyan et al., 2005). The possible sources of these parameters of pollution are diverse: originating from anthropogenic/ natural and point sources. Coal contains diverse amounts of trace elements in their overall composition. Certain trace elements such as lead, cadmium and chromium if present in high amount could preclude the coal from being used in environmentally sensitive situations. Ekulu River is the largest body of inland waters in Enugu Urban, which is of considerable importance industrially, culturally, and in agriculture. Ekulu coal mine is located by the bank of the Ekulu river. The coal mine station discharges its effluents directly into River Ekulu. Enugu coal mine occurs in the area where River Ekulu takes its source. Metal concentrations were generally higher in the coal samples than in the sediments. The metals (manganese, chromium, cadmium, nickel, and lead) analysed for were present throughout the period monitored in both the sediment and coal samples with some variations. Mean concentrations of Mn (0.256-0.389mg/kg) and Cr (0.214-0.267mg/kg) were high relative to concentrations of Cd (0.036-0.043mg/kg), Ni (0.064-0.067mg/kg) and Pb (0.013-0.017mg/kg). The presence of toxic metals in the area is established, calling for the assessment of their impact on the health of human and aquatic lives around the area (Adaikpoh et al., 2005). Other industrial effluents also contribute to the level of the heavy metals such as lead in the environment as reported by (Ayodele et al., 1996).

Commodity	**Greece**	**Japan and China**	**Nigeria**	**European Countries**
Rice	0.006	0.070	0.060	0.010
Cereal – other	0.002	0.023	0.075	0.016
Roots and tubers	0.022	0.015	0.103	0.025
Soya bean	–	0.041	0.200	0.021
Pulses – other	0.004	0.019	0.140	0.019
Sugars and honey	–	0.003	0.015	0.004
Groundnuts – shelled	–	–	0.370	0.050
Oilseeds – other	–	0.021	0.100	0.119
Vegetable oils – other	0.002	0.001	0.127	0.002
Stimulants – other	–	0.017	0.160	0.006
Spices	–	0.005	0.191	0.055
Leafy vegetables	0.054	0.025	0.155	0.034
Vegetables – other	0.024	0.020	0.343	0.013
Fish and other seafood – other	0.034	0.035	0.207	0.014
Eggs	0.001	0.003	0.500	0.003
Fruits	0.009	0.006	0.067	0.004
Milks	0.001	0.004	0.006	0.001
Milk products	0.004	0.004	0.375	0.005
Poultry meat	0.013	0.005	0.110	0.002
Meats – other	0.027	0.006	0.083	0.006

Source: Moriyama et al., (2002)

Table 1. Average concentrations of cadmium in foods (mg/kg)

Several works have been done to access the impact of improper waste management on the environment. The elevated level of heavy metal in the Niger Delta aquatic environment as a result of industrial discharges from refining operations has been elaborated by Spiff & Horsfall, (2004). Therefore it can be said that there is unregulated discharge of untreated effluents into natural receptors by industries in Nigeria. Samples of industrial effluents from Sharada industrial area Kano Nigeria were assessed for heavy metals. The study showed that about 60% of the industries discharge effluents with heavy metal concentration higher than 0.30 mg/L. Lead and chromium ions were the most prevalent with values above the minimum tolerable limit. The presence of these metal ions could pose a serious public health hazard. It is therefore recommended that these effluents be adequately treated before discharge. Table 2 shows the nickel content in naturally occurring waters.

WATER TYPE	LOCATION	CONCENTRATION RANGE (μg/L)
River water	Poland	2-75
	Germany-Rhine	8.9-24
	USA	0-71
Lake water	Poland-Lakes of Wielkopolska National Park *	2-11
	Poland-Lakes of the Golanieckie stream	1-8
Underground water	Poland-Poznzn	0.5-20
	Poland-Pozan voivodship	1-30
	Poland-Szczecin	1-15
Drinking water	USA	0.5-7
	Poland-Pozcan	0-5

Source: Barałkiewicz and Siepak (1999)

Table 2. Content of nickel in naturally occurring waters

6. Reports of research works done on heavy metals analysis in Nigerian environment

6.1 Blood

6.1.1 Methodology

3 ml of blood were collected directly from the select population comprised of 60 children, 114 women (pregnant, nursing mothers, others) and 66 men. This was carried out by venous puncture by a qualified nurse under contamination controlled conditions using pyrogen-free sterile disposable syringes and placed into 5 ml capacity EDTA plastic bottles containing K_3EDTA as anticoagulant. Each sample (3 ml) was transferred into 100 ml conical flasks. The EDTA bottle was rinsed with a little nitric acid and transferred into 100ml conical flask. Perchloric acid and nitric acid which were of analytical grade was added in the ratio 1:3 as follows: 2 ml perchloric acid and 6 ml nitric acid. The conical flask was covered with an evaporating dish and the mixture digested at low temperature using a thermostated Bitinett hot plate until a clear solution was obtained. The digest was made up to 20 ml with deionized water in a 20 ml standard flask (Rahman et al., 2006). The sample solutions were then analyzed for lead, cadmium, nickel, manganese and chromium using a GBC atomic absorption spectrophotometer, model A6600 AVANTA PM.

6.1.2 Results

As shown in Table 3, lead was detected in 235 of the 240 samples (97.92 %), the concentration range was from 0.039-0.881 ppm. Cadmium was detected in 205(85.42 %) samples, the concentration range was from 0.007-0.293 ppm. Nickel was detected in 137(57.08 %) samples while in 103(42.92 %) of the samples, the concentration range was from 0.007-0.849 ppm. Manganese was detected in 203(84.58 %) samples, the concentration range was from 0.006-0.861 ppm. Chromium was detected in 113(47.08 %) samples, the concentration range was from 0.006-0.829 ppm. Comparing the concentrations obtained from this study with the WHO (1996) guideline for heavy metals in blood, all the detectable samples had concentrations higher than the permissible levels stipulated for all the heavy metals except for 5 that were within the range stipulated for manganese i.e 0.008–0.012 ppm and 24 that were within the stipulated range for lead i.e 0.05-0.15 ppm. Thus there is a clear indication of high concentrations of the heavy metals in the general population in Nigeria especially the Southeast (Ibeto & Okoye, 2009; 2010a; 2010b).

Group		**Heavy metal**				
		Lead	**Cadmium**	**Nickel**	**Manganese**	**Chromium**
Men	N	66	65	40	63	26
	Mean ± sd	0.394 ± 0.126	0.093 ± 0.048	0.122 ± 0.079	0.119 ± 0.075	0.305 ± 0.228
	Range	0.07 - 0.76	0.01 - 0.21	0.01 - 0.33	0.01- 0.40	0.01- 0.83
Pregnant women/ Nursing mothers	N	56	48	31	52	35
	Mean ± sd	0.288 ± 0.198	0.099 ± 0.064	0.096 ± 0.061	0.088 ± 0.040	0.201 ± 0.150
	Range	0.04 - 0.72	0.01 - 0.28	0.01 0.21	0.01 - 0.21	0.01 - 0.57
Other women	N	54	50	20	53	28
	Mean ± sd	0.328 ± 0.121	0.080 ± 0.046	0.096 ± 0.067	0.121 ± 0.059	0.214 ± 0.167
	Range	0.12 - 0.67	0.01 - 0.29	0.01 - 0.25	0.02 - 0.31	0.01 - 0.67
Children	N	59	42	46	35	24
	Mean ± sd	0.488 ± 0.153	0.088 ± 0.056	0.411 ± 0.240	0.091 ± 0.147	0.267 ± 0.228
	Range	0.12 - 0.88	0.01 - 0.23	0.01 - 0.85	0.01 - 0.86	0.01 - 0.68
WHO guideline		0.05-0.15	0.0003-0.0012	0.001-0.005	0.008-0.012	<0.005

Source: Ibeto & Okoye, 2009; 2010a; 2010b

Table 3. Concentrations (ppm) of heavy metals in blood of different categories of the urban population in Enugu State Nigeria

However certain measures can be taken to reduce the effects of these heavy metals in the body. All of the currently available methods to obviate the toxic effects of the heavy metals are mainly by chelation. The chelating agents bind to the heavy metals, enhance its excretion by facilitating their transfer from soft tissues to where it can be excreted. Some of the standard chelating agents currently in use are meso-2,3- dimercaptosuccinic acid for cadmium, triethylenetetramine and cyclam (1,4,8,11-tetraazacyclotetradecane) for nickel, and ethylenediamine tetraacetic acid for lead, manganese and chromium. Also, through specific dietary supplementation, for example, sufficient iron or calcium stores, as opposed to a deficiency in these or other minerals, may reduce the heavy metals absorption, and thus reduce potential toxicity (Koplan, 2000a).

6.2 Fruit juices

6.2.1 Methodology

100ml of each of ten different brands of fruit juice was measured into a 200ml conical flask and heated till the volume reduced to 10ml. Perchloric acid and nitric acid was then added in a ratio of 1:2 with perchloric acid being 6ml and the nitric acid 12ml. The solution was then digested at low heat until a clear solution was obtained. It was then allowed to cool and made up to 25ml with distilled water using a standard flask. Heavy metals were then determined by atomic absorption spectrophotometry using Alpha 4 Serial no 4200 with air acetylene flame.

6.2.2 Results

As shown in Table 4, all the samples except one of guava brands contained lower concentration of copper than the 5ppm permissible limit set for the metal. All samples had concentrations of zinc well below the 5ppm maximum permissible level. The iron concentrations were below the limit of 15ppm in all the samples except for the pineapple brand, which showed a concentration of 50ppm. This could be due to many reasons such as the fact that the fruit juice brand was acidic and the fruit acids could pick up the metal from the equipment during processing or storage. As minerals are soil and species dependent, the fruit acids might also have picked up iron and other metals from the soil during growth. Iron could also be added for fortification.

Cadmium was more wide spread, occurring in seven brands with a range of 0.16 to 0.38ppm. Lead occurred in four brands with range 0.11 to 0.33 ppm. Only the foreign made apple juice brand with the lead content of 0.33ppm exceeded the maximum permissible level of 0.3ppm by FAO/W.H.O. The limit for cadmium was not stipulated but compared with the limit set for lead (since they are both non-nutritive elements), the foreign made guava brand and the pineapple brand may be considered to be high in cadmium (Okoye & Ibeto, 2009).

6.3 Soil

6.3.1 Methodology

Soil samples were collected from twenty different locations in three Local Government Areas in Enugu State. Soil samples were collected in duplicates at a dept of 15-20cm and transferred into a pre-washed polyethylene nylon bag to avoid contamination. Soil samples were dried at 105°C and sieved with 100mesh (152µm BS Screen 410). The samples were prepared for analysis by cold extraction. 1g of the dried soil sample was weighed into a labeled 100ml conical flask and 20ml of mixture of conc. HCl and conc. HNO_3 (1:1) were added and well shaken. The solution was kept overnight after which it was filtered through a whatman No 1 filter paper formerly leached by pouring cupious quantity of dilute HNO_3 on the filter paper while in the funnel. The clear solution obtained was made up to 50ml using a standard flask and transferred into a plastic bottle (Okoye, 2001). The sample solutions were analysed at various wavelengths for each metal using Buck Scientific Atomic Absorption Spectrophotometer 205.

SAMPLES	MEAN CONCENTRATIONS (mg/l) OF PARAMETERS				
	Cr	Mn	Ni	Cd	Pb
Lime	0.06	0.11	0.08	<0.002	<0.004
Mango	<0.002	0.67	<0.05	0.17	<0.004
Orange	<0.002	0.19	0.13	<0.002	0.11
Guava	<0.002	0.42	0.13	0.27	<0.004
Guava*	<0.002	0.67	0.03	0.37	<0.004
Black Currant	0.03	0.24	<0.05	0.16	0.13
Mixed fruit	<0.002	0.42	<0.05	<0.002	<0.004
Apple	<0.002	0.19	<0.05	0.26	0.33
Apple	<0.002	0.14	0.03	0.25	<0.004
Pine-apple	0.09	6.96	0.15	0.38	0.20

Source: Okoye & Ibeto, 2009

Table 4. Concentrations (ppm) of metals in fruit juice samples

6.3.2 Results

The ranges of concentrations were: Pb(30.3-235), Cr(9.0-15.5) and Cd(5.5-42.25) ppm in Igbo-Eze North. Pb (0.2-100), Cr(9.5-10.8) and Cd(0.51-44.8) ppm in Nsukka and Pb(14.8-165) and Cd(0.43-5.0) ppm in Udi. The order of abundance in the soil follow the order Pb>Cd> Cr. Compared with the work done in an automobile spare parts market the values for chromium and cadmium were relatively high. Compared with the Indian standard for heavy metals in soils, some of the samples exceeded the stipulated range of 3-6ppm for cadmium, indicating considerable cadmium contamination of some of the sampling points. However, the variations in the mean concentration of each metal in the three Local Government Areas were not significant (P>0.05) (Okoye & Ibeto, 2008).

6.4 Water

6.4.1 Methodology

Samples were collected from 17 different locations in Southeast Nigeria at various occasions covering the dry and wet seasons. In collecting samples from rivers, lakes and streams, the polyethylene sampling containers were dipped just below the surface to minimize the contamination of the water sample by surface films. For borehole samples, the mouth of the tap was cleaned with cotton wool and was left to run to waste for several minutes before collection while for spring water, samples were collected at different outlets of the spring. All the samples were collected with 2 L polyethylene cans which were leached with a 1:1 HCl and water and rinsed with distilled-de-ionized water (Okoye et al., 2010).

The samples were concentrated by evaporating 500 mL of water sample to about 100 mL followed by addition of 1 mL conc HCl and digesting until volume was about 15-20 mL. This was later made up to mark with distilled-de-ionized water in a 25 mL standard flask and later transferred into an acid-leached polyethylene bottle prior to analysis. Trace metals were determined with AAS (ALPHA Series 4200 CHEM TECH ANALYTICAL Ltd, UK) equipped with air-acetylene flame.

6.4.2 Results

The metal analysis gave values (mg/L) with ranges as follows: Pb (nd-13.5); Cd (nd-0.60); Ni (nd-0.075) and Cr (nd-0.10). Less than 40% had high levels of lead and cadmium which are indicative of the impact of indiscriminate discharge of untreated industrial effluents, domestic waste and inputs from other human activities on the pollution of the environment by trace metals. Concentrations of lead and cadmium in five locations were higher than the WHO limits of 0.01 mg/L and 0.003 mg/L respectively. Water containing high levels of lead and cadmium is not fit for drinking purposes. This study has created awareness concerning the risk of drinking from the identified water sources which have high concentrations of lead and cadmium (Okoye et al., 2010).

6.5 Chicken

6.5.1 Methodology

The samples of the liver, gizzards, muscles of chickens and also their feed were prepared by wet digestion. 10ml of nitric acid and 5ml of perchloric acid were added to 1g of each finely ground sample into different 100ml conical flasks covered with watch glasses for overnight predigestion. It was then heated on a hot plate until a clear solution was obtained. The contents were cooled and transferred to a 25ml standard flask and made up to mark with deionised water. These were then transferred to sample bottles until clear solutions were obtained. Each digested sample was transferred to prewashed sample bottles (Ibeto & Okoye, 2010a).

The sample solutions were then analyzed for the heavy metals: lead, cadmium, copper and zinc at required wavelength using a GBC atomic absorption spectrophotometer, model no A6600 AVANTA PM.

6.5.2 Results

The concentrations in ug/g of the heavy metals were in the range of 1.78 - 15.32, 9.7 - 147.07, 15.82 - 47.79 and 0.03 - 2.29 for cadmium, lead, copper and zinc respectively. Concentrations of cadmium were higher than the permissible limit of 0.5 ppm set by FAO/WHO and concentrations of lead were above the permissible limit of 1 ppm set by Australia New Zealand Food Authority. The high concentrations of the toxic metals obtained show a certain level of pollution of the environment. However, the low concentration of the essential metals in the feed shows there was no addition of nutritive supplements to the feed (Okoye et al., 2011).

7. Conclusion

Atomic absorption spectrophotometry was used to determine the heavy metal content of various samples from the environment and also human blood. The heavy metal content of the environmental samples indicated a certain level of heavy metal pollution in the Nigerian environment which can be attributed to fossil fuels combustion and indiscriminate disposal of wastes. This is also reflected in the level of heavy metals in the blood of the select population which on accumulation in the human system has led to low level of life expectancy globally. It is therefore recommended that utilization of alternative fuels be aggressively pursued and integrated into the energy mix of countries globally. These fuels include biogas, biodiesel and bioethanol and are becoming increasingly important not only because of the diminishing petroleum reserves but also because of the environmental consequences of exhaust gases from petroleum fuelled engines.

8. References

Adaikpoh, E. O., Nwajei, G.E. & Iogala, J. E. (2005). Heavy metals concentrations in coal and sediments from River Ekulu in Enugu, Coal City of Nigeria. J. Appl. Sci. Environ. Mgt. 9 (3) 5 - 8.

Ademoroti, C.M.A. (1986). Levels of heavy metals on bark and fruit of trees in Benin City, Nigeria. Environmental pollution. 11: 241-243.

Ademoroti, C.M.A. (1996). Environmental Chemistry and toxicology. March prints and Consultancy. Foludex Press Ltd. Ibadan. pp 177-195.

Agbo, S. (1997). Effects of lead poisoning in children. In: Proceeding at a workshop on vehicular emission and lead poisoning in Nigeria. Friends of the environment. pp 20-28.

Ajao, E.A. & Anurigwo, S. (2002). Land-based sources of pollution in the Niger Delta, Nigeria. AMBIO: Journal of the Human Environment. 31, (5) 442–445.

Akan, B.W. & Abiola, R.K. (2008): Assesment of trace metal levels in fish species of Lagos Lagoon. Conference Proceedings of Chemical Society of Nigeria. 31st Annual International Conference and Exhibition, Warri. 22nd-26th 2008. Delta State Nigeria. pp 394-399.

Akaninwor, J. O., Onyeike, E. N. & Ifemeje, J.C. (2005). Trace metal levels in raw and heat processed Nigerian staple foods from oil-producing areas of Rivers and Bayelsa States. Journal of Applied Sciences and Environmental Management. Vol. 10, No. 2, pp. 23-27.

Alamu, O. (2005). Watershed management to meet water quality standards and emerging TMDL (Total maximum daily load). Proceedings of the Third Conference 5-9 March 2005 (Atlanta, Georgia USA). American Society of Agricultural and Biological Engineers, St. Joseph, Michigan. www.asabe.org, 701P0105.

Amusan, A.A., Bada, S.B. & Salami, A.T. (2003). Effect of traffic density on heavy metal content of soil and vegetation along roadsides in Osun State, Nigeria. West African Journal of Applied Ecology. 4: 107-114.

Anderson, R.A. (1998). Effects of chromium on body composition and weight loss. Nutr. Rev. 56(9):266- 270.

ASTM (2007). International Standard Test Method for Determination of Additive Elements in Lubricating Oils by Inductively Coupled Plasma Atomic Emission Spectrometry. Annual Book of ASTM Standards, 2007, Vol. *05.03*, ASTM D4951-02.

Ayodele, J.T., Momoh, R.U. & Amm, M. (1996). Determination of heavy metals in Sharada Industrial effluents, in water quality monitoring and environmental status in Nigeria. Proceedings of the National Seminar on Water Quality Monitoring and Status in Nigeria, organized by Federal Environmental Protection Agency and National Water Resources Institute. October 16-18. pp 158-166.

Ayodele, R.I., Dawodu, M. & Akande, Y. (2007). Heavy metal contamination of topsoil and dispersion in the vicinities of reclaimed auto repair workshops in Iwo, Nigeria. Research Journal of Applied Sciences. 2(11): 1106-1115.

Barałkiewicz, D. & Siepak, J. (1999). Chromium, nickel and cobalt in environmental samples and existing legal norms. Polish Journal of Environmental Studies. Vol. 8, No. 4: 201-208.

Baumbach, G.U., Vogt, K.R.G., Hein, A.F., Oluwole, O.J., Ogunsola, H.B. & Akeredolu, F.A. (1995). Air pollution in large tropical city with high traffic density: results of measurements in Lagos, Nigeria. Sci. Total Environ. 169: 25-31.

Bayly, G.R., Braithwaite, R.A., Sheehan, T.M.T., Dyer, N.H., Grimley, C. & Ferner, R.E. (1995). Lead poisoning from Asian traditional remedies in the West Midlands-report of a series of five cases. Hum. Experiment Toxicol. 14: 24-28.

De Rosa, M., Zarrilli, S., Paesano, L., Carbone, U., Boggia, B., Petretta, M., Masto, A., Cimmino, F., Puca, G., Colao, A. & Lombardi, G. (2003). Traffic pollutants affect infertility in men. Human Reproduction. 18: 1055-1061.

Devenyi, A.G., Barron, T.F. & Mamourian, A.C. (1994). Dystonia, hyperintense basal ganglia, and whole blood manganese levels in Alagille's syndrome. Gastroenterology. 106:1068-1071.

EPA. 1984a. Health assessment document for chromium. Research Triangle Park, NC: Environmental Assessment and Criteria Office, U.S. Environmental Protection Agency. EPA 600/8-83-014F.

E.P.A. (1987). Toxic air pollutant/source crosswalk: A screening tool for locating possible sources emitting toxic air pollutants. Research Triangle Park, NC: U.S. Environmental Protection Agency, Office of Air Quality Planning and Standards. EPA-450/4-87-023a.

Elinder, C.G. (1985). Cadmium: uses, occurrence and intake. In: Friberg, L., Elinder, C.G., Kjellstrom, P. et al eds. Cadmium and health: A toxicological and epidermiological appraisal. Vol 1. Exposure, dose and metabolism. Effects and response. Boca Raton, FL. CRS. Press. pp 23-64.

Emoyan, O.O., Ogban, F.E. & Akarah, E. (2005). Evaluation of heavy metals loading in River Ijana in Ekpa-Warri, Nigeria. J. Applied Sciences and environmental management. Vol 10, No 2. pp 121-127.

Energy Ministry (2008). Business summary 21st to 28th 2008. Accessed from www.bpeng,org/NR/rdonlyres on 15/1/09.

Essoka, P.A., Ubogu, A.E. & Uzu, L. (2006). An overview of oil pollution and heavy metal concentration in Warri area, Nigeria. Management of environmental quality. Vol 10(2) 209-215.

European Committee for Standardization (ECS) (2006). Fat and oil derivatives -Fatty acid methyl ester (FAME) -Determination of Ca, K, Mg and Na content by optical emission spectral analysis with inductively coupled plasma (ICP-OES). EN 14538, 2006.

Fakayode, S. O. & Olu-Owolabi, B. I. (2003a). Heavy metal contamination of roadside topsoil in Osogbo, Nigeria: its relationship to traffic density and proximity to highways. Environmental Geology. Volume 44, (2) 150-157.

Fakayode, S. O. & Olu-Owolabi, B. I. (2003b). Trace metal content and estimated daily human intake from chicken eggs in Ibadan, Nigeria. Archives of Environmental Health. http:www.encyclopedia.com/beta/doc/IGI-111732614.

Flegal, A.R. & Smith, D.R. (1995). Measurements of environmental lead contamination and human exposure. Rev Environ Contam Toxicol. 143:1-45.

Forbes, G.M. & Forbes, A. (1997). Micronutrient status in patients receiving home parenteral nutrition. Nutrition. 13:941-944.

Gerbeding, J.L. (2005a). Toxicological profile for lead. Public health service, Agency for toxic substances and diseases. Atlanta Georgia. pp 3-5, 31, 113-130, 224-228 and 312-350.

Gerbeding, J.L. (2005b). Toxicological profile for nickel. Public health service. Agency for toxic substances and diseases. Atlanta Georgia. pp 27, 79, 134-144, 166-167.

Hazdat (2005). Hazdat data base. ASTDR's. Hazardous substance release and health effect data base. Atlanta. Agency for toxic substance and disease registry. www.astsdr.cdc.gov/hazdat-html. April 13, 2005.

Hughes, W.W. (1996). Essentials of environmental toxicology. The effects of environmental hazardous substances on human health. Loma, Lind Califonia. Tay and Francais Publishers. pp 3, 87-95.

IARC. (1990). IARC (International Agency for Research on Cancer) monographs on the evaluation of carcinogenic risks to humans. Chromium, nickel and welding. Lyon, France: International Agency for Research on Cancer. World Health Organization. Vol 49: 257-445.

Ibeto C. N. & Okoye C. O. B. (2009). Elevated Cadmium Levels in Blood of the Urban Population in Enugu State Nigeria. World Applied Sciences Journal 7 (10): 1255-1262, 2009. ISSN 1818-4952.

Ibeto, C.N. & Okoye, C.O.B. (2010a). High levels of heavy metals in blood of the urban population in Nigeria. Research Journal of Environmental Sciences. ISSN 1819-3412. 4 (4): 371-382.

Ibeto C. N. & Okoye C. O. B. (2010b). Elevated Levels of Lead in Blood of Different Groups in the Urban Population of Enugu State, Nigeria. International Journal of Human and Ecological Risk Assessment. Human and Ecological Risk Assessment: An International Journal, 16: 5, 1133 – 1144. DOI: 10.1080/10807039.2010.512257. http://dx.doi.org/10.1080/10807039.2010.512257.

Ikem, A., Osibanjo, O., Sridhar, M. K. C. & Sobande, A. (2002). Evaluation of groundwater quality characteristics near two waste sites in Ibadan and Lagos, Nigeria. Water, air and soil pollution. Vol 140, Nos 1-4. pp 307-333.

John, H., Cheryl, H., Richerd, S. & Christine, S. (1991). Toxics A-Z- A guide to everyday pollution hazards. University of Califonia, Press. Berkley. Angeles. Oxford. pp 47-104.

Jonathan, T. (2008). Leaded fuels update. Press information - Thursday 15th May 2008. Accessed from www.bayfordgroup.co.uk and www.leadedpetrol.co.uk. on the 24th of January 2008.

Kontturi, V., Hyvärinen, S., García, A., Carmona, R., Yu Murzin, D., Mikkola, J.P. & Peiponen, K.E. (2011). Simultaneous detection of the absorption spectrum and refractive index ratio with a spectrophotometer: monitoring contaminants in bioethanol. Measurement Science and Technology Vol. 22, No. 5, 055803 doi: 10.1088/0957-0233/22/5/055803.

Koplan, J.H. (2000a). Toxicological profile for manganese. Public health service. Agency for toxic substances and disease registry. Atlanta Georgia. pp 21-50, 175-207 and 295-400.

Koplan, J.H. (2000b). Toxicological profile for chromium. Public health service. Agency for toxic substances and disease registry. Atlanta Georgia. pp 1-9, 16-50, 122-157 and 301-315.

Koplan, J.P. (1999). Toxicological profile for cadmium. Public health service. Agency for Toxic Substance and Disease Registry (ATSDR). Atlanta Georgia. pp 126-140, 207 and 260- 270.

Lockitch, G. (1993). Prospective on lead toxicity. Clin Biochem. 26:371-81.

Lytle, C.M., Smith, B.N. & Mckinwu, C.Z. (1995). Manganese accumulation along the Utah roadways. A possible indication of motor exhaust pollution. Sci Total Environ. 162: 1056-109.

Maduabuchi, J.M.U., Nzegwu, C.N., Adigba, E.O., Aloke, R.U., Ezomike, C.N. Okocha, C.E., Obi, E. & Orisakwe, O.E. (2006): Lead and cadmium exposures from canned and non-canned beverages in Nigeria: A public health concern. Science of the Total Environment. Vol 366, Issues 2-3, pp 621-626.

Matte, T.D., Proops, D., Palazeulos, E., Graef, J. & Avila, H.A. (1994). Acute high dose lead exposure from beverage contaminated from traditional Mexican pottery. Lancet. 344: 1064-1065.

Moriyama, T., Taguchi, Y., Watanabe, H. & Joh, T. (2002). Changes in the cadmium content of wheat during the milling process (in Japanese). In: Report on Risk Evaluation of Cadmium in Food, Research on Environmental Health, Health Sciences Research Program, Ministry of Health, Labour and Welfare, pp. 153–160.

Nduka, J.K.C., Orisakwe, O.E., Ezenweke, L.O., Abiakam, C.A., Nwanguma, C.K. & Maduabuchi, U.J.M. (2006). Metal contamination and infiltration into the soil at refuse dump sites in Awka, Nigeria. Archives of Environmental and Occupational Health. Vol. 61, No 5, 197 - 204.

Nganje, T. N. Edet, A. E. & Ekwere, S. J. (2007). Concentrations of heavy metals and hydrocarbons in groundwater near petrol stations and mechanic workshops in Calabar metropolis, southeastern Nigeria. *Environmental Geosciences Vol.* 14; no. 1; p. 15-29; DOI:10.1306/eg.08230505005.

Nriagu, J.O. (1996): History of global metal pollution. Sci. 272:223-224.

Nriagu, J. O., Blankson, M. L. & Ocran, K. (1996). Childhood lead poisoning in Africa: a growing public health problem. Journal of Science of the Total Environment. 181(2, 15):93-100.

Nriagu, J.O. & Pacyna, J.M. (1988). Quantitative assessment of worldwide contamination of air, water and soils by trace metals. Nature. 333:134-139.

Ofoefule, A.U., Uzodinma, E.O. & Anyanwu C.N. (2010). Studies on the effect of Anaerobic digestion on the microbial flora of animal wastes 2: Digestion and modelling of process parameters. Trends in Appl. Sci. Res. 5(1) 39-47.

Okoye, C.O.B. (1991). Heavy metals and organisms in the Lagos lagoon. Inter. J. Environmental studies. Vol. 37, pp 285-292.

Okoye, C.O.B. (1994). Lead and other metals in dried fish from Nigerian markets. Bull. Environ. Contain. Toxicol. 52: 825 - 832.

Okoye, C.O.B. (2001). Trace metal concentrations in Nigerian fruits and vegetables. Intern. J. Environ. Studies. Vol. 58 pp 501-509.

Okoye C.O.B. & Ibeto C.N. (2008). Determination of Bioavailable Metals in Soils of Three Local Government Areas in Enugu State, Nigeria. Proceedings of the 31st Annual International Conference and Exhibition, Delta Chem, 2008. Petroleum Training Institute (PTI), Conference Centre Complex, Effurun-Warri, Delta State, Nigeria, pp 767-771.

Okoye C.O.B. & Ibeto C.N. (2009). Analysis of different brands of fruit juice with emphasis on their sugar and trace metal content. Bioresearch Journal. 7 (2): 493-495.

Okoye, C. O. B., Aneke A.U., Ibeto, C. N. & Ihedioha, J. N. (2011). Heavy Metals Analysis of Local and Exotic Poultry Meat. International Journal of Applied Environmental Sciences. ISSN 0973-6077 Vol. 6, No 1. pp. 49-55.

Okoye, C.O.B., Ugwu, J.N. & Ibeto, C.N. (2010). Characterization of rural water resources for potable water supply in some parts of South-eastern Nigeria. J. Chem. Soc. Nig. Vol. 35. No.1. pp 83-88.

Okuo, J.M., Okonji, E.I. & Omeyerere, F.R. (2007). Hydrophysico-chemical assessment of the Warri coastal aquifer, Southern Nigeria. J. Chem Soc. Nig. Vol 32, No 2. 53-64.

Onianwa, P. C.; Jaiyeola, O. M. & Egekenze, R. N. (2001). Heavy metals contamination of topsoils in the vicinities of auto-repair workshops, gas stations and motor parks in a Nigeria city. Toxicol. and Environ. Chem. 84(1-4), 33 -39.

Osibanjo, O. & Ajayi, S.O. (1989). Trace metal analysis of petroleum products by flame atomic absorption spectrophotometry. Nigeria Journal of Nutritional Health. 4: 33-40.

Rahman, S., Khalid, N. Zaidi, J.H., Ahmad, S. & Iqbal, M. Z. (2006): Non occupational lead exposure and hypertension in Pakistani adults. J.Zhepang University Science B. 9: 732-737.

Ramakrishnaiah, H. & Somashekar, R.K. (2002). Heavy metal contamination in roadside soil and their mobility in relations to pH and organic Carbon. Soil and sediment contamination: An International Journal, Vol.11, Issue 5, pp 643 - 654.

Reeves, P.G. & Vanderpool, R.A. (1997). Cadmium burden in men and women who report regular consumption of confectionery sunflower kernels containing a natural abundance of cadmium. Environ. Health. Perspect. 105 (10), 98-104.

Shy, C. M. (1990). Lead in petrol. The mistake of the 20th century. World health statistics, Quaterly. 43: 168-176.

Sidnei, G. S. & Fábio, P.R. (2010). A flow injection procedure based on solenoid micro-pumps for spectrophotometric determination of free glycerol in biodiesel. Talanta. Volume 83, Issue 2, 15 December 2010, Pages 559-564. doi:10.1016/j.talanta.2010.09.061.

Spiff, A. I. & Horsfall. M. Jnr. (2004). Trace metal concentrations in inter-tidal flate sediments of the upper new Calabar River in the Niger Delta area of Nigeria. Scientia African. Vol 3. 19-28.

Taylor, A., Branch, S. Halls, D.J., Owen, L.M.W. & White, M. (2000). Atomic Spectrometry update: Clinical and biological material, food and beverages. J. Anal. At. Spectrom. 15, 451-487.

Thornton, I. (1992). Sources and pathways of cadmium in the environment. IARC Sci Publ. 118:149-162.

Tinggi, U., Reilly, C., & Patterson, C. (1997). Determination of manganese and chromium in food by atomic absorption spectromety after wet digestion. Food Chem 60:123-128.

Todorovska, N., Karadjova, I. & Stafilov, T. (2002). ETAAS determination of nickel in serum and urine. Anal. Bioanal. Chem. 373(4-5):310-313.

United Nations (2006). Interim review of scientific information on lead. Overview of existing and future national actions, including legistation, relevant to lead. Appendix. Acessed from http://www.chem.unep.ch/pb_and_cd/SR/Files/Interim_reviews/UNEP-Lead-review-Interim-APPENDIX-Oct 2006.doc on December 20th 2008.

Uzodinma, E.O. Ofoefule, A.U. & Enwere N.J. (2011). Optimization of biogas fuel production from blending maize bract with biogenic wastes. Amer. J. Food and Nutr. 1 (1): 1-6.

WHO (1996). Trace elements in human nutrition and health. International atomic energy agency. WHO Library Publication Data. Geneva. pp 194-215, 256-259.

Woods, G.D. & Fryer, F.I. (2007). Direct elemental analysis of biodiesel by inductively coupled plasma-mass spectrometry. Anal Bioanal Chem. 389(3):753-761.

Wu, T.N., Yang, G.Y., Shen, C.Y. & Liou, S.H. (1995). Lead contamination of candy: an example of crisis management in public health. *Lancet*. 346: 1437-1442.

4

An Assay for Determination of Hepatic Zinc by AAS – Comparison of Fresh and Deparaffinized Tissue

Raquel Borges Pinto[1], Pedro Eduardo Fröehlich[2], Ana Cláudia Reis Schneider[3], André Castagna Wortmann[3], Tiago Muller Weber[2] and Themis Reverbel da Silveira[1,*]

[1]Post Graduate Program in Medicine: Pediatrics,
[2]Post Graduate Program in Pharmaceutical Sciences, UFRGS, Porto Alegre,
[3]Post Graduate Program in Medical Sciences: Gastroenterology and Hepatology,
Universidade Federal do Rio Grande do Sul (UFRGS),
Hospital de Clínicas de Porto Alegre,
Brazil

1. Introduction

Atomic absorption spectroscopy (AAS) is a reliable method to determine metal concentrations. Zinc is a fundamental trace element because of its role in several essential biochemical functions. It is a component or co-factor of several enzymes, such as alcohol dehydrogenase and superoxide dismutase. It is of fundamental importance in cell division, genetic expression, and physiological processes, such as growth and development, immunity and wound healing; also, it plays a structural role in stabilizing biomembranes (Hambidge, 2000; Kruse-Jarres, 2001).

The importance of the determination of the hepatic concentration of certain metals is clearly established in the investigation of hereditary hemochromatosis and Wilson's disease (Pietrangelo, 2003, Roberts et al; 2003). As well, some interesting studies were published on the importance of zinc related to hepatic diseases. Decreases in plasma (Halifeoglu et al., 2004; Schneider et al., 2009; Pereira et al., 2011) or serum (Hamed et al., 2008, Matsuoka et al., 2009) zinc concentrations have been described in patients with chronic liver diseases. Authors that measured zinc in the liver parenchyma of adults and children with liver cirrhosis found low zinc levels (Milman et al., 1986; Göksu & Özsoylu, 1986; Sharda & Bhandari, 1986; Kollmeier et al., 1992; Adams et al., 1994). In patients with alcoholic cirrhosis, studies found abnormal zinc concentrations not only in the liver parenchyma (Rodriguez-Moreno et al., 1997), but also in subcellular fractions of the liver (Bode et al., 1988). In other diseases, such as biliary atresia (Bayliss et al., 1995; Sato et al., 2005), Indian childhood cirrhosis (Bhardwaj et al., 1980; Sharda & Bhandari, 1986), and chronic hepatitis B (Gür et al., 1998), also were found low liver zinc concentrations.

* Corresponding Author

Paraffin-embedded liver tissue usually stored in Pathology Laboratories may be used for analysis when fresh tissue is not available. Some hepatic diseases present a severe imbalance in the metabolism of metals, for example the excess of copper in Wilson's disease and iron in hemochromatosis. Recently, Wortmann and colleagues have validated an analytical method similar to ours for hepatic iron quantification, following the guidelines recommended by the International Conference on Harmonisation (ICH) of Technical Requirements for Registration of Pharmaceuticals for Human Use (Wortmann et al., 2007). Due to the fact that zinc is a protective metal to human health, the assessment of zinc status has a great importance in clinical investigation. Since there is only a few data available in the literature about methods to determine zinc concentration in hepatic tissue, the assay proposed can be helpful to this analysis. Therefore, the purpose of this study was to compare zinc concentrations in fresh and deparaffinized tissues and to determine whether the concentration of this metal in liver specimens is influenced by tissue processing in paraffin blocks.

2. Material and methods

This study was conducted after the validation of the graphite furnace AAS method to determine zinc concentration in bovine liver tissue (Fröehlich et al., 2006), in accordance with the guidelines established by ICH, FDA and ANVISA (ICH, 2005; FDA, 2001; ANVISA, 2003). We used a standard zinc solution (1 mg/ml) and standard bovine liver material from the National Institute of Standards and Technology (NIST, SRM 1577b) with known zinc concentrations (127 ± 16μg/g dry weight).

A steel scalpel was used to obtain 29 wedge biopsies from the same bovine liver obtained from a local market. Each specimen measured about 4 x 2 cm, and 2 were excluded from each group due to contamination and loss of material. Each of the 27 remaining specimens was divided in half, and two groups of samples were formed.

This study was approved by the Ethics in Research Committee of the Research and Graduate Studies, Hospital de Clínicas de Porto Alegre, Porto Alegre, Brazil.

2.1 Group 1

The samples in group 1 were placed in eppendorf tubes previously decontaminated with 10% nitric acid (HNO_3). Samples were lyophilized, using a lyophilizer (Micro moduli 97, Edwards®) for 72 hours. After lyophilization, 500 μL of concentrated HNO_3 (twice distilled, Zn concentration < 0.5 μg/g) was added to each sample, and sonication was applied for 1 hour. The samples were then placed in an incubator (303, Biomatic®) at 60° C for 1 hour to complete digestion of organic matter. From each of the 27 solutions prepared, 50 μl were poured into automatic sampler vials, and 950 μl of pure water (Milli-Q Plus, Millipore®) was added. Concentrations were then determined with a graphite furnace (HGA 800, Perkin-Elmer®) AAS (AAnalyst-3000, Perkin-Elmer®) and an automatic sampler (AS-72).

2.2 Group 2

Samples in the second group were fixed in formalin, embedded in paraffin, and later deparaffinized. The material was kept in paraffin blocks for about 7 days. The samples were deparaffinized by placing them in an incubator at 60° C for about 30 minutes to dissolve the paraffin block, and then in an average of two xylene baths (about 30 minutes each) alternating with 4 baths of alcohol (99%) and distilled water until the paraffin was totally removed. The samples were lyophilized for 72 h and concentrations were determined using AAS, following the same procedure described for samples in group 1. Mean dry weight of the 27 samples after lyophilization was 39.9 mg, and, after deparaffinization, 38 mg. Reagents were analyzed to assess contamination by zinc, which was negligible.

2.3 Statistical analysis

The software Microsoft Excel for Windows® and the Statistical Package for Social Sciences® 12.0 (SPSS) were used to create a database and to conduct statistical analysis. Measures of central tendency and dispersion were used to describe data, with means and standard deviations for quantitative variables. The Student t test for paired samples was used for comparisons between groups. The level of significance was established at $p<0.05$.

3. Results

3.1 Group 1 – Analysis of fresh liver tissue

Zinc concentration in group 1 was 173.6 ± 37.9 µg/g dry tissue (mean ± standard deviation). Table 1 shows the results for the 27 fresh bovine liver samples analyzed after lyophilization.

3.2 Group 2 – Liver tissue analyzed after deparaffinization

Zinc concentration in group 2 was 220.2 ± 127.0 µg/g dry tissue (mean ± standard deviation). Table 1 shows the results of 27 bovine liver samples that were embedded in paraffin, deparaffinized, lyophilized and then analyzed.

3.3 Comparison between analysis of fresh and deparaffinized samples

Liver zinc concentrations in fresh and deparaffinized samples were compared by Student t test for paired samples, and there were no statistically significant differences between these two groups ($p = 0.057$).

Samples	*Concentration fresh liver (µg/g dry tissue)*	*Concentration deparaffinized liver (µg/g dry tissue)*
1	213.3	246.6
2	178.5	148.0
3	169.5	163.3
4	241.2	153.8
5	154.3	201.8
6	154.0	201.4
7	179.4	171.6
8	169.2	193.7
9	170.8	185.1

Samples	*Concentration fresh liver (µg/g dry tissue)*	*Concentration deparaffinized liver (µg/g dry tissue)*
10	168.0	200.2
11	188.2	215.4
12	221.6	212.9
13	136.6	216.4
14	299.0	744.7
15	170.0	523.7
16	179.8	166.5
17	160.5	260.8
18	151.1	153.1
19	155.9	220.2
20	163.8	175.0
21	124.1	141.9
22	145.6	161.4
23	140.0	164.4
24	190.3	179.9
25	174.2	149.0
26	103.7	169.3
27	184.3	225.2
Mean	173.6	220.2
SD[a]	37.9	127.0
RSD[b]	21.9	57.5

[a] SD = standard deviation,[b] RSD = relative standard deviation (%).

Table 1. Liver zinc concentration in samples of fresh and deparaffinized bovine liver tissue.

4. Discussion

Zinc is an essential trace element to human health. It plays an important role in membrane stabilization and in cell protection against oxidative stress because it is part of structure of superoxide dismutase, the main enzyme in endogenous control of some types of free oxygen radicals. It also inhibits transition metals, such as copper and iron, from producing reactive types of oxygen (Powell, 2000). This metal is also essential for DNA and RNA polymerase, which has an important effect in hepatic regeneration (Sato et al., 2005).

The liver is one of the main organs in the metabolism of zinc. Disorders in zinc metabolism have been described in patients with chronic liver disease, and several studies found a decrease in plasma, serum or liver zinc concentrations (Loguercio et al., 2001; Halifeoglu et al., Schneider et al., 2009; Matsuoka et al., 2009, Milman et al., 1986; Göksu & Özsoylu, 1986; Sharda & Bhandari, 1986; Bode et al., 1988; Kollmeier et al., 1992). The decrease of zinc in liver disease seems to be associated with decreased intake, poor absorption associated with portal hypertension, and greater urinary excretion (Loguercio et al., 2001). Collagenase is a zinc-metalloenzyme and zinc is the most effective inhibitor for prolylhydroxylase, an enzyme which plays a key role in collagen synthesis. These two assumptions could explain the role of zinc in collagen deposition and reabsorption in liver disease, the role played by zinc in liver fibrosis and in the evolution of chronic hepatitis toward cirrhosis (Faa et al., 2008).

Some reports found an increase in liver zinc concentrations in chronic liver disease. An increase in copper and zinc liver concentrations was found in Canadian children with chronic cholestasis (Phillips et al., 1996). Another case report described the increase in zinc concentration in hepatic tissue of a child with hepatosplenomegaly and symptoms of zinc deficiency, and the authors speculated about the existence of a zinc metabolism disorder (Sampson et al., 1997). A study that investigated the concentration of metals in liver tissue of adults with hereditary hemochromatosis found an increase in zinc in the liver parenchyma. The authors suggested that the concurrent increase in iron and zinc might be explained by the greater intestinal absorption of these metals (Adams et al., 1991).

The test usually conducted to determine body zinc is the measurement of plasma zinc concentration. However, plasma zinc concentrations do not seem to reflect the concentration found in the liver parenchyma (Göksu & Özsoylu, 1986; Sato et al., 2005). This may be explained by the fact that there are very efficient homeostatic mechanisms to correct plasma or serum zinc deficiencies, which makes it difficult to diagnose marginal deficiency by using this method. Therefore, the investigation of zinc concentration in liver tissue is important.

Studies report a great variation in liver zinc concentrations, maybe due to the different techniques used (Table 2). Kollmeier et al. (1992) studied the distribution of zinc in adult liver parenchyma from necropsy material, and found a small variation in intraorgan metal concentrations. They reported that zinc concentrations in the liver do not seem to be associated with sex or age (Kollmeier et al., 1992). Another study, conducted with children by Coni et al. (1996), confirmed these findings. They measured the concentration of metal in necropsy material from infants that died of sudden infant death syndrome and from pediatric control subjects, and found that a small liver sample is representative of liver concentration in the whole liver.

Author and year	Subjects and technique	Zinc concentration in liver tissues (μg/g dry tissue)
Adams et al. (1991)	Healthy controls (n=21) Flame AAS[a]	326.3 ± 65.4
Kollmeier et al. (1992)	Unselected necropsies (n=58) Flameless AAS[a]	280.0 ± 178.0
Bush et al. (1995)	Unselected necropsies (n=30) ICP-ES[b] AAS[a]	191.0 ± 56.3
Treble et al. (1998)	Unselected necropsies (n=73) Graphite furnace AAS[a]	118.3 ± 44.4
Hatano et al. (2000)	Healthy controls (n=21) Particle-induced X-ray emission	281.0 ± 25.5

[a] AAS = atomic absorption spectrophotometry.
[b] ICP-ES = inductively coupled plasma emission spectroscopy.

Table 2. Zinc concentration in hepatic tissue according to the literature.

Fresh tissue is not always available for chemical analysis, but formalin-fixed tissue often is. In our study, zinc concentration measurements in fresh liver tissue and deparaffinized tissue have been shown to be concordant. We have found in fresh e deparaffinized tissues respectively, 173,6 ± 37,9 μg/g dry weight and 220,2 ± 127 μg/g dry weight. Two deparaffinized tissue samples showed higher values of zinc than the others samples, causing a higher standard deviation. There was no evidence of mineral contamination during the embedding process to account for these divergent values.

We found only one study in the literature, conducted at the Mayo Clinic by Bush et al. (1995) that compared zinc concentration in fresh and deparaffinized liver tissue. Their study investigated the concentration of metals in several organs using material obtained from autopsy of 30 presumably healthy individuals. Zinc concentration found in fresh liver tissue was 191 ± 56.3 μg/g dry weight, and, in formalin-fixed tissue, 204 ± 63.2 μg/g dry weight. They concluded that formalin fixation long-term storage has little effect on zinc concentrations in tissue and that zinc was homogeneously distributed in liver.

Due to the clinical importance of zinc in liver diseases, the use of paraffin-embedded specimens for analysis is extremely useful when fresh tissue is not available. Stored material for analysis may be available even years after the biopsy or autopsy sample was obtained.

4.1 Conclusion

More than the results themselves, the proposed protocol for paraffinization/deparaffinization as well as for sample preparation for zinc determination by atomic spectroscopy in paraffinized samples were adequately established. According to the results of this study, paraffin embedding and deparaffinization do not significantly affect the determination of zinc concentrations in liver tissue, and, therefore, stored material can be used for analysis.

5. Glossary

AAS - Atomic Absorption Spectrophotometry.
ANVISA - Agência Nacional de Vigilância Sanitária (Sanitary Surveillance National Agency).
FDA - Food and Drug Administration.
ICH - International Conference on Harmonisation of Technical Requirements for Registration of Pharmaceuticals for Human Use.

6. Acknowledgements

We are grateful to Coordenação de Aperfeiçoamento de Pessoal de Nível Superior (CAPES), Conselho Nacional de Desenvolvimento Científico e Tecnológico (CNPq) and Fundo de Incentivo a Pesquisas - Hospital de Clínicas de Porto Alegre (FIPE-HCPA).

7. References

Adams PC, Bradley C, Frei JV. Hepatic zinc in hemochromatosis. Clin Invest Med.1991;14:16-20.

Adams PC, Bradley C, Frei JV. Hepatic iron and zinc after portocaval shunting for nonalcoholic cirrhosis. Hepatology. 1994; 19(1):101-105.
ANVISA - Agência Nacional de Vigilância Sanitária. Guidelines for validation of analytical and bioanalytical methods. Diário Oficial da União, Brasília, 2003.
Bayliss EA, Hambidge KM, Sokol RJ, et al. Hepatic concentrations of zinc, copper and manganese in infants with extrahepatic biliary atresia. J Trace Elem Med Biol 1995; 9:40-43.
Beilby JP, Prins AW, Swanson NR. Determination of hepatic iron concentration in fresh and paraffin-embedded tissue. Clin Chem 1999; 45(4):573-574.
Bhardwaj S, Miglani N, Gupta BD, et al. Hepatic zinc levels in Indian childhood cirrhosis. Indian J Med Res 1980; 71:278-281.
Bode JC, Hanisch P, Henning H, et al. Hepatic zinc content in patients with various stages of alcoholic liver disease and in patients with chronic active and chronic hepatitis. Hepatology 1988; 8:1605-1609.
Bush VJ, Moyer TP, Batts KP, et al. Essential and toxic element concentrations in fresh and formalin-fixed human autopsy tissue. Clin Chem 1995; 41:284-294.
Coni P, Ravarino A, Farci AM, et al. Zinc content and distribution in the newborn liver. J Pediatr Gastroenterol Nutr 1996; 23:125-129.
Faa G, Nurchi VM, Ravamino A, et al. Zinc in gastrointestinal and liver disease. Coordination Chem Rev 2008; 252:1257-1269
Fröelich PE, Pinto RB, Wortmann AC, et al. Full validation of an electrothermal atomic absorption for zinc in hepatic tissue using a fast sample preparation procedure. Spectroscopy 2006; 20:81-87.
Göksu, N, Özsoylu S. Hepatic and serum levels of zinc, copper and magnesium in childhood cirrhosis J Pediatr Gastroenterol Nutr 1986; 5:459-462.
Gür G, Bayraktar Y, Ozer D, et al. Determination of hepatic zinc content in chronic liver disease due to hepatitis B virus. Hepatogastroenterology 1998; 45:472-476.
Hambidge M. Human zinc deficiency. J Nutr 2000; 130:1344-1349.
Hamed SA, Hamed EA, Farghaly MH, et al. Trace elements and flapping tremors in patients with liver cirrhosis. Is there a relationship? Saudi Med J. 2008; 29(3):345-351.
Halifeoglu I, Gur B, Aydin S, et al. Plasma trace elements, vitamin B12, folate, and homocysteine levels in cirrhotic patients compared to healthy controls. Biochemistry (Mosc) 2004; 69(6):693-696.
Hatano R, Ebara M, Fukuda H, et al. Accumulation of copper in the liver and hepatic injury in chronic hepatitis C. J Gastroenterol Hepatol 2000; 15:786-791.
ICH - Harmonized Tripartite Guideline, Test on Validation of Analytical Procedures - Q2(R1). In: International Conference on Harmonisation of Technical Requirements for Registration of Pharmaceuticals for Human Use, 2005.
Kollmeier H, Seemann J, Wittig P, et al. Zinc concentrations in human tissues. Liver zinc in carcinoma and severe liver disease. Pathol Res Pract 1992; 188:942-945.
Kruse-Jarres JD. Pathogenesis and symptoms of zinc deficiency. Am Clin Lab 2001; 20:17-22.
Loguercio C, De Girolamo V, Federico A, et al. Trace elements and chronic liver diseases. J Trace Elem Med Biol 2001; 11:158-161.
Matsuoka S, Matsumura H, Nakamura H, et al. Zinc supplementation improves the outcome of chronic hepatitis C and liver cirrhosis. J Clin Biochem Nutr 2009; 25:292-303.

Milman N, Laursen J, Podenphant, et al. Trace elements in normal and cirrhotic human liver tissue. I. Iron, copper, zinc, selenium, manganese, titanium and lead measured by X-ray fluorescence spectrometry. Liver 1986; 6:111-117.

Olynyk JK, O'Neill R, Britton RS, et al. Determination of hepatic iron concentration in fresh and paraffin-embedded tissue: diagnostic implications. Gastroenterology 1994; 106:674-677.

Pereira TC, Saron ML. Carvalho WA, et al. Research on zinc blood levels and nutritional status in adolescents with autoimmune hepatitis. Arq Gastroenterol 2011; 48(1):62-65.

Pietrangelo A. Haemochromatosis. Gut 2003; 52 Suppl 2:ii23-ii30

Phillips MJ, Ackerley CA, Superina RA, et al. Excess zinc associated with severe progressive cholestasis in Cree and Ojibwa-Cree children. Lancet 1996; 347:866-868.

Powell SR. The antioxidant properties of zinc. J Nutr 2000; 130:1447-54.

Roberts EA, Schilsky ML. A practice guideline on Wilson disease. Hepatology 2003; 37:1475-1492.

Rodriguez-Moreno F, González-Reimers E, Santolaria-Fernandez F, et al. Zinc, copper, manganese and iron in chronic alcoholic liver disease. Alcohol 1997; 14:39-44.

Sampson B, Kovar IZ, Rauscher A, et al. A case of hyperzincemia with functional zinc depletion: a new disorder? Pediatr Res 1997; 42:219-225.

Sato C, Koyama H, Satoh H, et al. Concentrations of copper and zinc in liver and serum samples in biliary atresia patients at different stages of traditional surgeries. Tohoku J Exp Med 2005; 207:271-277.

Schneider AC, Pinto RB, Fröelich PE et al. Low plasma zinc concentrations in patients with cirrhosis. J Pediatr (Rio J). 2009; 85(4):359-364.

Sharda B, Bhandari B. Studies of trace elements in childhood cirrhosis. Acta Pharmacol Toxicol (Copenh) 1986; 59 Suppl 7:206-10.

Treble RG, Thompson TS, Lynch HR. Determination of copper, manganese and zinc in human liver. Biometals 1998; 11:49-53.

U.S., Department of Health and Human Services. Food and Drug Administration. Guidance for industry: bioanalytical method validation. Available at: http://www.fda.gov/CDER/GUIDANCE/4252fnl.htm. Accessed: June 30, 2011.

Wortmann AC, Froehlich PE, Pinto RB, et al. Hepatic iron quantification by atomic absorption spectrophotometry: Full validation of an analytical method using a fast sample preparation. Spectroscopy 2007; 21:161-167.

Section 2

UV-VIS Spectroscopy

5

The Use of Spectrophotometry UV-Vis for the Study of Porphyrins

Rita Giovannetti
University of Camerino, Chemistry Section of School of Environmental Sciences, Camerino Italy

1. Introduction

The porphyrins (Fig. 1) are an important class of naturally occurring macrocyclic compounds found in biological compounds that play a very important role in the metabolism of living organisms. They have a universal biological distribution and were involved in the oldest metabolic phenomena on earth. Some of the best examples are the iron-containing porphyrins found as heme (of haemoglobin) and the magnesium-containing reduced porphyrin (or chlorine) found in chlorophyll. Without porphyrins and their relative compounds, life as we know it would be impossible and therefore the knowledge of these systems and their excited states is essential in understanding a wide variety of biological processes, including oxygen binding, electron transfer, catalysis, and the initial photochemical step in photosynthesis.

The word porphyrin is derived from the Greek porphura meaning purple. They are in fact a large class of deeply coloured pigment, of natural or synthetic origin, having in common a substituted aromatic macrocycle ring and consists of four pyrrole rings linked by four methine bridges (Milgrom, 1997; D. Dolphin, 1978).

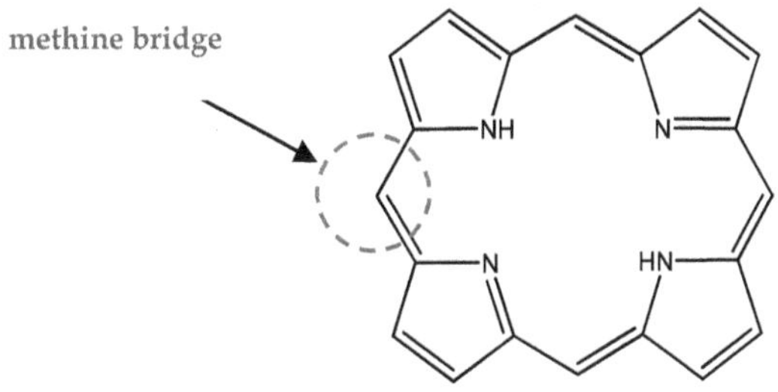

methine bridge
NH
N
N
HN

Fig. 1. The structure of porphyrin.

The porphyrins have attracted considerable attention because are ubiquitous in natural systems and have prospective applications in mimicking enzymes, catalytic reactions, photodynamic therapy, molecular electronic devices and conversion of solar energy. In particular, numerous porphyrins based artificial light-harvesting antennae, and donor acceptor dyads and triads have been prepared and tested to improve our understanding of the photochemical aspect of natural photosynthesis.

The porphyrins play important roles in the nature, due to their special absorption, emission, charge transfer and complexing properties as a result of their characteristic ring structure of conjugated double bonds (Rest et al.,1982).

As to their electronic absorption, they display extreme intense bands, the so-called Soret or B-bands in the 380–500 nm range with molar extinction coefficients of 10^5 M^{-1} cm^{-1}. Moreover, at longer wavelengths, in the 500–750-nm range, their spectra contain a set of weaker, but still considerably intense Q bands with molar extinction coefficients of 10^4 M^{-1} cm^{-1}. Thus, their absorption bands significantly overlap with the emission spectrum of the solar radiation reaching the biosphere, resulting in efficient tools for conversion of radiation to chemical energy. In such a conversion, the favourable emission and energy transfer properties of porphyrin derivatives are indispensable as in the case of chlorophylls, which contain magnesium ion in the core of the macrocycle. Also, metalloporphyrins can be utilized in artificial photosynthetic systems, modelling the most important function of the green plants (Harriman et al., 1996).

The studies of the wavelength shift of their adsorption band and the absorbance changes as function of pH, temperature, solvent change, reaction with metal ions and other parameters permits to obtained accurate information about equilibrium, complexation, kinetic and aggregation of porphyrins.

This review, resumes the best successes in the use of spectrophotometer UV-Vis for explained the chemical characteristics of this extraordinary group of natural occurring molecules and clarifies the potential of these molecules in many fields of application.

2. The chemical characteristics of porphyrins

The synthetic world of porphyrins is extremely rich and its history began in the middle of 1930s. An enormous number of synthetic procedures have been reported until now, and the reason can be easily understood analysing the porphyrin skeleton. In principle, there are many chemical strategies to synthesized porphyrins, involving different building blocks, like pyrroles, aldehydes, dipyrromethanes, dipyrromethenes, tripyrranes and linear tetrapyrroles.

The most famous monopyrrole polymerization route to obtain porphyrins involves the synthesis of tetraphenyl porphyrins, from reaction between pyrrole and benzaldehyde (Atwood et al., 1996). This procedure was first developed by Rothemund (Rothemund, 1935) and, after modification by Adler, Longo and colleagues (Adler et al., 1967), was finally optimized by Lindsey's group (Lindsey et al.,1987). In the Rothemund and Adler/Longo methodology the crude product contains between 5 and 10% of a byproduct, discovered later to be the *meso*-tetraphenylchlorin which is converted in the product under oxidative conditions (Fig. 2).

Rothemund in 1935 set up the synthesis of porphyrins in one step by reaction of benzaldehyde and pyrrole in pyridine in a sealed flask at 150 °C for 24 h but the yields were low, and the experimental conditions so severe that few benzaldehydes could be converted to the corresponding substituted porphyrin (Rothemund, 1936; Menotti, 1941). The reason in the low yield is that the main by-product of reaction was *meso*substituted chlorin and in understanding the nature of its formation, Calvin and coworkers (Calvin et al., 1946) discovered that the addition of metal salts to the reaction mixture, such as zinc acetate, increases the yield of porphyrin from 4-5% for the free-base derivative, and decreases the amount of chlorin compound. Others improvement were obtained by changing opportunely the reaction conditions and substituents in benzaldehyde molecule framework.

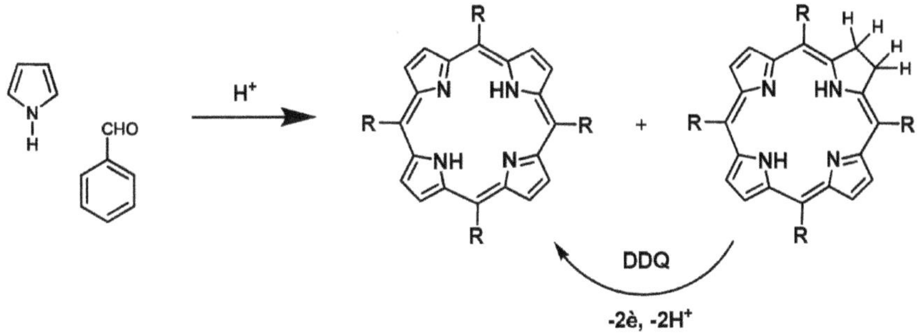

Fig. 2. Synthesis of 5,10,15,20-tetraphenyl porphyrin.

Adler, Longo and coworkers, in the 1960s (Adler et al. 1967), re-examined the synthesis of *meso*substituted porphyrins and developed an alternative approach (Fig. 3) with a method that involves an acid catalyzed pyrrole aldehyde condensation in glassware open to the atmosphere in the presence of air. The reactions were carried out at high temperature, in different solvents and concentrations range of reactants with a yields of 30-40%, and with chlorin contamination lower than that obtained with the Rothemund synthesis.

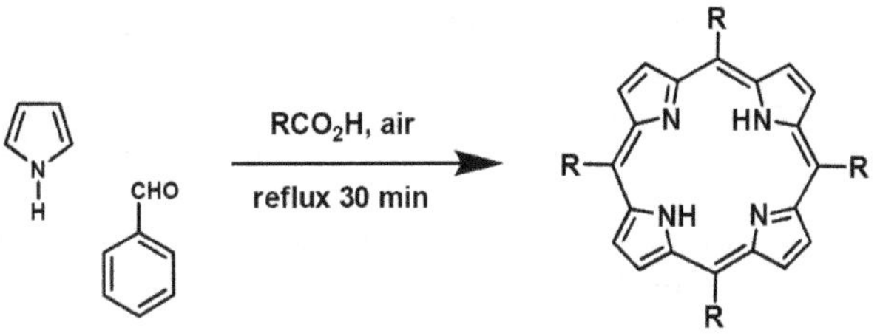

Fig. 3. Adler-Longo method for preparing *meso*-substituted porphyrins.

Over the period 1979-1986, Lindsey developed a new and innovative two-step room temperature method to synthesize porphyrins, motivated by the need for more gentle conditions for the condensation of aldehydes and pyrrole, in order to enlarge the number of the aldehydes utilizable and then the porphyrins available (Anderson et al., 1990; Acheson et al., 1976; Dailey, 1990; Porra, 1997; Mauzarall, 1960). The method has been a new strategy for the synthesis of porphyrins, using a sequential process of condensation and oxidation steps. The reactions were carried out under mild conditions in an attempt to achieve equilibrium during condensation, and to avoid side reactions in all steps of the porphyrin-forming process (Fig. 4)

The porphyrin macrocycle is a highly-conjugated molecule containing 22 π-electrons, but only 18 of them are delocalized according to the Hückel's rule of aromaticity (4n+2 delocalized π-electrons, where n = 4).

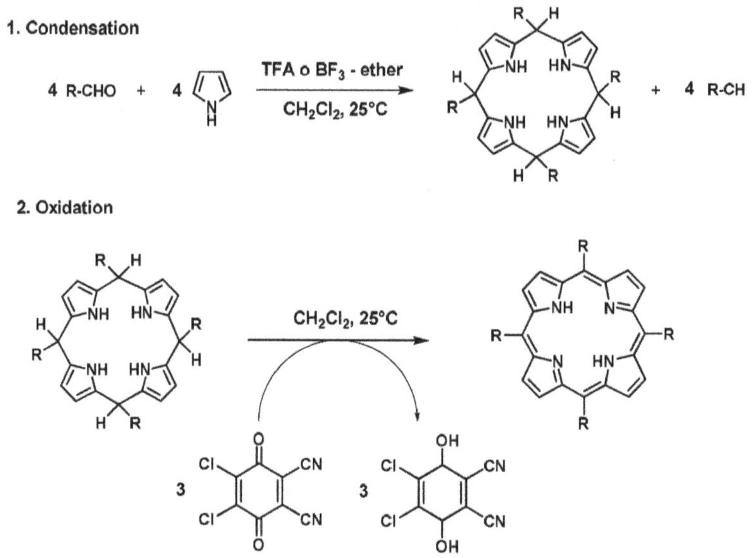

1. Condensation

4 R-CHO + 4 [pyrrole] —TFA o BF_3 - ether / CH_2Cl_2, 25°C→ [porphyrinogen] + 4 R-CH

2. Oxidation

[porphyrinogen] —CH_2Cl_2, 25°C→ [porphyrin]

3 [DDQ] → 3 [DDQ-H_2]

Fig. 4. Two-step one-flask room-temperature synthesis of porphyrins.

Its structure supports a highly stable configuration of single and double bonds with aromatic characteristics that permit the electrophilic substitution reactions typical of aromatic compounds such as halogenation, nitration, sulphonation, acylation, deuteration, formylation. Although this, in the porphyrins there are two different sites on the macrocycle where electrophylic substitution can take place with different reactivity (Milgrom, 1997): positions 5, 10, 15 e 20, called *meso* and positions 2, 3, 7, 8, 12, 13, 17 and 18, called β-pyrrole positions (Fig. 5). The first kind of compounds are widely present in natural products, while the second have no counterpart in nature and were developed as functional artificial models. The activation of these sites depends of the porphyrins electronegativity that can be controlled by the choice of the metal to coordinate to the central nitrogen atoms. For this, the introduction of divalent central metals produces electronegative porphyrin ligands and these complexes can be substituted on their *meso*-carbon. On the other hand, metal ions in electrophylic oxidation states (e.g. Sn IV) tend to deactivate the *meso*-position and activate the β pirrole to electrophylic attack. The chemical characteristics of substituents in β-pyrrole and meso-position determine the water or solvent solubility of porphyrins.

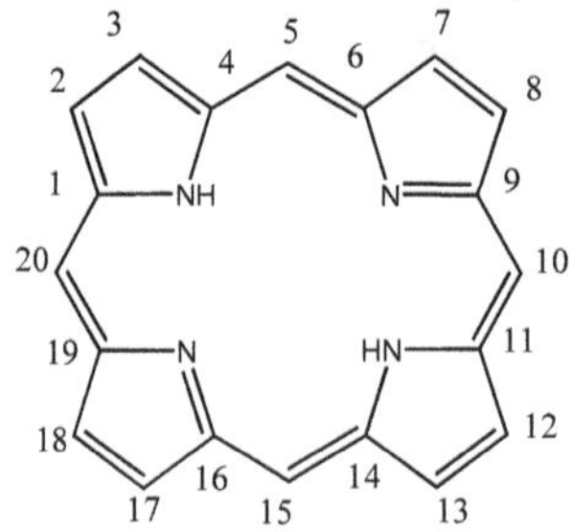

Fig. 5. Porphyrin numeration.

3. Uv-vis spectra of porphyrins

It was recognized early that the intensity and colour of porphyrins are derived from the highly conjugated π-electron systems and the most fascinating feature of porphyrins is their characteristic UV-visible spectra that consist of two distinct region regions: in the near ultraviolet and in the visible region (Fig. 6).

It has been well documented that changes in the conjugation pathway and symmetry of a porphyrin can affect its UV/Vis absorption spectrum (Gouterman, 1961; Whitten et al. 1968; Smith, 1976; Dolphin, 1978; Nappa & Valentine, 1978; Wang et al. 1984; Rubio et al. 1999).

The absorption spectrum of porphyrins has long been understood in terms of the highly successful "four-orbital" (two highest occupied π orbitals and two lowest unoccupied π^* orbitals) model first applied in 1959 by Martin Gouterman that has discussed the importance of charge localization on electronic spectroscopic properties and has proposed the four-orbital model in the 1960s to explain the absorption spectra of porphyrins (Gouterman, 1959; Gouterman, 1961).

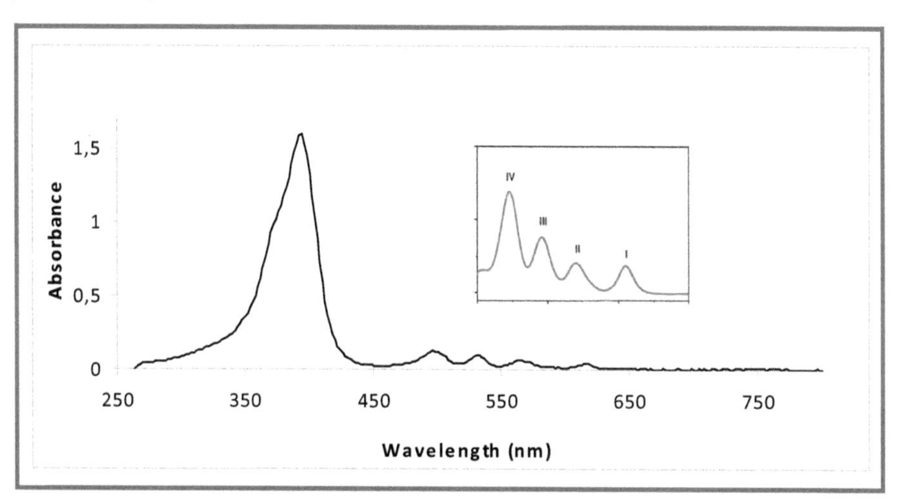

Fig. 6. UV-vis spectrum of porphyrin with in insert the enlargement of Q region between 480-720 nm.

According to this theory, as reported in Figure 7, the absorption bands in porphyrin systems arise from transitions between two HOMOs and two LUMOs, and it is the identities of the metal center and the substituents on the ring that affect the relative energies of these transitions. The HOMOs were calculated to be an $a1_u$ and an $a2_u$ orbital, while the LUMOs were calculated to be a degenerate set of e_g orbitals. Transitions between these orbitals gave rise to two excited states. Orbital mixing splits these two states in energy, creating a higher energy state with greater oscillator strength, giving rise to the Soret band, and a lower energy state with less oscillator strength, giving rise to the Q-bands.

The electronic absorption spectrum of a typical porphyrin (Fig. 6) consists therefore of two distinct regions. The first involve the transition from the ground state to the second excited state (S0 → S2) and the corresponding band is called the Soret or B band. The range of absorption is between 380-500 nm depending on whether the porphyrin is β- or *meso*-substituted. The second region consists of a weak transition to the first excited state (S0 → S1) in the range between 500-750 nm (the Q bands). These favourable spectroscopic features of porphyrins are due to the conjugation of 18 π- electrons and provide the advantage of easy and precise monitoring of guest-binding processes by UV-visible spectroscopic methods (Yang et al. 2002; Gulino et al., 2005; Di Natale et al. 2000; Paolesse & D'Amico, 2007) CD, (Scolaro et al. 2004; Balaz et al. , 2005) fluorescence, (Zhang et al., 2004; Zhou et al., 2006) and NMR spectroscopy (Shundo et al., 2009; Tong et al., 1999).

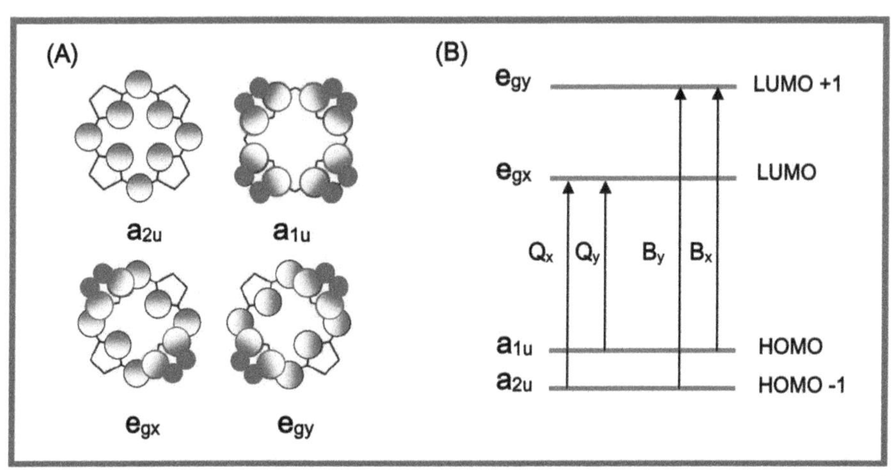

Fig. 7. Porphyrin HOMOs and LUMOs. (A) Representation of the four Gouterman orbitals in porphyrins. (B) Drawing of the energy levels of the four Gouterman orbitals upon symmetry lowering from *D4h* to C2V. The set of e_g orbitals gives rise to Q and B bands.

The relative intensity of Q bands is due to the kind and the position of substituents on the macrocycle ring. Basing on this latter consideration, porphyrins could be classified as *etio, rhodo, oxo-rhodo* e *phyllo* (Prins et al 2001).

When the relative intensities of Q bands are such that IV > III > II > I, the spectrum is said *etio-type* and porphyrins called *etioporphyrins*. This kind of spectrum is found in all porphyrins in which six or more of the β-positions are substituted with groups without π-electrons, *e.g.*, alkyl groups. Substituent with π-electrons, as carbonyl or vinyl groups, attached directly to the β-positions gave a change in the relative intensities of the Q bands, such that III > IV > II > I. This is called *rhodo-type* spectrum (*rhodoporphyrin*) because these groups have a "reddening" effect on the spectrum by shifting it to longer wavelengths. However, when these groups are on opposite pyrrole units, the reddening is intensified to give an *oxo-rhodo-type* spectrum in which III > II > IV > I. On the other hand, when *meso*-positions are occupied, the *phyllo-type* spectrum is obtained, in which the intensity of Q bands is IV > II > III > I (Milgron 1997).

While variations of the peripheral substituents on the porphyrin ring often cause minor changes to the intensity and wavelength of the absorption features, protonation of two of the inner nitrogen atoms or the insertion/change of metal atoms into the macrocycle usually strongly change the visible absorption spectrum.

When porphyrinic macrocycle is protonated or coordinated with any metal, there is a more symmetrical situation than in the porphyrin free base and this produces a simplification of Q bands pattern for the formation of two Q bands.

4. The equilibrium of porphyrins

Neglecting the overall charge of the macrocycle, a monomeric free-base porphyrin H_2-P in aqueous solution can add protons to produce mono H_3-P^+ and dications H_4-P^{2+} at very low pHs, or loose protons to form the centrally monoprotic H-P^- at pH about 6 or aprotic P^{2-} species at pH ≥ 10 (Fig. 8). These chemical forms of porphyrin may exist in equilibrium, depending upon the pH of the solution and can be characterized from the change of the electronic absorption spectrum. The change in spectra upon addition of acid or basic substances can generally be attributed to the attachment or the loss of protons to the two imino nitrogen atoms of the pyrrolenine-like ring in the free-base (Gouterman, 1979; Giovannetti et al, 2010). The N-protonation induced a red-shifts that are consistent with frontier molecular orbital calculations for protonated porphyrins (Daniel et al., 1996).

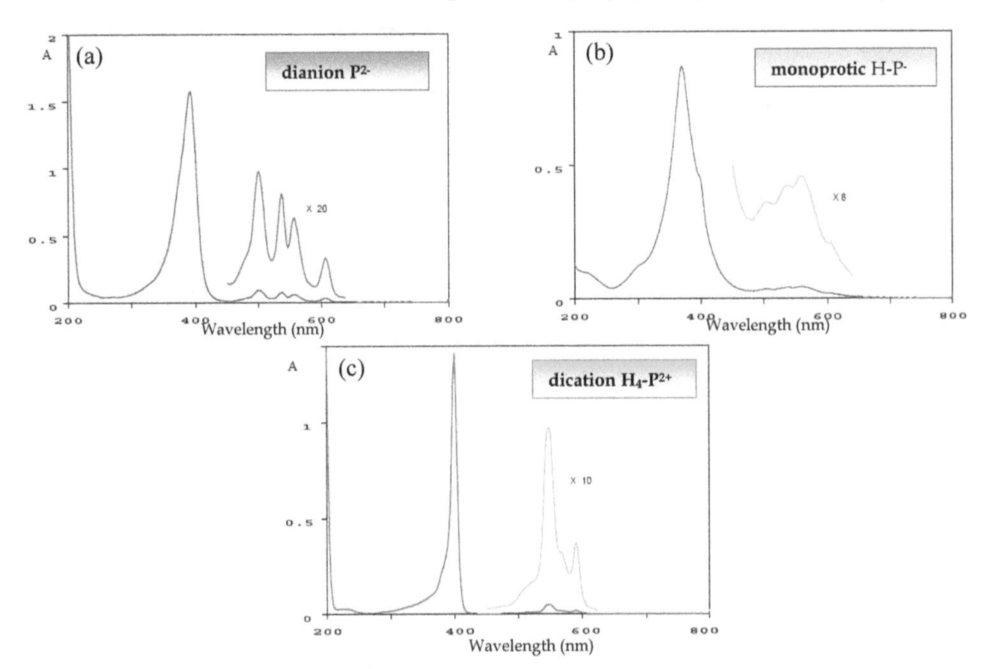

Fig. 8. Typical Uv-vis specrtum of dianion P^{2-} (pH about 10) monoprotic H-P^- (pH about 6) and dication H_4-P^{2+} porphyrin (pH about 1).

Spectrophotometric titration was employed for determining the acid dissociation constants over the inter pH range and change in absorbance with pH can be attributed to the following acid dissociation reactions of porphyrins. Upon addition of acid the spectral pattern of porphyrins changes from the four Q-band spectrum, indicating *D2h* symmetry for free-base porphine, to a two Q-band spectrum for the formation of dications H_4-P^{2+} (Fig. 8 c), indicating *D4h* symmetry, characteristic of porphyrin coordinated to a metal ion through the four N-heteronuclei. In addition, in all cases, the intense Soret band is red-shifted (to an extent dependent on the particular meso-substituents).

5. The reaction of porphyrins with metal ions: Regular and sitting-atop complexes

The metalloporphyrin formation reaction is one of the important processes from both analytical and bioinorganic points of view. The large molar absorption coefficient and the very high stability of porphyrins is valuable for the separation of various kinds of metal ions (Tabata et al.,1998). A variety of metalloporphyrin formation rates are also applicable for the kinetic analysis of metal ions (Tabata & Tanaka, 1991). Also, kinetic studies of metalloporphyrin formation are indispensable in order to understand in vivo metal incorporation processes leading to the natural metalloporphyrins. Generally porphyrins are synthesized in a metal-free form and metal ions are successively inserted.

When the metal ion M^{n+} is incorporated into the porphyrin H_2P to form $MP^{(n-2)+}$, the two amine protons in H_2P are dissociated from the two pyrrole groups as reported in equation (1):

$$M^{n+} + H_2P \leftrightarrow MP^{(n-2)+} + 2H^{+} \quad (1)$$

In the formation of metalloporphyrins an marked colour changes with trasformation of the Uv-Vis spectrum especially in the Q zone has been observed. The two Q band obtained are called α and β (Fig. 9). The relative intensities of these bands can be correlated with the stability of the metal complex; in fact when α > β, the metal forms a stable square-planar complex with the porphyrin, in the other case when β > α (e.g. Ni(II), Pd(II), Cd(II)), the metals are easily displaced by protons (Milgron, 1997).

Studies on water soluble and insoluble porphyrins have elucidated aspects of the mechanisms of metal ion incorporation into porphyrins to form metalloporphyrins (Bailey & Hambright, 2003; Hambright et al., 2001; Lavallee, 1987; Funahashi et al., 2001).

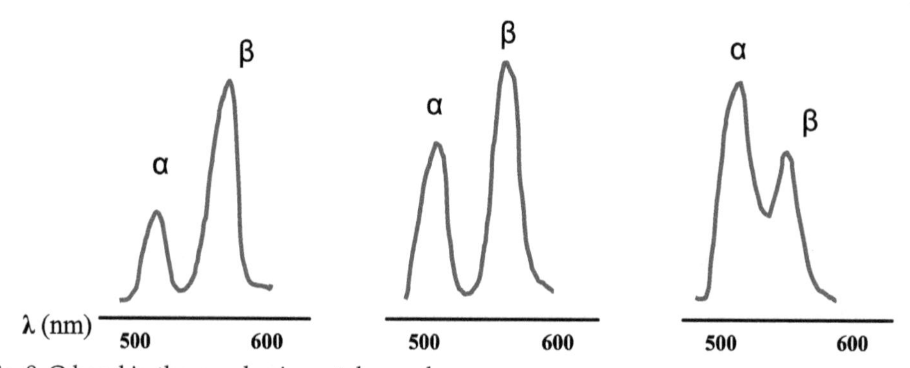

Fig. 9. Q band in the porphyrin metal complexes

The size of the porphyrin-macrocycle is perfectly suited to bind almost all metal ions and indeed a large number of metals can be inserted in the center of the macrocycle forming metalloporphyrins that play key roles in several biochemical processes, due to their central role in photosynthesis, oxygen transport and in various redox reactions (Mathews et al., 2000; Garret & Grisham, 1999; Knör & Strasser, 2005; Lim et al., 2005; Martirosyan et al., 2004; Tovmasyan et al., 2008; Ren et al., 2010; Kawamura et al., 2011).

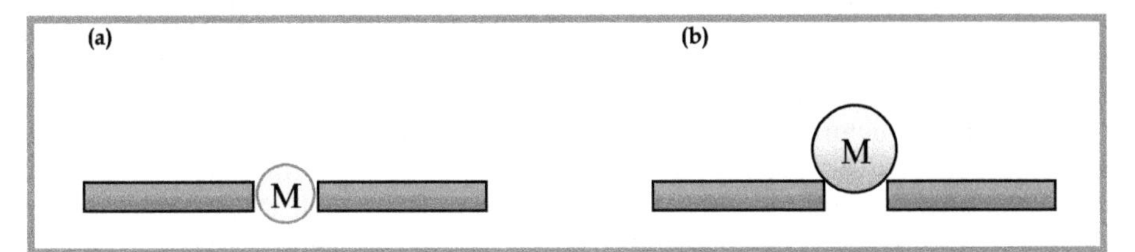

Fig. 10. Schematic representation of (a) regular and (b) SAT metalloporphyrins.

Depending on their size, charge, and spin multiplicity, metal ions (*e.g.* Zn, Cu, Ni, Co, *etc.*) can fit into the center of the planar tetrapyrrolic ring system forming *regular metalloporphyrins* resulting in a kinetically inert complexes (Fig. 10a).

When divalent metal ions (e.g. Co(II), Ni(II), Cu(II)) are chelated, the resulting tetracoordinate chelate has no residual charge. While Cu(II) and Ni(II) in their porphyrin complexes have generally low affinity for additional ligands, the chelates with Mg(II), Cd(II) and Zn(II) readily combine with one more ligand to form pentacoordinated complexes with square-pyramidal structure (Fig. 11a). Some metalloporphyrins (Fe(II), Co(II), Mn(II)) are able to form distorted octahedral (Fig. 11b) with two extra ligand molecules (Biesaga et al., 2000).

Most of the natural metalloporphyrins are of regular type, i.e. their metal centres are located within the plane of the macrocyclic ligand as a consequence of their fitting size. The cationic radii are in the range of 55–80 pm corresponding to the sphere in the porphyrin core surrounded by the four pyrrolic nitrogens. While the symmetry group of the free-base porphyrins is D2h due to the two hydrogen atoms on the diagonally located pyrrolic nitrogens, the coplanar (regular) metalloporphyrins (without these protons) are of higher symmetry (Khan & Bruice, 2003).

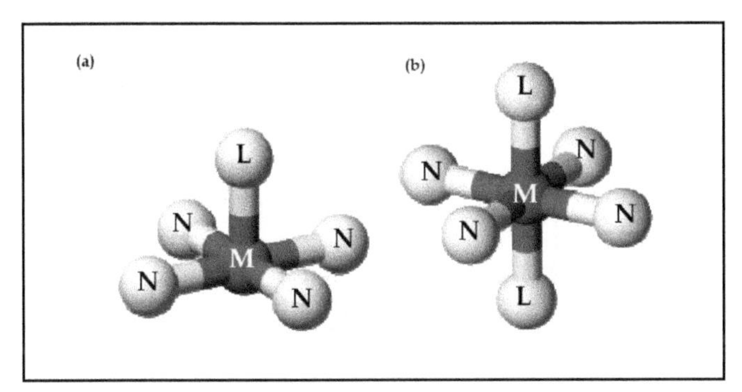

Fig. 11. Schematic pictures of square-pyramidal (a) and octahedral structures (b) (only enclose nitrogen N, metal M and extra ligands L).

If, however, the ionic radius of the metal ions is too large (over ca. 80-90 pm) to fit into the hole in the centre of the macrocycle, they are located out of the ligand plane, distorting it forming *sitting-atop (SAT) metalloporphyrins* (Fig. 10b) that are characterized by special properties (Fleischer & Wang 1960; Barkigia et al., 1980; Liao et al., 2006; Walker et al., 2010) originating from the non-planar structure caused by, first of all, the size of the metal center.

These complexes are kinetically labile and display characteristic structural and photoinduced properties that strongly deviates from those of the regular metalloporphyrins. The latter kind of structure induces special photophysical and photochemical features that are characteristic for all SAT complexes. The symmetry of this structures is lower (generally C_{4v}-C_1) than that of both the free-base porphyrin (D2h) and the regular, coplanar metalloporphyrins (D4h), in which the metal center fits into the ligand cavity.

The rate of formation of in-plane (or normal) metalloporphyrins is much slower than that of the *SAT complexes* because of the inflexibility of porphyrins. In fact, in an SAT complex the distortion of the porphyrin caused by the out-of-plane location of the metal center makes two diagonal pyrrolic nitrogens more accessible on the other side of the ligand due to the increase of their sp^3 hybridization (Tung & Chen, 2000).

Deviating from the regular metalloporphyrins, the SAT complexes, on account of their distorted structure and kinetic lability, display peculiar photochemical properties, such as photoinduced charge transfer from the porphyrin ligand to the metal center, leading to irreversible ring opening of the ligand and dissociation on excitation at both the Soret- and the Q-bands (Horváth et al., 2006). Moreover, the absorption and emission characteristics of these complexes are also significantly deviating from those of the normal (in-plane) metalloporphyrins (Horváth et al., 2006). Also the formation of bi and even trinuclear (bis-porphyrin) complexes has been observed (Lehn, 2002).

In Figure 12 is shown a schematic Energy-level diagram of the frontier orbital of a porphyrin in free-base state (H_2P), in a regular and in a SAT metalloporphyrin.

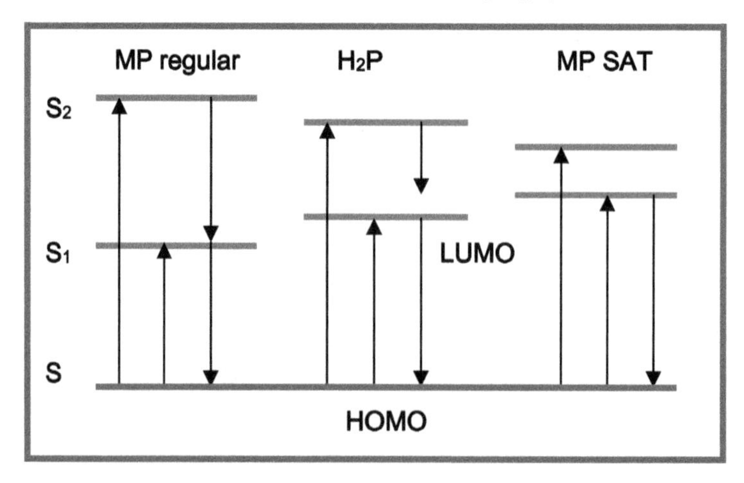

Fig. 12. Simplified energy-level diagram of the frontier orbital of a porphyrin in free-base state H_2P, in a regular and in a SAT metalloporphyrin.

The photoinduced behavior of normal metalloporphyrins have been thoroughly studied for several decades, while the investigation of SAT complexes started in this respect only in the past 8–10 years (Horváth et al., 2004; Valicsek et al., 2008; Valicsek et al., 2009; Valicsek et al., 2007; Huszánk et al., 2005; Huszánk et al., 2007; Valicsek et al., 2011).

Interestingly, in the case of lanthanide ions as metal centers, triple decker porphyrin sandwich complexes were also synthesized and studied (Wittmer &. Holten, 1996).

While the natural porphyrin derivatives are exclusively hydrophobic, some artificial porphyrins having ionic substituents made it possible to prepare water-soluble metalloporphyrins of both regular and SAT type. Kinetically labile complexes are mostly examined in the excess of the ligand.

In the case of metalloporphyrins, however, metal ions are applied generally in excess, especially for spectrophotometric measurements, partly because of the extremely high molar absorbances (mainly at the Soret-bands) of the porphyrins. The formation of kinetically labile SAT complexes, deviating from the regular metalloporphyrins, is an equilibrium process. It can be spectrophotometrically monitored because the absorption and emission bands assigned to ligand-centered electron transitions undergo significant shift and intensity change upon coordination of metal ions.

Special attention was devoted to the reaction of porphyrins with essential metal ions as manganese, iron and chromium show that the most important properties of manganese in complex biological systems is the highly variable oxidation states of the metal from +2 to +5 (Kadish et al. 1999). All these compounds can be easily spectrophotometrically distinguished among them; this is because they have different absorption spectra (Spasojevic & Batinic-Haberle, 2001,) from which is possible to know the oxidation state. Manganese–porphyrin complexes have more extensively studied because were found to be similar to the biologically active compounds (Nakanishi et al. 2000; Meunier, 1992; Perie & Barbe, 1996; Balahura & Kirby, 1994; Haber et al., 2000; Cuzzocrea et al., 2001), and because were also used as catalysts for the oxygenation of alkanes, alkenes and compounds containing nitrogen and sulphur (Mansuy & Momenteau, 1982; Fontecave & Mansuy, 1984). The very important properties that influence the reactivity of the Mn(III)-porphyrin concerns the changes in the oxidation states of Mn in the complexes for its high reactivity with O_2 (Cuzzocrea et al., 2001). Manganese, in the complexes obtained by the reaction of Mn(II) with the porphyrins, has oxidation number +3, so the complex of Mn(II) can be obtained only by reduction, while those of Mn(IV) and Mn(V) for the oxidation of Mn(III)-complexes. Interesting is the reactions of a natural porphyrin, the acid 2,7,12,17 tetrapropionic of 3,8,13,18 tetramethyl-21H, 23H-porphyrin called Coproporphyrin- I (CPI), with manganese (III) that, with different pH and solvent compositions, show the formation of $[Mn^{III}CPI(H_2O)_2]$, $[Mn^{III}CPI(OH)_2]$, $[Mn^{IV}(O)CPI(OH)]$, $[Mn^{V}(O)CPI(OH)]$, $[Mn^{II}CPI(OH)]$ (Fig. 13) with specific Uv-Vis adsorptions as reported in Table 1. (Giovannetti et al., 2010).

5.1 Complexation kinetics

Rates of the complexation of porphyrins with metal ions are very much slower by several orders of magnitude than those of acyclic ligands (Funahashi, S. et al., 1984). Such very slow rates have been discussed in terms of the rigidity of the planar porphyrin framework. The electronic nature of porphyrins, and also the steric accessibility of the bound metal center, can be varied by using electron-donating or with drawing substituents at the meso carbon or in the pyrrolic positions. While such substituent-based changes have been seen to influence the extent of apical ligand binding, as well as the stability of the metal complexes, there is a relatively small effect on the ability to insert cations into the nitrogen core (Lin & Lash, 1995).

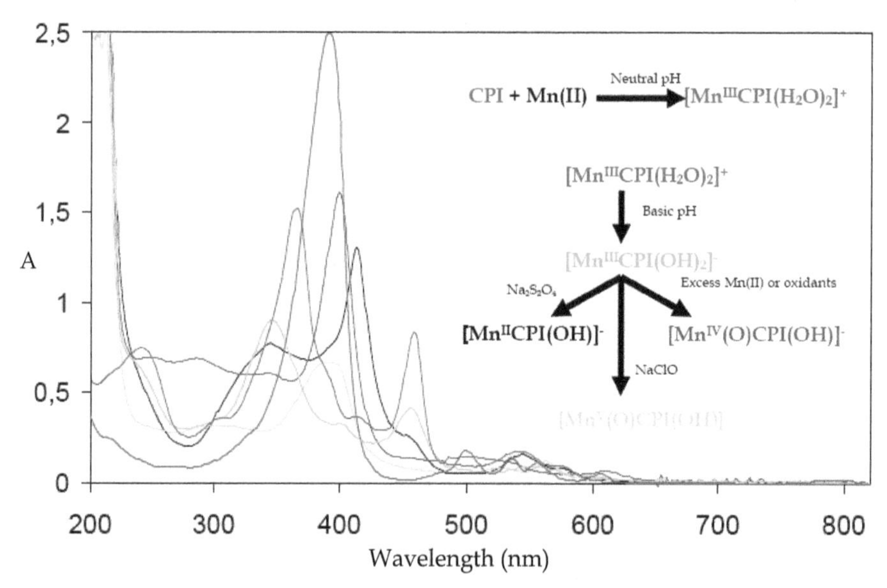

Fig. 13. Uv-Vis adsorption spectra of CPI, [Mn^{III}CPI(H$_2$O)$_2$] [Mn^{III}CPI(OH)$_2$], [Mn^{IV}(O)CPI(OH)], [Mn^{V}(O)CPI(OH)], [Mn^{II}CPI(OH)] with in insert, the experimental conditions for the preparation of all complexes obtained with several reagents.

Complex	Soret band λ (nm)	Q band (β, α) λ (nm)
[Mn^{II}CPIOH]-	414	544, 574
[Mn^{III}CPI(H$_2$O)$_2$]$^+$	366, 458	542, 572
[Mn^{III}CPI(OH)$_2$]	348, 458	556
[Mn^{IV}OCPI(OH)]-	400	502, 610
[Mn^{V}OCPI(OH)]	384	526, 562

Table 1 Spectral characteristics of Mn–CPI complexes.

In the case of N-substituted porphyrins, in which one of the two hydrogen atoms bound to the pyrrole nitrogen atoms is substituted by an alkyl or aryl group, displacement of the substituent from the porphyrin plane due to its bulkiness causes tilting distortion of the pyrrole rings of the porphyrin as shown by X-ray crystallography (Lavallee & Anderson, 1982; Aizawa et al., 1993). The investigation of the kinetics of the complexation of N-substituted porphyrins with several metal ions, showing that N-alkylporphyrins form metal complexes much faster than corresponding non-N-alkylated porphyrins (Aizawa et al., 1993; Shah et al., 1971; Shah et al., 1971; Anderson & Lavallee, 1977; Lavalle et al., 1978; Anderson et al., 1980; Kuila & Lavallee, 1984; Schauer et al., 1987; Balch et al., 1990; McLaughlin, 1974;

Bain-Ackerman & Lavallee, 1979; Funahashi et al., 1984; Funahashi et al., 1986;Lavallee et al., 1986; Bartczak et al., 1990; Shimizu et al., 1992).

The formation of metalloporphyrins can be accelerated substantially by the use of an auxiliary complexing agent. For example, the rate of complexation of TMPyP with Cu(II) and Mg(II) was accelerated by L-cysteine (Watanabe & Ohmori 1981) and 8-quinolinol (Makino & Itoh, 1981), respectively. The organic ligands with an extended π-electron structure, such as imidazole or bipyridine (Giovannetti et al., 1995; Kawamura et al., 1988; Ishi & Tsuchiai 1987; Tabata & Kajhara, 1989) and L-tryptophan (Tabata & Tanaka, 1988) also show a tendency to accelerate the formation of metalloporphyrins because form intermediate molecular complexes with metal ions and porphyrin reagents. In the presence of tryptophan, the rate of incorporation of Zn(II) to TPPS4 is about 100 times greater than in its absence (Tabata & Tanaka, 1988).

Hence, larger metal ions such as Pb^{2+}, Hg^{2+}, or Cd^{2+} can catalyse the formation of regular metalloporphyrins via generation of SAT complexes as intermediates. This because in the SAT complexes, the distortion caused by the out-of-plane location of the larger metal center, makes two diagonal pyrrolic nitrogens more accessible to another metal ion, even with smaller ionic radius, on the other side of the porphyrin ligand that can easily coordinate to them (Stinson & Hambright, 1977; Inamo et al 2001; Wittmer & Holten, 1996; Tabata & Tanaka 1985; Tung & Chen, 2000; Tabata et al., 1995; Grant & Hambright 1969; .R. Robinson & Hambright, 1992; C. Stinson & Hambright, 1977; Barkigia et al., 1990; Giovannetti et al., 1998).

Since the deformation of the porphyrin ring proved to be the main factor governing the acceleration of the metalloporphyrin formation (Lavallee, 1985; Tabata & Tanaka, 1991), this can also be achieved by substituents at the porphyrin core or at the peripheral ring. Thus, e.g., the peripheral or substituted octabromoporphyrins display a buckled structure due to steric hindrance between the substituents (Bhyrappa & Krishnan, 1991; Mandon et al., 1992; Henling et al., 1993; Brinbaum et al., 1995). Such a deformation profoundly enhanced the reactivity of the porphyrin even towards Hg^{2+} (Nahar & Tabata, 1998), the ionic radius of which is rather large anyway.

6. Aggregation of porphyrins

An increasing interest in recent years is due to supramolecular assemblies of π-conjugated systems for their potential applications in optoelectronic and photovoltaic devices (Schenning & Meijer, 2005).

Molecular aggregates of several dyes have been studied as organic photoconductors (Borsenberger et al.,1978), as markers for biological and artificial membrane systems (Waggoner, 1976), as materials with high non-linear optical properties suitable for optical devices ([Hanamura,1988; Sasaki & Kobayashi, 1993; Wang, 1986; Wang, 1991). Some properties of molecular structure of aggregates permit their use in superconductivity, and other processes (Kobayashi,1992; Schouten et al., 1991; Collman, 1986). Aggregation of small organic molecules to form large clusters is of large interest in chemistry, physics and biology. In nature, particularly in living systems, self-association of molecules plays a very important role; an example is given by molecular aggregates of chlorophyll that have been found to mediate the primary light harvesting and charge-transfer processes in photosynthetic complexes (Creightonet al., 1988; Kuhlbrandt, 1995). In fact, light-harvesting and the primary charge-separation steps in photosynthesis are facilitated by aggregated species, i.e., chlorophylls.

Self-assembly of molecules, driven by not-covalent intermolecular interactions, is a convenient route for manufacturing of new functional materials (Lidzey at al., 2000; Van der Boom et al., 2002; Fudickar et al. 2002; Lagoudakis et al., 2004; Li et al., 2003).

Recently, porphyrin assembly has been used for light-driven energy transduction systems, copying the photophysical processes of photosynthetic organisms (Choi et al.,2004); Choi et al., 2003; Choi et al., 2002; Choi et al., 2001; Luo et al., 2005).

The aggregation and dimerization of porphyrins and metalloporphyrins in aqueous solution have been widely investigated (Borissevitch & Gandini, 1998; Pasternack et al, 1985) and it has been deduced that it is dependent strictly on physical-chemical characteristics, such as, ionic strength, pH and solvent composition; the combination of these factors can facilitate the aggregation processes (Kubat et al., 2003; Giovannetti et al., 2010).

The aggregation of porphyrins, changing their spectral and energetic characteristics, influences their efficacy in several applications thus, it is very important take on detailed informations about the formation dynamic and on the typology of aggregates. In the metal complexation of porphyrins the efficiency reaction is affected by their aggregation (Yusmanov et al, 1996). Several authors have observed that in the photogeneration of H_2O_2 by porphyrins, the efficiency of production was highly dependent on their aggregation state (Komagoe et al, 2006).

The diverse chemical and photophysical properties of porphyrins are in many cases due to their different aggregation mode and, as a result of interchromophoric interactions, perturbations in the electronic absorption spectra of dyes occur. Deviations from Beer's law are often used to investigate the porphyrin aggregation in solution.

Because the aggregates of porphyrins show peculiar spectroscopic properties, the molecular associations of porphyrins were generally investigated using UV–vis absorption and fluorescence spectroscopy (Ohmo, 1993).

The characteristic of porphyrin molecule with 22 π-electrons causes a strong π –π interaction (Van de Craats, & Warman, 2001), facilitating the formation of two structure types: "H-type" with bathochromic shift of B and Q bands and "J-type" with blue shift of B band and red-shift of Q band, with respect to those of monomer.

The J-type aggregates (side-by-side) were formed for transitions polarized parallel to the long axis of the aggregate, while H-type (face-to-face) for transitions polarized perpendicular to it (Fig. 14).

J-aggregates are formed with the monomeric molecules arranged in one dimension such that the transition moment of the monomers are parallel and the angle between the transition moment and the line joining the molecular centers is zero (Bohn, 1993). The strong coupling of monomers results in a coherent excitation with a red-shift relative to the monomer band. *H-aggregates* are again a one-dimensional arrangement of strongly coupled monomers, but the transition moments of the monomers are perpendicular (ideal case) to the line of centers. On the contrary of J-aggregates, the arrangement in H-aggregates is face-to-face. The dipolar coupling between monomers leads to a blue shift of the absorption band (Czikklely et al.,

1970; Nuesch et al., 1995). The H-aggregates are not known to have sharp spectra like the J-aggregates; nevertheless, there are many examples where the spectroscopic blue shift, evident for formation of H-aggregates, was observed.

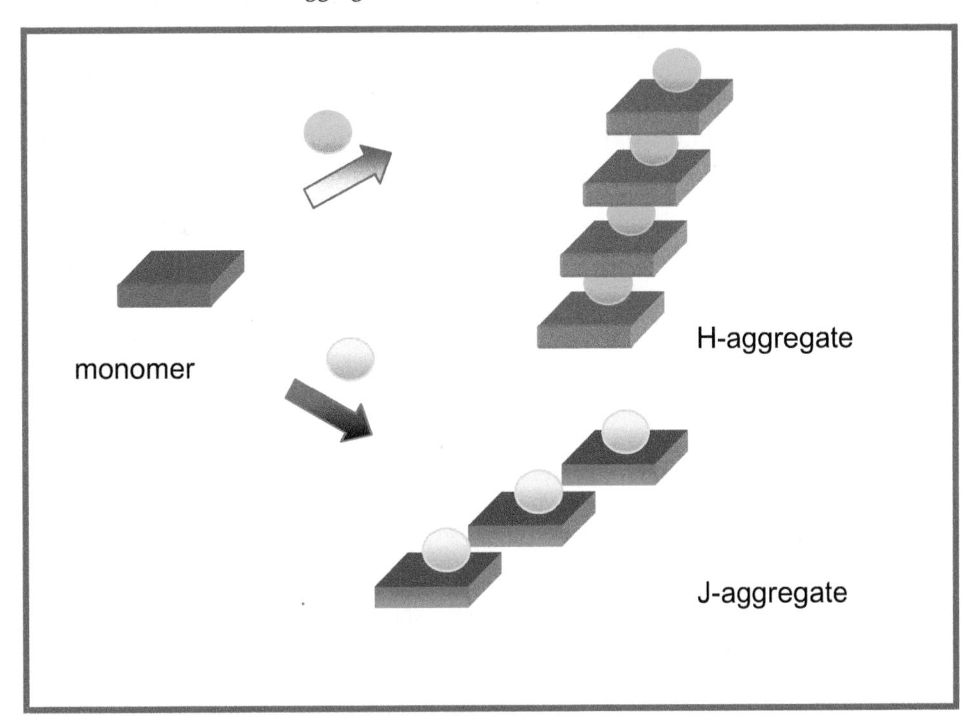

Fig. 14. Schematic representation of H and J aggregates.

These aggregated are of particular interest because the highly ordered molecular arrangement present unique electronic and spectroscopic properties that can be predicted (Fidder et al., 1991; West & Carroll,1966; Furuki et al., 1989; Spano & Mukamel, 1989; Bohn, 1993).

H and J aggregates can be obtained under specific conditions (Misawa & Kobayashi, 1999; Maiti et al., 1995; Kanojk & Kobayashi, 2002; Pasternack et al., 1994; Akins et al., 1994; Ohno et al., 1993; Luca et al., 2005; Luca et al., 2006; Luca et al., 2006). Due to the distinct optical properties, also the control of the formation of H- and J-aggregated states of dyes has attracted much research interest (Maiti et al., 1998; Shirakawa et al., 2003; De Luca et al., 2006; Egawa et al., 2007; Yagai et al., 2008; Gadde et al., 2008; Ghosh et al., 2008; Zhao et al., 2008; Delbosc et al., 2010).

As a result, a wide variety of self-assembled porphyrin structures are highly desirable for practical use, which can be applied to nonlinear optical materials (Collini et al., 2006; Liu et al., 2006; Matsuzaki et al., 2006; Terazima et al., 1997), organic solar cells (Hasobe et al., 2003) and sensor devices (Fujii et al., 2005; Lucaet al., 2007).

In general, the aggregate formation of porphyrins have been studied in solution, and their physicochemical properties can be affected by the ionic strength, nature of the titrating acid, temperature, pH, peripheral substitution and presence of surfactants of ions (Choi et al., 2003; Ohno et al., 1993; Napoli et al., 2004; Kubat et al., 2003; Siskova et al., 2005).

H- and J-aggregates was formed by simply mixing aqueous solutions of two kinds of porphyrins with opposite charges (Xiangqing et al., 2007).

Moreover, self-assembly can be mediated by templates that allows for obtaining aggregates with additional properties, e.g., chiral templates (Koti & Periasamy, 2003; Mammana et al., 2007). In these systems, coupling of strong transition dipoles can result in a perturbations to the electronic absorption spectra of monomer with hyposochromic and bathochromic shift of the monomer Soret band, for H and J aggregates formation respectively (Gourterman et al., 1977; Zimmermann et al., 2003; Scherz & Parson, 1984). The produced splitting is proportional to the magnitude of transition dipole coupling between adjacent molecules.

Although much has been studied about the spectroscopic features and excitonic interactions in molecular aggregates, the detailed information of geometrical structure, especially the molecular orientation, are still the subjects of continuing interests (Nikiforov et al., 2008; Jeukens et al., 2004).

The aggregates of porphyrins have been formed in solutions in the form of fibers, ribbons and tubules by the self-association or aggregation method (Rotomskis, 2004; Fuhrhop, 1993; Giovannetti et al., 2010).

Some H- or J-type aggregates of porphyrins play a role as light harvesting assemblies to gather and transfer energy to the assembled devices, and to obtain a higher incident photon-to-photocurrent generation efficiency (Kamat et al.,2000; Sudeep et al., 2002).

The structural, kinetic, and spectroscopic studies on J- and H-aggregates provide useful information for understanding molecular interactions in aggregation processes.

The kinetics of the formation of the porphyrin aggregate and its structure are sensitive of experimental conditions (Giovannetti et al., 2010). The monomer – aggregated species is a system of multiple equilibria. Spectrophotometric monitoring in the time of the Uv- Vis absorbance permit to obtain information of intermediate species, of type of the aggregate, and of their transformation. For this, for evaluated the polymerization kinetic constants, the concentrations of monomeric [M] and dimeric form [D], can be calculated from the relative absorption maxima at each time. If k_{pol} is the polymerisation kinetic constant, C_M and C_D denoted the initial monomer and dimer concentrations, the reaction rate can be expressed as in equation (2):

$$k_{pol}t = \frac{1}{C_M - C_D} \ln \frac{C_D[M]}{C_M[D]} \tag{2}$$

The plot of the right term of this equation versus t gave good straight lines the slopes of which represented the values of k_{pol}.

The information derived from such studies can help in achieving appropriate design of photoactive aggregates for mimicking light-harvesting natural photosynthetic pigments, photodynamic therapeutic use, and advanced nonlinear optical materials.

7. Conclusions

The porphyrins represent a fascinating world of molecules with sensational properties. Many results have been obtained by careful observation and with detailed studies of their chemical and physical properties due to the use of UV-Vis spectrophotometry between the interpretation of Soret and Q band transformations.

In this contest, the light absorbing power of porphyrins and related compounds should be used in the near future for other many applications and much more can still be studied in the future.

8. References

Acheson, R.M. (1976). *An Introduction to the Chemistry of Heterocyclic Compounds*, 3rd ed; John Wiley & Sons: New York.

Adler, A. D. Longo, F. R. Finarelli, J. D. Goldmacher, J. Assour, J. Korsakoff, L. (1967). *J. Org. Chem.* 32, 476.

Aizawa, S. Tsuda, Y. Ito, Y. Hatano K. Funahashi, S. (1993). *Inorg. Chem.*, 32, 1119.

Akins, D. L. Zhu, H.-R. Guo, C. (1996). *J. Phys. Chem.*, 100, 5420.

Akins, D. L. Zhu, H.-R. Guo, C. (1994). *J. Phys. Chem.*, 98, 3612.

Anderson, H.J. Loader, C.E. Jackson, A.H., (1990) In *Pyrrole*, Jones, R. A. Ed. *The Chemistry of Heterocyclic Compounds*, Vol. 48, John Wiley & Sons: New York 295-397.

Anderson O.P. Lavallee, D.K. (1976). *J. Am. Chem. Soc.*, 98, 4670.

Anderson O.P. Lavallee, D.K. (1977). *J. Am. Chem. Soc.*, 99, 1404.

Anderson O.P. Lavallee, D.K. (1977). *Inorg. Chem.*, 16, 1634.

Anderson, O.P. Kopelove A.B. Lavallee, D.K. (1980). *Inorg. Chem.*, 19, 2101.

Arai Y. & Segawa H. (2011). *J. Phys. Chem. B*, 115, 7773.

Atwood, J.L. Davies, J.E.D. MacNicol, D.D. Vogtle, F. Lehn, J.-M., Eds., *Handbook : Comprehensive Supramolecular Chemistry*, (1996). Pergamon, Oxford.

Bailey, S. L. , Hambright P. (2003). *Inorganica Chimica Acta* , 344, 43.

Bain-Ackerman M.J. Lavallee, D.K. (1979). *Inorg. Chem.*, 18, 3358.

Balaban, T. S. (2005). *Acc.Chem. Res.*, 38, 612.

Balahura, R.J. Kirby, R.A. (1994). *Inorg. Chem.*, 33, 1021.

Balaz, M. Holmes, A.E. Benedetti, M. Rodriguez, P.C. Berova, N. Nakanishi, K. Proni, G., (2005). *J. Am. Chem. Soc., 127*, 4172.

Balch, A.L. Cornman, C.R. Latos-Grazynski L. Olmstead, M.M. (1990), *J. Am. Chem. Soc.*, 112, 7552.

Barkigia, K.M. Berber, M.D. Fajer J., Medforth, C.J. Renner M.W., Smith, K.M. (1990). *J. Am. Chem. Soc.*, 112,

Barkigia, K.M. Fajer, J. Adler, A.D. Williams, G.J.B. (1980). *Inorg. Chem.*, 19, 2057.

Bartczak, T.J. Latos-Grazynski L. Wyslouch, A. (1990). *Inorg. Chim. Acta*, 171, 205.

Beletskaya, I. Tyurin, V. S. Tsivadze, A. Y. Guilard, R. Stern, C. (2009). *Chem. Rev.*, 109, 1659.

Bhyrappa, P. Krishnan, V. (1991) *Inorg. Chem.*, 30, 239.

Biesaga, M. Pyrzynska, K. Trojanowicz, M. (2000). *Talanta*, 51, 209.

Bohn, P. W. (1993). *Annu. Rev. Phys. Chem.*, 44, 37.

Borissevitch, I.E. Gandini, S.C. (1998). *J. Photochem. Photobiol. B: Biol.*, 43, 112.

Borsenberger, P. Chowdry, A. Hoesterey, D. Mey, W. (1978). *J. Appl. Phys.*, 44, 5555.

Brinbaum, E.R. Hodge, J.A. Grinstaff, M.W. Schaefer, W.P. Henling, L. Labinger, J.A. Bercaw, J.E. Gray, H.B. (1995). *Inorg. Chem*,. 34, 3625.
Calvin, M. Ball, R. H. Arnoff, S. (1943). *J. Am. Chem. Soc.*, 65, 2259.
Choi, M. Y. Pollard, J. A. Webb, M. A. McHale, J. L. (003). *J. Am. Chem. Soc.*, 125, 810.
Choi, M.-S. Aida, T. Luo, H. Araki, Y. Ito, O. (2003). *Angew. Chem., Int. Ed.*, 42, 4060.
Choi, M.-S. Aida, T. Yamazaki, I. Yamazaki, T. (2001). *Angew. Chem., Int. Ed.*, 40, 194.
Choi, M.-S. Aida, T. Yamazaki, I. Yamazaki, T. (2002). *Chem. Eur. J.*, 8, 2667.
Choi, M.-S. Aida, T. Yamazaki, I. Yamazaki, T. (2004). *Angew. Chem., Int. Ed.* ,43, 150.
Collini, E. Ferrante, C. Bozio, R. Lodi, A. Ponterini, G. (2006). *J. Mater. Chem.* 16, 1573.
Collman, J.P. McDeevitt, J.T. Yee, G.T. Leidner, C.R. McCullough, L.G. Little, W.A. Torrance, J.B. (1986). *Proc. Natl. Acad. Sci. U.S.A.*, 83, 44581.
Creighton, S. Hwang, J. Warshel, A. Parson, W. Norris, J. (1988). *Biochemistry*, 27, 774.
Cuzzocrea, S. Riley D.P., Caputi A.P., Salvemini D. (2001). *Pharmacol. Rev.* , 53, 135.
Czikklely, V. Forsterling, H. D. Kuhn, H. (1970). *Chem. Phys. Lett.*, 6, 207.
Dailey, H.A. Ed. (1990) *Biosynthesis of Heme and Chlorophylls*, McGrawHill, Inc.: New York.
De Luca, G. Pollicino, G. Romeo, A. Scolaro, L. M. (2006). *Chem. Mater.*, 8, 18.
De Luca, G. Romeo, A. Scolaro, L. M. (2005). *J. Phys. Chem. B*, 109, 7149.
De Luca, G.; Romeo, A.; Scolaro, L. M. *(2006). J. Phys. Chem. B*, 110, 14135.
De Luca, G.; Romeo, A.; Scolaro, L. M. (2006). *J. Phys. Chem. B*, 110, 7309.
Delbosc, N. Reynes, M. Dautel, O. J. Wantz, G. Lere-Porte, Fidder, H. Terpstra, J. Wiersma, D. A. (1991). *J. Chem. Phys, 94*, 6895.
Delbosc, N. Reynes, M. Dautel, O. J. Wantz, G. Lere-Porte, J.-P. Moreau, J. J. E. (2010). *Chem. Mater.*, 22, 5258.
Di Natale, C. Salimbeni, D. Paolesse, R. Macagnano, A. D'Amico, A., (2000). *Sens. Actuators, B, 65*, 220
Di Natale,C. Paolesse, R. D'Amico, A., (2007). *Sens. Actuators, B, 121*, 238.
Dolphin, D. Ed *The Porphyrins*, (1978).Academic, New York,.
Drain, C. M. Alessandro, V. Radivojevic, I. (2009). *Chem. Rev.*, 109, 1630.
Egawa, Y. Hayashida, R. Anzai, J. (2007) .*Langmuir*, 23, 13146.
Elemans, J. A. A. W. Van Hameren, R. Nolte, R. J. M. Rowan, A. E. (2006). *Adv. Mater.*, 18, 1251.
Fleischer, E.B. Wang, J.H. (1960). *J. Am. Chem. Soc.*, 82, 3498.
Fontecave, M. Mansuy, D. (1984). *Tetrahedron Lett.*, 21, 4297.
Fuhrhop, J.H. Bindig, U. Siggel, U. (1993). *J. Am. Chem. Soc.*, 115, 11036.
Fujii, Y. Hasegawa, Y. Yanagida, S. Wada, Y. (2005). *Chem. Commun.* 3065.
Funahashi S., Inada, Y. Inamo, (2001). M. *Anal. Sci.* 917.
Funahashi, S. Ito, Y. Kakito, H. Inamo, M. Hamada Y. Tanaka, M. (1986). *Mikrochim. Acta*, 33.
Funahashi, S. Yamaguchi, Y. Tanaka, M. (1984). *Bull. Chem. Soc. Jpn.*, 57, 204.
Funahashi, S. Yamaguchi, Y. Tanaka, M. (1984). *Inorg. Chem.*, 23, 2249
Furuki, M. Ageishi, K. Kim, S. Ando, I. Pu, L. S. (1989).*Thin Solid Films, 180*, 193.
Gadde, S. Batchelor, E. K. Weiss, J. P. Ling, Y. Kaifer, A. E. (2008). *J. Am. Chem. Soc.*, 130, 17114.
Ghosh, S. Li, X. Stepanenko, V. Wrthner, F. (2008).*Chem. Eur. J.*, 14, 11343.
Giovannetti R., Alibabaei, L. Petetta, L. (2010).*Journal of Photochemistry and Photobiology A: Chemistry*, 211,108.
Giovannetti, R. Bartocci, V. Ferraro, S. Gusteri, M. Passamonti, P. (1995). *Talanta*, 42, 1913.

Giovannetti, R. Alibabaei, L. Pucciarelli, F. (2010). *Inorganica Chimica Acta*,363, 1561.
Goldberg D.E. Thomas, K.M. (1976). *J. Am. Chem. Soc.*, 98, 913.
Gourterman, M. Holten, D. Lieberman, E. (1977). *Chem. Phys.*, 25, 139.
Gouterman, M. (1961). *J. Mol. Spectroscopy*, 6, 138.
Gouterman, M. (1959). *J. Chem. Phys.*, *30*, 1139.
Grant, C. Hambright, P. (1969). *J. Am. Chem. Soc.*, 91, 4195.
Gulino, A. Mineo, P. Bazzano, S. Vitalini, D. Fragalà, I., (2005). *Chem. Mater.*, *17*, 4043.
Haber, J. Matachowski, L. Pamin, K. Potowicz, J. (2000). *J. Mol. Catal. A*, 162, 105.
Hambright, P. inK.H. Kadish, K.M. Smith, R. Guilard (Eds.), *The Porphyrin Handbook*, (2001). vol. 3 (chap. 18), Academic Press, NewYork,.
Hanamura, E. (1988). *Phys. Rev. B*, 37,1273.
Hasobe, T. Imahori, H. Fukuzumi, S. Kamat P.V., (2003). *J. Mater. Chem.*,13, 2515.
Henling, L.M. Schaeffer, W.P. Hodge, J.A. Hughes, M.E. Gray, H.B. (1993). *Acta Crystallogr.* , 49, 1743.
Hoeben, F. J. M. Jonkheijm, P. Meijer, E. W. Schenning, A. P. H. (2005). *J. Chem. Rev.*, 105, 1491.
Horváth, O. Huszánk, R. Valicsek, Z. Lendvay, G. (2006). *Coord. Chem.Rev.*, 250, 1792.
Horváth, O. Valicsek, Z. Vogler, A. (2004). *Inorg. Chem. Commun.*, 7, 854.
Huszánk, R. Horváth, O. (2005). *Chem. Commun.* 224.
Huszánk, R. Lendvay, G. Horváth, O. (2007). *J. Bioinorg. Chem.*, 12, 681.
Inamo, M. Kamiya, N. Inada, Y. Nomura, M. Funahashi, S. (2001). *Inorg. Chem.*, 40, 5636.
Ishi, H. Tsuchiai, H. (1987) *Anal. Sci.*, 3, 229.
Jeukens, C.R.L.P.N. Lensen, M.C. Wijnen, F.J.P. Elemans, J.A.A.W. Christianen, P.C.M.
Kadish, K.M. Smith, K.M. Guilard, R. (1999). *The Porphyrin Handbook*, Academic Press, S. Diego, CA.
Kahn, K. Bruice, T.C., (2003). *J. Phys. Chem. B*, *107*, 6876.
Kamat, P.V. Barazzouk, S. Hotchandani, S. Thomas, K.G. (2000). *Chem. Eur. J.*, 6, 3914.
Kamat, P.V. Barazzouk, S. Thomas, K.G. Hotchandani, S. (2000).*J. Phys. Chem. B*, 104, 4014.
Kano, H. Kobayashi, T. J. (2002). *Chem. Phys.*, 116, 184.
Kawamura, K. Igarashi, S. Yotsuyanagi, T. (2011). *Microchim. Acta* ,172, 319.
Kawamura, K. Igrashi, S. Yotsuyanagi, T. (1988). *Anal. Sci.*, 4, 175.
Kilian, K. Pyrzynska, K. (2003). *Talanta*, 60, 669.
Knör, G. Strasser, A. (2005). *Inorg. Chem. Commun.* , 9, 471.
Kobayashi, (1992). S. *Mol. Cryst. Liq. Cryst.*, 217, 77.
Komagoe, K. Katsu, T. (2006). *Anal. Sci.*, 22, 255.
Koti, A.S.R. Periasamy, N. (2003). *Chem. Mater.*, 15, 369.
Kubat, P. Lang, K. Prochazkovà, K. Anzenbacher Jr. , P. (2003). *Langmuir*, 19, 422.
Kuhlbrandt, W. (1995). *Nature* , 374, 497.
Kuila D. Lavallee, D.K. (1984). *J. Am. Chem. Soc.*, 106, 448.
Lagoudakis, P.G. de Souza M.M., Schindler F., Lupton J.M., Feldmann, J. Wenus, J. Lidzey, D.G. (2004). *Phys. Rev. Lett.* ,93, 257401.
Lavallee D.K. and Anderson, O.P. (1982). *J. Am. Chem. Soc.*, 104, 4707.
Lavallee, D.K. (1987).*The Chemistry and Biochemistry of N-Substituted Porphyrins, VCH, New York.*
Lavallee, D.K. (1985). *Coord. Chem. Rev.*, 61, 55.
Lavallee, D.K. Kopelove A.B. Anderson, O.P. (1978). *J. Am. Chem. Soc.*, 100, 3025.

Lavallee, D.K. Wite, A. Diaz, ABattioni . J.-P. Mansuy, D. (1986). *Tetrahedron Lett.*, 27, 3521.

Lehn, J.-M., (2002). *Science, 295,* 2400.

Li, G. Fudickar, W. Skupin, M. Klyszcz, A. Draeger, C. Lauer, M. Fuhrhop, J.-H. (2002). *Angew. Chem. Int.* Edit 41, 1828.

Li, L.-L. Yang, C.-J. Chen, W.-H. Lin, K.-J. (2003). *Angew. Chem. Int.* Edit 42, 1505.

Liao, M.S. Watts, J.D. Huang, M.J. (2006). *J. Phys. Chem. A,* 110, 13089.

Lindsey, J. S. Schreiman, I.C. Hsu, H.C. Kearney, P.C. Marguerettaz, A.M. (1987). *J. Org. Chem.* 52, 827.

Lidzey, D.G. Bradley, D.D.C. Armitage, A. Walker, S. Skolnick, M.S. (2000). *Science,* 288, 1620.

Lim, M.D. Lorkovic, I.M. Ford, P.C. (2005). *J. Inorg. Biochem.*, 99, 151.

Lin, Y. Lash, T. (1995).*Tetrahedron Lett.*, 36, 9441

Liu, Z.-B. Zhu, Y.-Z. Chen, S.-Q. Zheng, J.-Y. Tian, J.-G. (2006). *J. Phys. Chem. B,* 110, 15140.

Luca, G.D. Pollicino, G. Romeo, A. Scolaro, L.M. (2007). *Chem. Mater.*, 18, 2005.

Luca, G.D. Romeo, A. Scolaro, L.M. (2005). *J. Phys. Chem. B,* 109, 7149.

Luca, G.D. Romeo, A. Scolaro, L.M. (2006). *J. Phys. Chem. B,* 110, 14135.

Luca, G.D. Romeo, A. Scolaro, L.M. (2006). *J. Phys. Chem. B,* 110, 7309.

Luo, H. Choi, M.-S. Araki, Y. Ito, O. Aida, T. (2005). *Bull. Chem. Soc. Jpn.*, 78, 405.

Ma, S.Y. (2000). *J. Chem. Phys. Lett*, 332, 603,

Maiti, N. C. Mazumdar, S. Periasamy, N. (1998). *J. Phys. Chem. B* 102, 1528.

Maiti, N.C. Ravikanth, M. Mazumdar, S. Periasamy N., (1995). *J. Phys. Chem.*, 99, 17192.

Makino, T. Itoh, J. (1981) *Clin. Chim. Acta,* 111, 1.

Mammana, A. Durso, A. Lauceri, R. Purrello, R. (2007). *J. Am. Chem. Soc.*, 129, 8062.

Mandon, D. Ochsenbein, P. Fischer, J. Weiss, R. Jayaraj, K. Austin, R.N. Gold, A. White, P.S. Brigaud, O. Battioni, P. Mansuy, D. (1992). *Inorg. Chem.*, 31, 2044.

Mansuy, D. Momenteau, (1982). M. *Tetrahedron Lett.*, 2781.

Martirosyan, G.G. Azizyan, A.S. Kurtikyan, T.S. Ford, P.C., (2004). *Chem. Commun.*, 1488.

Matsuzaki, Y. Nogami, A. Tsuda, A. Osuka, A. Tanaka K., (2006). *J. Phys. Chem. A* ,110, 4888.

McLaughlin, G.M. (1974). *J. Chem. Soc., Perkin Trans.* 2, 136.

Medforth, C. J. Wang, Z. Martin, K. E. Song, Y. Jacobsenc, J. L. Shelnutt, (2009). *J. A. Chem. Commun.*, 7261.

Menotti, A. R. (1941). *J. Am. Chem. Soc.*, 63, 267.

Meunier, B. (1992). *Chem. Rev.* , 92, 1411.

Milgrom, L.R. (1997) *The Colours of Life*, OUP, Oxford,;

Misawa, K. Kobayashi, T. (1999). *J. Chem. Phys.*, 110, 5844.

Nahar, N. Tabata, M. (1998). *J. Porphyr. Phthalocyan.*, 2, 397.

Nakanishi, I. Fukuzumi, S. Barbe, J.M. Guilard, R. Kadish, K.M. (2000). *Eur. J. Inorg.Chem.*, 1557.

Napoli, M.D. Nardis, S. Paolesse, R. Graca, M. Vicente, H. Lauceri, R. Purrello, R. (2004). *J. Am. Chem. Soc.* 126, 5934.

Nappa, M. J. S. Valentine, (1978). *J. Am. Chem. Soc.*, 100, 5075

Nikiforov, M.P. Zerweck, U. Milde, P. Loppacher, C. Park, T.-H. Tetsuo Uyeda, H. Therien, M.J. Eng, L. Bonnell, D. (2008). *Nano Lett.*, 8,110.

Nuesch, F.; Gratzel, M. (1995). *Chem. Phys.*, 193, 1.

Ohmo, O. Kaizu, Y. Kobayashi, H. (1993). *J. Chem. Phys.* 99, 4128.

Okada, S.; Segawa, H. (2003). *J. Am. Chem. Soc.*, 125, 2792.

Pasternack, R. F. Schaefer, K. F. Hambright, P. (1994). *Inorg. Chem.*, 33, 2062.

Pasternack, R.F. Gibbs, E.J. Antebi, A. Bassner, S. Depoy, L. Turner, D.H. Williams, A. Laplace, F. Lansard, M.H. Merienne, C. Perrée-Fauvet, M. (1985). *J. Am. Chem. Soc.*, 107, 8179.

Perie, K. Barbe J.M., Cocolios P., (1996). *Soc. Chim. Fr.*, 133, 697.

Porra, R.J. (1997) *Photochem. Photobiol.*, *65*, 492.

Prins, L. J. Reinhoudt, D. N. Timmerman, P., (2001).*Angew. Chem. Int. Ed.*, *40*, 2382

Ren, Q.G. Zhou, X.T. Ji, H.B. (2010). *Chin. J. Org. Chem.* , 30, 1605.

Ribo, J. M. Crusats, J. Farrera, J. A. Valero, M. L. (1994). *J. Chem. Soc., Chem. Commun.*, 681.

Robinson, L.R. Hambright, P. (1992). *Inorg. Chem.*, 31, 652.

Rothemund, P. (1935). *J. Am. Chem. Soc.* 57, 2010.

Rothemund, P. (1936). *J. Am. Chem. Soc.* 58, 625.

Rotomskis, R. Augulis, R. Snitka, V. Valiokas, R. Liedberg, B. (2004). *J. Phys. Chem., B* 108, 2833.

Rowan, A.E. Gerritsen, J.W. Nolte, R.J.M. Maan, J.C. (2004). *Nano Lett.*, 4, 1401.

Rubio, M. Rios, B. O. Serrano-Andres, L. Merchan, M. (1999). *J. Chem. Phys.*, 110, 7202

Sasaki, F. Kobayashi, (1993). S. *Appl. Phys. Lett.*, 63, 2887.

Schauer, C.K. Anderson, O.P. Lavallee, D.K. Battioni J.-P. Mansuy, D. (1987). *J. Am. Chem. Soc.*, 109, 3922.

Schenning, A.P.H.J. Meijer, E.W. (2005). *Chem. Commun.*, 3245.

Scherz, A. Parson, W.W. (1984). *Biochim. Biophys. Acta* , 766, 653.

Schouten, P. Warman, J. De Haas, M. Fox, M. Pan, H. (1991). *Nature*, 353, 736.

Scolaro, L.M. Andrea Romeo, A. Pasternack, R.F., (2004). *J. Am. Chem. Soc.*, *126*, 7178.

Shah, B. Shears B. Hambright, P. (1971). *Inorg. Chem.*, 10, 1828.

Shen, Y. Ryde, U. (2005). *Chem. Eur.* J. ,11 1549.

Shimizu, Y. Taniguchi, K. Inada, Y. Funahashi, S. Tsuda, Y. Ito, Y. Inamo, M. Tanaka, M. (1992). *Bull. Chem. Soc. Jpn.*, 65, 771.

Shirakawa, M. Kawano, S. Fujita, N. Sada, K. Shinkai, S. (2003). *J. Org. Chem.*, 68, 5037.

Shundo, A. Labuta, J. Hill, J.P. Ishihara, S. Ariga, K., (2009). *J. Am. Chem. Soc.*, *131*, 9494.

Siskova, K. Vlckova, B. Mojzes, P. (2005). *J. Mol. Struct.* 744.

Spano, F. C. Mukamel, (1989). S. *Phys. Rev. A*, *40*, 5783.

Spasojevic´ I., Batinic´-Haberle, (2001). I. *Inorg. Chim. Acta*, 317, 230.

Stinson, C. Hambright, P. (1977). *J. Am. Chem. Soc.*, 99, 2357.

Sudeep, P.K. Ipe, B.I. Thomas, K.G. George, M.V. Barazzouk, S. Hotchandai, S. Kamat, P.V. (2002). *Nano Lett.*, 2, 29.

Tabata, M. Kajhara, N. (1989) *Anal. Sci.*, 5, 719.

Tabata, M. Miyata, W. Nahar, N. (1995). *Inorg. Chem.*, 34, 6492.

Tabata, M. Tanaka, M. (1985). *J. Chem. Soc., Chem. Commun.* 42.

Tabata, M. Tanaka, M. (1988).*Inorg. Chem.*, 27, 203.

Tabata, M. Tanaka, M. (1991).*Trends Anal. Chem.*, 10, 128.

Tabata, M. Nishimoto, J. Kusano, K. (1998). *Talanta* 46, 703.

Terazima, M. Shimizu, H. Osuka, A. (1997). *J. Appl. Phys.*, 81, 2946.

Tong,Y. Hamilton, D.G. Meillon, J.-C. Sanders, J.K.M., (1999) *Org. Lett.*, *1*, 1343.

Tovmasyan, A.G. Babayan, N.S. Sahakyan, L.A. Shahkhatuni, A.G. Gasparyan G.H., Aroutiounian, R.M. Ghazaryarn, R.K. (2008). *J. Porphyr. Phthalocya.*, 12, 1100.

Tung, J.Y. Chen, J.-H. (2000). *Inorg. Chem.*, 39, 2120.

Valicsek, Z. Horváth, O. (2007). *J. Photoch. Photobio. A*, 186, 1.
Valicsek, Z. Horváth, O. Lendvay, G. Kikas, Skoric, I.I. (2011). *J. Photoch. Photobio. A*, 218, 143.
Valicsek, Z. Lendvay, G. Horváth, O. (2008). *J. Phys. Chem. B*, 112, 14509.
Valicsek, Z. Lendvay, G. Horváth, O. (2009). *J. Porphyr. Phthalocya.*, 13, 910.
Van de Craats, A.M. Warman, J.M. (2001). *Adv. Mater.*, 12, 130.
Van der Boom, T. Hayes, R.T. Zhao, Y. Bushard, P.J. Weiss, E.A. Wasielewski, M.R. (2002). *J. Am. Chem. Soc.* 124 9582.
Waggoner, A. Membr. J. (1976). *Biol.* , 27, 317.
Walker, V.E.J. Castillo, N. Matta, C.F. Boyd, R.J. (2010). *J. Phys. Chem. A.*, 114, 10315.
Wang, M.-Y. R. Hoffman, B. M. (1984) *J. Am. Chem. Soc.*, 106, 4235.
Wang, Y. (1986). *Chem. Phys. Lett.*, 126, 209.
Wang, Y. (1991). *J. Opt. Soc. Am. B*, 8, 981.
Watanabe, H. Ohmori, H. (1981). *Talanta* 28 774.
West, W. Carroll, B. H. (1966). In *The Theory of Photographic Processes*, 3rd ed. James, T. H., Ed.; The McMillan Company: New York,; Chapter 12.
Whitten, D. G. Lopp, I. G. Wildes, P. D. (1968). *J. Am. Chem. Soc.*, 90, 7196
Wittmer, L.L. Holten, D. (1996), *J. Phys. Chem.*, 100, 860.
Wittmer, L.L. Holten, D. (1996). *J. Phys. Chem.*, 100, 860.
Xiangqing Li, Line Zhang, Jin Mu, (2007). *Colloids and Surfaces A: Physicochem. Eng. Aspects* ., 311, 187.
Yagai, S. Seki, T. Karatsu, T. Kitamura, A. Wrthner, F. (2008). *Angew. Chem.*, 47, 3367.
Yamamoto, S. Watarai, H. (2008). *J. Phys. Chem. C*, 112, 12417.
Yang, R. Wang, K. Long, L. Xiao, D. Xiaohai Yang, X. Tan, W. (2002). *Anal. Chem.*, , *74*, 1088.
Yusmanov, V.E. Tominaga, T.T. Borissevich, L.E. Imasato, H. Tabak, M. (1996). *Magn. Reson. Imaging*, 14, 255.
Zhang, Y. Yang, R. H. Liu, F. Li, K.A., (2004) *Anal. Chem.*, *76*, 7336.
Zhang, Y. Chen, P. Liu, M. (2008). *Chem. Eur. J.*, 14, 1793.
Zhang, Y. Chen, P. Ma, Y. He, S. Liu, M. (2009). *ACS Appl. Mater. Interfaces*, 1, 2036.
Zheng, W. Shan, N. Yu, L. Eang, X. (2008). *Dyes and Pigments* 77, 153.
Zhou, H. Baldini, L. Hong, J. Wilson, A. J. Hamilton, A.D., (2006). ,*J. Am. Chem. Soc.*, *128*, 2421.
Zimmermann, J. Siggel, U. Fuhrhop, J.-H. Roder, B. (2003). *J. Phys. Chem. B*. 107, 6019.

Synthesis and Characterization of CdSe Quantum Dots by UV-Vis Spectroscopy

Petero Kwizera[1,*], Alleyne Angela[1], Moses Wekesa[2,*], Md. Jamal Uddin[2,*] and M. Mobin Shaikh[3]
[1]*Department of Mathematics, Edward Waters College, Jacksonville, Florida,*
[2]*Department of Natural Sciences, Coppin State University, Baltimore, Maryland,*
[3]*Sophisticated Instrument Centre (SIC), School of Basic Science, Indian Institute of Technology Indore, Indore,*
[1,2]*USA*
[3]*India*

1. Introduction

CdSe nanocrystals are effective visual aid to demonstrate quantum mechanics, since their transition energies can be explained as a Particle in a Box, where a delocalized electron is the particle and the nanocrystal is the box. Kippeny and co-workers[1] have provided more background information and theoretical discussion. Additionally, Ellis et al.[2] have stated that modern science is becoming increasingly interdisciplinary. One example is material science, a broad, chemically oriented view of solids that results from the combined viewpoints of chemistry, physics, engineering, and for biotechnology, the biological sciences. Schulz[3] has suggested that nanotechnology is an exciting emerging field that involves the manipulation of the atoms and molecules at the nano scale. It is projected that important advances in engineering will come from understanding of the properties of matter constructed from building blocks whose size and shape is uniform and on the 1-100 nm scale. These consequences include technologies to be used in medicine[4], advances in computer technologies[5], defense[6] and everyday applications[3].

Several methods exist for synthesizing Cd-Se Quantum Dots. The Molecular Beam Epitax (MBE) is expensive and not readily accessible. Kippeny et al. have used dimethyl cadmium, which is expensive, explosive, and pyrophonic making the system difficult to control and reproduce. Peng and others[7-9] have pioneered the kinetic synthesis of Cd-Se nanocrystals from CdO and elemental Se. Boatman et al.[10] have prepared Cd-Se nanocrystals using a kinetic synthesis with a quenching technique where the temperature was 225°C. The visible absorption and emission spectra of individual samples collected at various time intervals during the experimental run were recorded and the maximum wavelength peak were determined. In this paper we report a modified technique of kinetic synthesis of Cd-Se nanocrystals that is safer, simple and can easily be carried out by students in the normal chemistry lab.

---

* Corresponding Authors

2. Experimental

UV-Vis spectrophotometer (Perkin Elmer Lambda 950) was used for spectroscopic measurements. The scan speed was 11.54 nm / min, integration time was 0.52 s and the data interval was 0.10 nm. The hotplate was Labcongo (115 V, 12 A) and the heating was set at level 3. All chemicals used were bought from Sigma-Aldrich and were of analytical grade. 60 mg of Se, 10 cm^3 of 1-octadecene and 0.8 cm^3 of trioctylphospine were mixed together in a round-bottomed flask. The solution was then continuously stirred with a magnetic stirrer on a hot-plate and warmed for a few minutes in a fume-hood. Separately, 26 mg of CdO was added to a 25 cm^3 round-bottomed flask and clamped in a heating mantle. 1.2 cm^3 of acid and 20 cm^3 of octadecene were added and mixed together. The solution was heated until CdO dissolved. The CdO solution was then sub-divided into 5 Erlenmeyer flasks each containing 4 cm^3 of the stock solution. 0.5 cm^3 of Se stock solution was then transferred into the CdO solution with pipette. The samples were heated for 50 s, 60 s, 70 s, 80 s and 120 s, respectively.

3. Results and discussion

The colloidal suspensions of Cd-Se quantum dots of increasing size from left to right are shown in Figure 1.

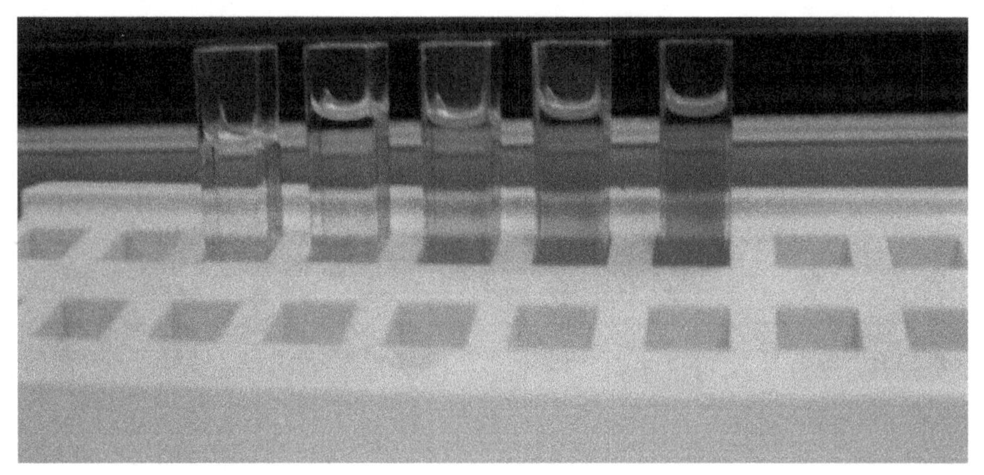

Fig. 1. Colloidal suspensions of Cd-Se quantum dots

The samples viewed in ambient light vary from green-yellow to orange-red. These changes in color which have been noted by other workers[1, 10] are attributed to the increasing size of the Cd-Se nano-crystals.

Figure 2 presents the calculated diameter of nanocrystal with time[1]. The diameter of nanocrystals increases with increasing time. As the nano-crystal size increases, the energy of the first excited state decreases.

The Cd-Se nano-crystals stay suspended in solution and cannot be filtered out. The oleic acid acts as a surfactant, binding to the exterior of the crystal lattice and allowing for the crystal to remain soluble in the octadecene[10]. The diameter of the nanocrystal was calculated using Kippeny[1] method and was found to be in the range found by other workers[10]. The Cd-Se crystal growth has been found to be temperature dependent. Transmission electron microscope (TEM) measurements of Cd-Se nanocrystals by others suggest that such wavelengths correspond to 2- 4 nm diameter crystals[10] with at most a few hundred atoms.

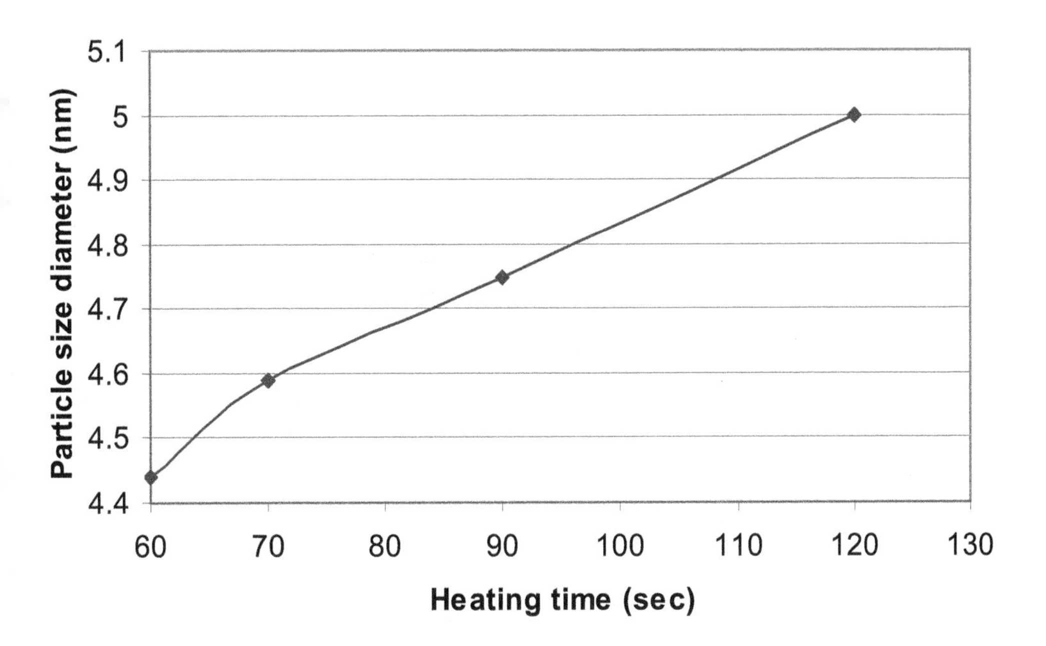

Fig. 2. The effect of particle size growth with time

Figure 3 presents ground state peak wavelengths as a function of reaction time. As the reaction progresses, the peak wavelength decreases. As nanocrystals grow, it has been suggested that their peak emission quickly approaches the band gap of bulk Cd-Se (730 nm).

The observable peak maximum shifts from violet to green with increasing crystal size. The absorption shows peak maxima with additional absorption at lower wavelengths due to the starting materials and oleic acid polymerization. Heating oleic acid and octadecene alone yields increasing visible absorption at increasing wavelengths over time as the effects of oleic acid polymerization become noticeable.

Figure 4 presents UV-Vis spectra of Cd-Se colloidal nanocrystals. The scan range was between 400 nm to 600 nm. The maximum peak shifted toward the longer wavelength. This observation is expected because as the crystal size increases, the energy absorbed or emitted decreases. The sample heated for only 50 sec did not show any peak.

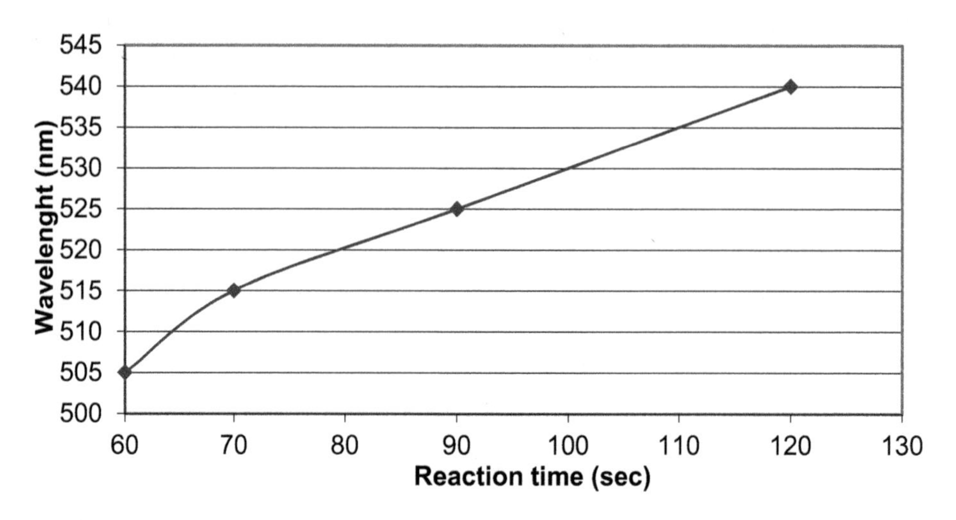

Fig. 3. The change of wavelength with reaction time

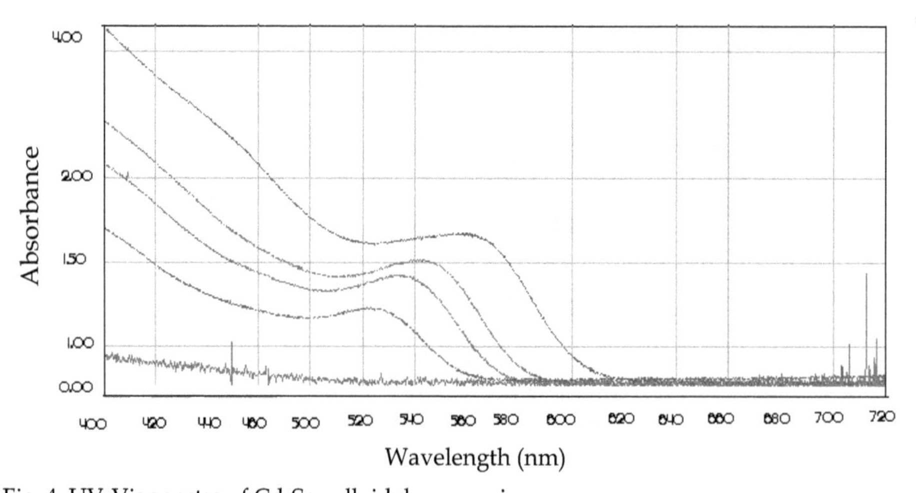

Fig. 4. UV-Vis spectra of Cd-Se colloidal suspension

4. Discussion of the optical measurements and results

As the nanocrystal size increases, the energy of the first excited state decreases qualitatively following particle in a box behavior[1]. The optical absorption results using Perkin Elmer Lambda 950 spectrometer are indicated in Figure 4.

a. Energy Shift and Nano- Crystal Size.

Using the L E. Brus [1,11,12] model we assume the following:

1. The nanocrystal is spherical with a radius R.
2. The interior of the nanocrystal consists of uniform medium and the excited electron and hole pair.
3. The potential energy outside the radius R is infinite—the radius R defines the confining boundary of the box.

The solution to the spherical Schrödinger equation leads to the energy of the exciton–electron hole pair as[1]:

$$E_{ex} = \frac{h^2}{8R^2}\left(\frac{1}{m_e} + \frac{1}{m_h}\right) - \frac{1.8e^2}{4\pi\varepsilon_{CdSe}\varepsilon_0 R} + \frac{e^2}{R}\left\langle \sum_{k=1}^{\infty} \alpha_k \left(\frac{S}{R}\right)^{2k} \right\rangle$$

The first term is the kinetic energy and the second term the Coulomb potential attractive energy; and the third term is the polarization energy.

b. Using the first term to calculate the exciton energy

At small R the predominant term is the first term (because of the inverse square R dependence since R<1 for a simple example $\frac{1}{(0.5)^2} > \frac{1}{0.5}$).

We can therefore use the first term to approximate R the radius of the nanoparticles as follows:

The energy needed to create the first peak - corresponding to the peak position in the spectra is $E_u = E_g + E_{ex}$ this energy corresponds to 500 nm from Figure 4.

The energy then converts to 2.48 eV (using the well known conversion formula $\frac{1.24}{1\mu m}$ for photon energy to eV.

The energy gap of bulk CdSe corresponds to 730 nm (0.73 μm) and is 1.70 eV [1,10]

This leads to the exciton energy of 0.78 eV

Using the formula:

$$E_{ex} = \frac{h^2}{8R^2}\left(\frac{1}{m_e} + \frac{1}{m_h}\right) - \frac{1.8e^2}{4\pi\varepsilon_{CdSe}\varepsilon_0 R} + \frac{e^2}{R}\left\langle \sum_{k=1}^{\infty} \alpha_k \left(\frac{S}{R}\right)^{2k} \right\rangle$$

and the first term alone as the approximation for small R

$$E_{ex} = \frac{h^2}{8R^2}\left(\frac{1}{m_e} + \frac{1}{m_h}\right) = 0.78eV$$

Using h as Planck's constant ; the electron effective mass m_e = 0.13 mass of a free electron and m_h equals 0.45 times the free electron mass

R can then be calculated to be the following:

$$R^2 = \frac{h^2}{8E_{ex}}\left(\frac{1}{m_e} + \frac{1}{m_h}\right)$$

$$R = \sqrt{\frac{(6.626)^2 x10^{-68} x9.91452}{9.1095x10^{-31} x8x0.78x1.602x10^{-19}}} = 2.18x10^{-9} m.$$

This leads to diameter of about 4 nm

5. Conclusion

We have demonstrated a more convenient synthesis method for colloidal CdSe quantum dots. This method does not involve quenching. This makes it easier for students to make the semiconductor nanoparticles. This synthesis method depends on different heating times for premixed CdO and Se solutions.

6. Acknowledgements

We would like to thank;

- Congresswoman Corrine Brown who was instrumental in procuring EWC grant to purchase the optical laboratory equipment.
- Army Research Laboratory and Dr N. Sundaralingam, Chair Department of Math and Sciences, Edward Waters College for their assistance.
- Dr. Elias Towe, ECE, MSE and Director-CNXT, Carnegie Mellon University for collaborative advice.
- Finally EWC who made the time available and provided the necessary resources to conduct this research possible.

7. References

[1] Tadd, K.; Laura, A. S.; Sandra, J. R. *J. of Chem. Edu.* 2002, 79, 9.
[2] Ellis, A. B.; Geselbracht, M. J.; Johnson, B.J.; Lisensky, G.C.; Robinson, W.R.; *American Chemical Society*, Washington, D.C., 1993.
[3] Schulz, W.G.; *Chem. Eng. News*, 2000, 78, 41.
[4] Rawls, R.L.; *Chem. Eng. News*, 2003, 81, 39.
[5] Halford, B.; *Chem. Eng. News*, 2004, 82, 5.
[6] Wilson, E.K.; *Chem. Eng. News*, 2003, 81, 29.
[7] Peng, Z.A.; Peng, X,; *J. Am. Chem. Soc.*, 2001, 123, 183-184.
[8] Yu, W.W.; Peng, X.; *Angew. Chem. Int. Ed. Engl.*, 2002, 41, 2368-2371.
[9] Peng, Z.A.; Peng, X.J.; *J. Am. Chem. Soc.*, 2002, 124, 3343-3353.
[10] Elizabeth M. B.; George C. L.; *J. of Chem. Edu.*, 2005, 82, 1697-1699.
[11] Brus, L.E.; *J. Chem. Phys.*, 1983, 79, 5566-5571.
[12] Brus, L.E.; *J. Chem. Phys.*, 1984, 80, 4403-4409.

7

Spectrophotometric Methods as Solutions to Pharmaceutical Analysis of β-Lactam Antibiotics

Judyta Cielecka-Piontek[1], Przemysław Zalewski[1],
Anna Krause[2] and Marek Milewski[2]
[1]*Poznan University of Medical Sciences, Department of Pharmaceutical Chemistry*
[2]*PozLab Contract Research Organization at Centre of Transfer of Medical Technologies*
Poland

1. Introduction

Following the discovery of the first analog of penicillin by A. Fleming (1929), the β-lactam antibiotics are still a developing group of chemotherapeutics and are used in treatment of majority of diseases with bacterial etiology. β-lactam antibiotics have a broad spectrum of antibacterial activity, favourable pharmacokinetic parameters and low side effects. In β-lactam therapy two main problems are still current. The increasing resistance of some bacterial strains which implicates necessity to combine the therapy with inhibitors of β-lactamases and other chemotherapeutics. The second problem of therapy of β-lactam antibiotics is their significant instability [1-3]. The analogs from that group are easily degraded in aqueous solutions and in solid state. They are a special group of drugs because parallel to losing the antibacterial efficiency, the strong allergic properties can also appear as a results of their degradation. Therefore in terms of quality control, the stability of β-lactam antibiotics in solutions was widely studied. The evaluation of stability concerned also the studies of their metabolites and intravenous solutions after preparations of pharmaceutical dosage forms. Moreover, the evaluation of concentration changes during storage of substance in solid state was also conducted. As problem of the instability of some β-lactam analogs has been solved their oral administration is possible. An intake of oral formulations is connected with appearance of excipients, which can influence rate of degradation and cause formation of different degradation products.

The common element of chemical structure of all β-lactam antibiotics is five-membered β-lactam ring. Currently, higher significance in treatment have derivatives in which the β-lactam ring is fused to:

- thiazolidine ring in penam analogs,
- 2,3-dihydro-2*H*-1,3-tiazine ring in cephem analogs,
- 2,3-dihydro-1*H*-pyrrole in carbapenem analogs,
- 2,3-dihydrotiazole in penem analogs (Fig .1).

These connections implicate the different intra-ring stress. The presence of sulphur atom and/or double bonds influence on length of bond and intra-molecular angle in molecule of β-lactam analog. Finally for some derivatives, the differences in stability are noticeable. Additionally, the factor distinguishing a stability of derivatives of β-lactam analogs are

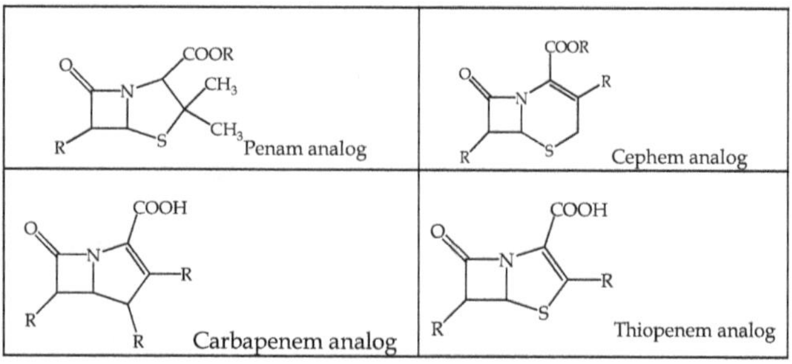

Fig. 1. Chemical structure of penam, cephem, carbapenem and thiopenem nuclei.

chemical structures of substituents at C2, C3, C5, C6, and C7. The amount and type of degradation products of β-lactam antibiotics often depend on affecting factors (solvents, concentration of substance and hydrogen ions, temperature). Moreover, most of the β-lactam antibiotics obtained by chemical synthesis or fermentation contain impurities being remnants of the process. In the development of analytical methods for the determination of β-lactam antibiotics, selectivity is a fundamental validation parameter. A reliable, selective method is expected to allow separation and determination of parental substance in the presence of related ones. Current International Conference on Harmonization (ICH) guidelines require the development of analytical methods permitting analysis in the presence of related products (Q1A–R2) [4]. These requirements are restrictions but also challenges during the development of analytical methods for the determination of β-lactam antibiotics. The problem of the overlapping of the "background" originating from related products (impurities, degradation products and metabolites) and/or the presence of other active substances in a sample (inhibitors of β-lactamases, other drugs) was solved during the determination of β-lactam antibiotics by using chromatographic techniques (high-performance liquid chromatography, thin layer chromatography). On the other hand, search of new solutions and analytical methods, especially being in accordance with the "*green chemistry*" concept, is very important and up-to-date. Analytical methods based on determination of spectrophotometric properties of β-lactam analogs are a developing tools in their analysis. Non-destructive investigations of β-lactam analogs, did not producing residues, were reported in fields of many spectrophotometric methods. A few methods of determination of β-lactam analogs were developed by using infrared spectrophotometry enriched by chemometric procedures [5-6]. Most of all analytical methods for the determination of β-lactam analogs were developed in range of visible and ultraviolet radiations. Desired, selective signals were possible to obtain by application of following techniques:

- as spectrophotometric methods
  - direct spectrophotometry
  - direct spectrophotometry enriched by chemometric procedures
  - derivative spetrophotometry
  - derivative spectrotometry enriched by chemometric procedures
- as visible spectrophotometric methods
  - measurement of absorption of species being a result of reaction between analyte and derivatizating reagent
  - measurement of absorption of species being a result of reaction between degradation products of analyte and derivatizating reagent (Table 1).

Derivative	Ultraviolet region			Visible region	
	1	2	3	1*	2*
Analog of penam	✓	✓	✓	✓	✓
Analog of cephem					
I generation	✓	✓	✓	✓	✓
II generation		✓		✓	✓
III generation	✓	✓	✓	✓	✓
IV generation		✓			
Analog of carbapenem	✓	✓	✓		
Analog of penam		✓			

1. direct spectrophotometry enriched by chemometric procedures
2. derivative spetrophotometry
3. derivative spectrotometry enriched by chemometric procedures
1.* measurement of absorption of species being a result of reaction between analyte and derivatizating reagent
2.* measurement of absorption of species being a result of reaction between degradation products of analyte and derivatizating reagent

Table 1. Possibilities of application of visible and ultraviolet spectrophotometric determinations for analysis of β-lactam antibiotics in the period of time 1994–2011.

2. Spectrophotometric methods for determination of β-lactam antibiotics

2.1 Direct spectrophotometry

Spectra of β-lactam antibiotics recorded by using direct spectrophotometry do not have desired selectivity due to the presence of related products. A comparison of sharp zero-order spectra and/or value of absorption maxima for some β-lactam analogs with ones obtained for CRS (*chemical reference substance*) is recommended by pharmacopeias for an their identification [7] . Lack of desired absorbing species in chemical structure of penam analog often do not allow to apply direct spectrophotometry even for qualitative studies of substance of high purity.

Paradoxically, the significant instability of analogs can sometimes solve this problem due to formation of degradation products that can absorb ultraviolet radiation permitting determination of parental substance.

Significant susceptibility of β-lactam analogs to degradation in basic medium was reported during analysis of cephem analogs. It was confirmed that formation of piperazine-2,5-dione derivative, peak at 340 nm, was possible via intra-molecular nucleophilic attack of the primary amine from the side chain on β-lactam ring (pH = 11 was required) (Fig. 2) [8].

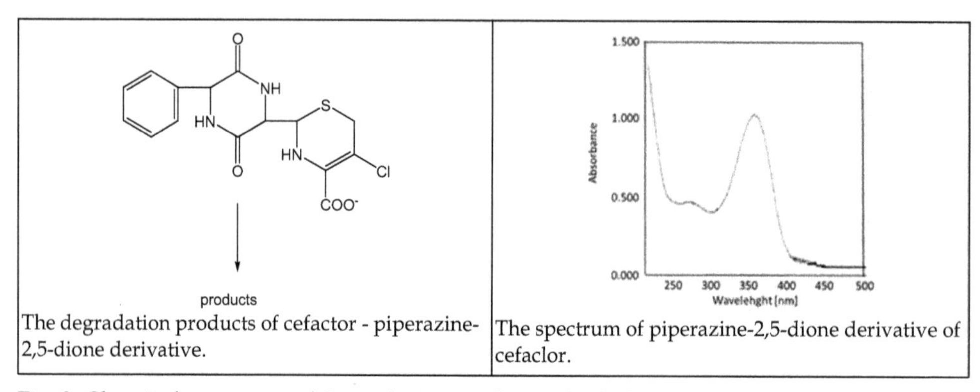

The degradation products of cefactor - piperazine-2,5-dione derivative.	The spectrum of piperazine-2,5-dione derivative of cefaclor.

Fig. 2. Chemical structures of degradation products of cefaclor (1.0 mmol/l) formed at pH 11.0 and its spectrum [8].

The degradation of penam analogs in acidic conditions was also a base for spectrophotometric determination. As it is shown in Fig. 3, different pathways of degradation (including enzymatic one) can lead to obtaining absorbing species in the range of ultraviolet radiation. As a results of chemical degradation of penam analog in acidic conditions, the penicilloic acid, penillic acid and penicillenic acid are formed and absorb the ultraviolet radiation in the range 320–360 nm, respectively [9]. While during the enzymatic degradation under the influence of penicillin acylase, D-4-hydroxyphenylglycine (D-HPhG) and 6-aminopenicillanic acid are formed. Then the D-HPhG was catalyzed by D-phenylglycine aminotransferase to form L-glutamate and hydroxybenzoylformate which strongly absorb UV light at 335 nm [10].

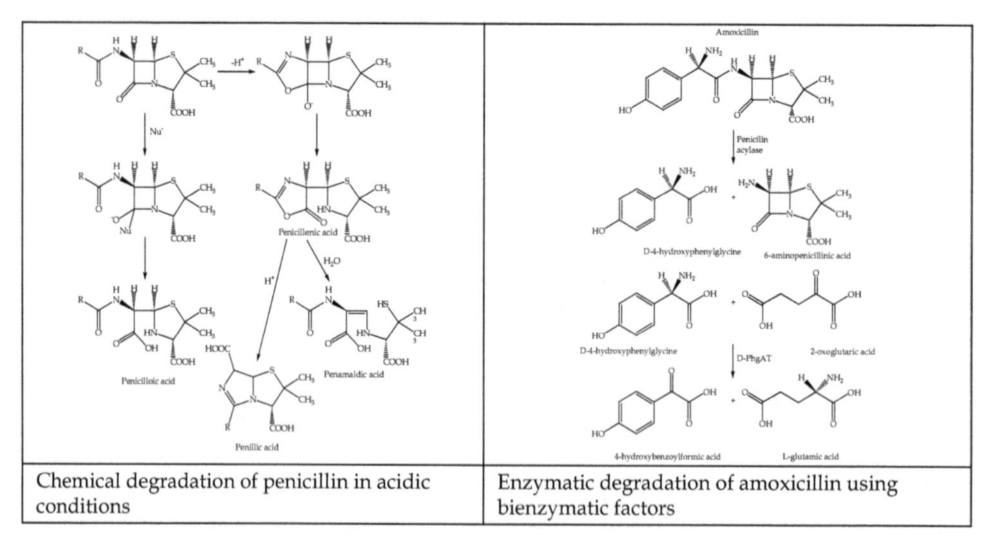

Chemical degradation of penicillin in acidic conditions	Enzymatic degradation of amoxicillin using bienzymatic factors

Fig. 3. The pathways of obtaining of absorbing degradation products of penam analog [9-10].

2.2 Direct spectrophotometry enriched by chemometric procedures

The other way of improving the selectivity of direct spectrophotometry for the determination of β-lactam antibiotics is the enrichment of data analysis by chemometric procedures. A literature review revealed the application of the following determinations of β-lactam antibiotics enriched by chemometric procedures that solved the problem of spectral overlap without additional separation techniques at the stage of sample preparation, were used:

- a separation of analog of cephem in the presence of impurities originating from synthesis (e.g., cephalexin in the presence of 7-aminocephalosporanic acid and acid-induced degradation products) using H-point standard additions method (HPSAM) [11]
- determination of analog of penam in the presence of other drugs (e.g., amoxicillin in the presence of diclofenac) using partial least squares (PLS) regression analysis [12]
- determination of analog of cephem in the presence of alkali-induced degradation products using full spectrum quantitation (FSQ) (e.g., cefotaxime, ceftazidine, ceftiaxome, in the presence of degradation products) [13].

Each chemometric method relies on different tools of regression analysis of multicomponent system permitting simultaneous determination of two or more components.

The determination of β-lactam analyte in the presence of known and unknown inferences was possible by the application of HPSAM procedure, where analyte concentration is calculated from the following equation:

$$\frac{(A_0-b_0)+(A'-b)}{M(\lambda_1)-M(\lambda_2)} = -C_X + \frac{(A'-b)}{M(\lambda_1)-M(\lambda_2)} \quad (1)$$

where b_0 and A_0 are the absorbance values for β-lactam analyte, b and A' ones for the interferent, at λ_1 and λ_2 and $M(\lambda_1)$, $M(\lambda_2)$ are slopes of plots at selected wavelengths.

In PLS technique, analytical sensibility was defined as $\gamma = \frac{SEN_k}{\|\sigma_r\|}$ where $SEN_k = \frac{1}{\|b_k\|}$, σ_r is a value estimated from standard deviation of blank samples, b_k value is a vector of the regression coefficient for the *k* analytes and *k* is a number of components in a mixture.

The FSQ technique during a determination of β-lactam antibiotics applies Fourier pre-processing of the entire absorption spectra of the individual β-lactam analogs with their degradation products at variable concentration to calculate matrix calibration coefficients.

2.3 Derivative spectrophotometry

A derivative spectrophotometry using derivatives of absorbance with respect to wavelength (first $\frac{dA}{d\lambda} = f(\lambda)'$, second $\frac{d^2A}{d^2\lambda} = f(\lambda)''$; third $\frac{d^3A}{d^3\lambda} = f(\lambda)'''$; respectively) is a suitable tool for overcoming the overlapping spectra problem in analysis of many β-lactam analogs. Possibility of application of derivative spectrophotometry with zero-crossing point is widely used in analysis of all β-lactam analogs. The direct correlation between order of used derivative spectrophotometry and similarities of chemical structures of nuclei of β-lactam analogs has not been observed, e.g., both second-derivative and first-derivative were developed for cephem analogs including the same nuclei (Fig. 4).

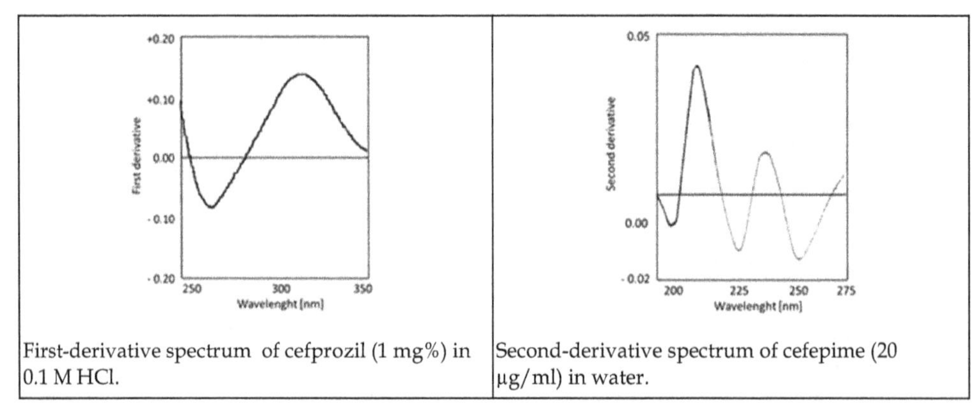

First-derivative spectrum of cefprozil (1 mg%) in 0.1 M HCl.

Second-derivative spectrum of cefepime (20 μg/ml) in water.

Fig. 4. The application of derivative spectrophotometry for analysis of cephem analogs [15-16].

The application of derivative spectrophotometry for determination of β-lactam antibiotics was used in the following areas:

- a separation and determination of penam/cephem analogs and inhibitors of β-lactamase in aqueous solution (e.g., determination of ampicillin sodium in the presence of sulbactam sodium; determination of cefsulodin in the presence of clavulanic acid) [14-15]
- a separation and determination of cephem/carbapenem analog and excipients used in parenteral pharmaceutical dosage forms (e.g., determination of cefepime in the presence of *L*-argininie) [16]
- a separation and determination of cephem analog and its degradation products (e.g., determination of cefprozil in the presence of its degradation products) [17]
- a separation and determination of cephem analog and related compounds from the synthesis (e.g., determination of triethylammonium salt of cefotaxime in the presence of 2-mercaptobenzothiazole) [18]
- a separation and determination of penam/cephem/carbapenem analogs in biological matrix (e.g., determination of amoxicillin, cefuroxime, imipenem in urine) [19].

The separation of often structurally very similar species (e.g., two analogs of cephem, cephem analog and its impurities from synthesis or carbapenem analog and its degradation products) was possible by using derivative spectrophotometry (Fig. 5).

It was proved that the derivative spectrophotometry can be recommended as a method for routine control analysis of pharmaceutical preparation of β-lactam antibiotics. Derivative spectrophotometry ensured the rapid analysis of parenteral dosage forms and also removed a "*background*" excipients in oral pharmaceutical dosage forms.

The special potency of derivative spectrophotometry was possibility of its usage in determination of β-lactam analogs in biological matrix. In this case, to meet the requirements of analytical methods, the selectivity had to be extended in regard with interference of biological endogenous components. It was noticed that the selective determination of penam/cephem/carbapenem in the presence of metabolites (open-ring degradation product) and endogenous substance of urine was possible to achieve.

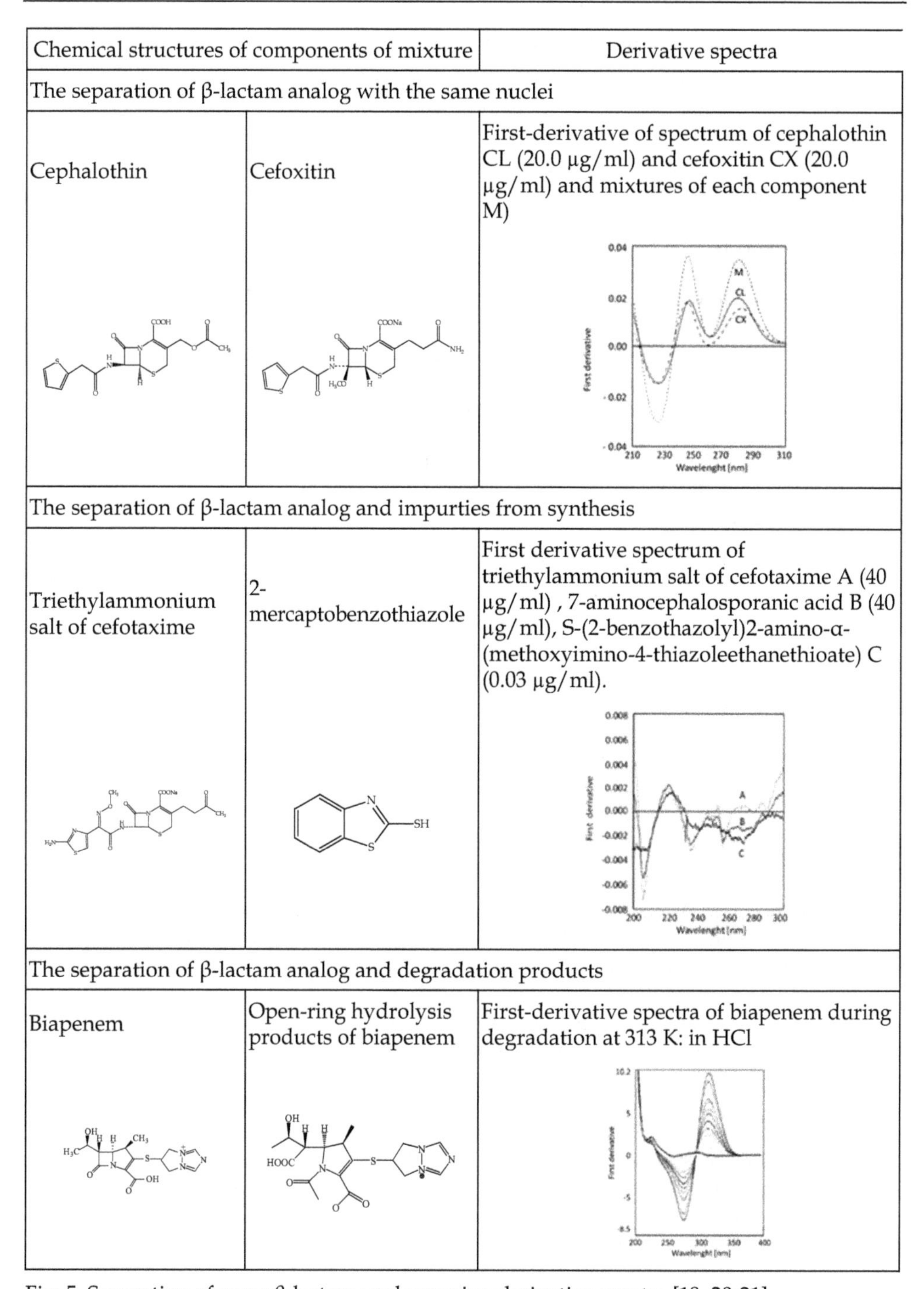

Chemical structures of components of mixture		Derivative spectra
The separation of β-lactam analog with the same nuclei		
Cephalothin	Cefoxitin	First-derivative of spectrum of cephalothin CL (20.0 μg/ml) and cefoxitin CX (20.0 μg/ml) and mixtures of each component M)
The separation of β-lactam analog and impurties from synthesis		
Triethylammonium salt of cefotaxime	2-mercaptobenzothiazole	First derivative spectrum of triethylammonium salt of cefotaxime A (40 μg/ml) , 7-aminocephalosporanic acid B (40 μg/ml), S-(2-benzothazolyl)2-amino-α-(methoxyimino-4-thiazoleethanethioate) C (0.03 μg/ml).
The separation of β-lactam analog and degradation products		
Biapenem	Open-ring hydrolysis products of biapenem	First-derivative spectra of biapenem during degradation at 313 K: in HCl

Fig. 5. Separation of some β-lactam analogs using derivative spectra [18, 20-21].

2.4 Derivative spectrophotometry enriched by chemometric procedures

The application of chemometric procedures coupled with derivative spectroscopy permits achievement of higher selectivity in determination of β-lactam antibiotics. Currently, chemometric procedures based on the estimated ratio of spectra derivative for the selective determination of β-lactam analogs are the most common. It was proved that the application of the ratio of different-order spectra derivatives permitted the separation of binary and tertiary mixtures of β-lactam antibiotics [22]. During the determination of concentrations of three components (e.g., penicillin-G sodium, penicillin-G procain and dihydrostreptomycin sulphate salts) in a mixture the equation describing the ratio spectra derivative spetrophotometry is as follows:

$$\frac{d(A_{a+b,\lambda}/A_{a,\lambda^0})}{d\lambda} = C_b\frac{d(k_{b,\lambda}/A_{a,\lambda^0})}{d\lambda} + C_c\frac{d(k_{c,\lambda}/A_{a,\lambda^0})}{d\lambda} \quad (2)$$

where $A_{a+b+c,\lambda}$ is the absorbance of the ternary mixture of *a, b* and *c* at wavelength λ, A_{a,λ^0} is the absorbance of pure component at wavelength λ, C_b and C_c – are the concentrations of *b* and *c*, $k_{b,\lambda}$ and $k_{c,\lambda}$ are the products of the molar absorption coefficient of *b* at wavelength λ and the thickness of the absorption cell. Equation 2 is divided by C_b while divisor can be any component of ternary mixture (Fig. 6):

$$\frac{d(A_{a+b,\lambda}/A_{a,\lambda^0})}{d(A_{b,\lambda^0}/A_{a,\lambda^0})} = \frac{C_b}{C_b^0} + (C_c d\frac{d(k_{c,\lambda}/A_{a,\lambda^0})}{d(A_{b,\lambda^0}/A_{a,\lambda^0})}) \quad (3)$$

Equation 3 is drawn:

$$J = C_c\, d(d(\frac{d(k_{c,\lambda}/A_{a,\lambda^0})}{d(A_{b,\lambda^0}/A_{a,\lambda^0})}))/d\lambda \quad (4)$$

Finally, after the next derivation J (as the left side of equation 3), is proportional to the C_c value and can be used to determine concentration of component in the ternary mixture (when A_{a,λ^0} and A_{b,λ^0} are fixed) [23].

Depending on chemometric procedure, the selective determination of following analogs was possible:

- a separation and determination of carbapenem and degradant (e.g., the determination of ertapenem and its degradant) when the substraction technique was used [24-25]
- a separation and determination of penam and cephem analogs (e.g., the determination of penicillin-G, penicillin-G procain in the presence of dihydrostreptomycin sulphate salts or the determination of cefotaxime and cefadroxil), cephem analog and inhibitor of β-lactamases (e.g., the determination of cephradine and clavulanic acid) and carbapenem analog and degradation products (e.g., the determination of meropenem and its degradant) when ratio spectra of derivative with all orders were used [26-29]
- a separation and determination of carbapenem and degradation products (e.g., the determination of ertapenem) when the substraction technique was used [30]
- a separation and determination of carbapenem and degradation products (e.g., the determination of ertapenem) when the Kraise's method was used [31]
- a separation and determination of penam analogs (e.g., the determination of ampicillin and flucloxacillin) when multivariate methods (classical least squares and principle component regression) were used.

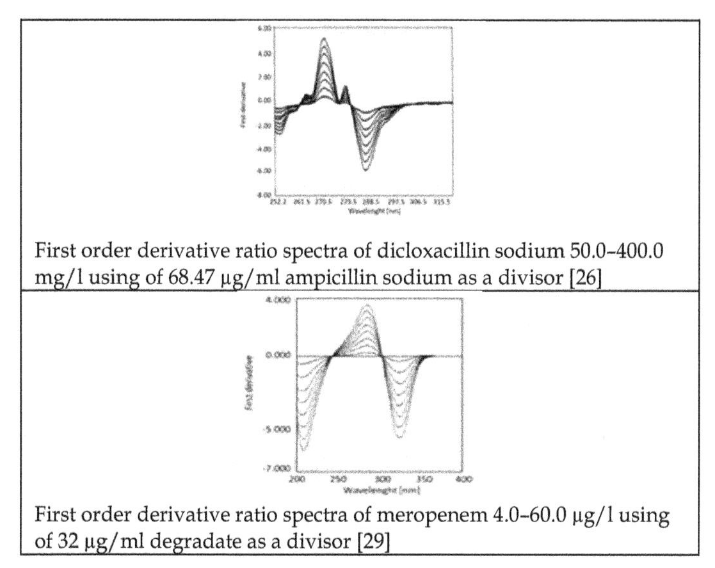

First order derivative ratio spectra of dicloxacillin sodium 50.0–400.0 mg/l using of 68.47 μg/ml ampicillin sodium as a divisor [26]

First order derivative ratio spectra of meropenem 4.0–60.0 μg/l using of 32 μg/ml degradate as a divisor [29]

Fig. 6. The application of ratio spectra of derivative spectrophotometry in analysis of β-lactam antibiotics.

3. Visible spectrophotometric methods for determination of β-lactam antibiotics

The β-lactam analogs itselfs do not absorb in visible region of radiation. However, many visible spectrophotometric methods were developed for the determination of β-lactam antibiotics using the effect of formation of "species" giving signals in visible region as the result of chemical derivatization (Fig. 7).

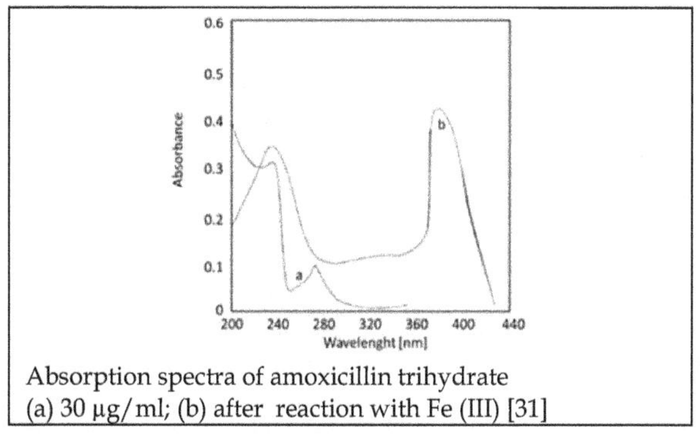

Absorption spectra of amoxicillin trihydrate
(a) 30 μg/ml; (b) after reaction with Fe (III) [31]

Fig. 7. The application of derivatization for detetmination of penam analog.

Formation of "species" absorbing visible radiation can be a result of reactions of chemical reagents with:

- β-lactam analog

- degradation product of some β-lactam analog.

Above-mentioned methods were described only for the determination of penam and cephem analogs. The species, absorbing visible radiation, used in analysis of β-lactam analogs were formed as a result of the following reactions:

- redox
- complexation of metals
- complexation on the base of charge-tranfer process
- formation of ion pairs
- coupling with specified reagents.

3.1 Visible spectrophotometric methods based on redox reactions

The methods based on the selective oxidation were reported for penam and cephem analogs containing phenolic substitutes at C6 and C7, respectively. These methods permitted also selective determination of β-lactam analogs in the presence of excipients being in their pharmaceutical preparations. The application of oxidation properties of iron ions was used in analysis of a huge number of β-lactam analogs:

- when, as the result of direct reaction with Fe(III) in acidic medium, yellow coloured products (λ_{max} = 397 nm) were produced (cefoperazone sodium, cefadroxil monohydrate, cefprozil anhydrous, amoxicillin trihydrate). A possible mechanism of reaction is presented in Fig. 8 [32].

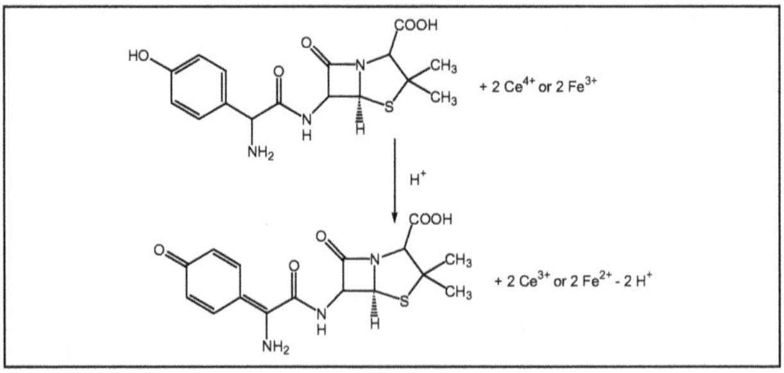

Fig. 8. The mechanism of reaction of amoxicillin and Fe(III)[32].

- when, as the result of indirect reaction, red complex Fe-(o-phen)$_{2/3}$ (λ_{max} = 510 nm) was produced. This complex is formed between *o*-phenanthroline and Fe(II) which previously was reduced from Fe(III) as the result of oxidation of β-lactam analogs in alkali medium [33].

The reduction of oxidized quercetin by cephem analogs was used in development of visible spectrophotmetric method for determination of β-lactam analogs. Quercetin is a flavonol (3,5,7,3′,4′- pentahydroxyflavone) which is oxidized by *N*-bromosuccinimide giving reddish green colour (λ_{max} = 510 nm). As the result of reduction of oxidized form of quercetin by cephem analog fade colour was observed (Fig. 9). This colour is the result of formation of *o*-quinone derivative of quercetin under the mild oxidants [34].

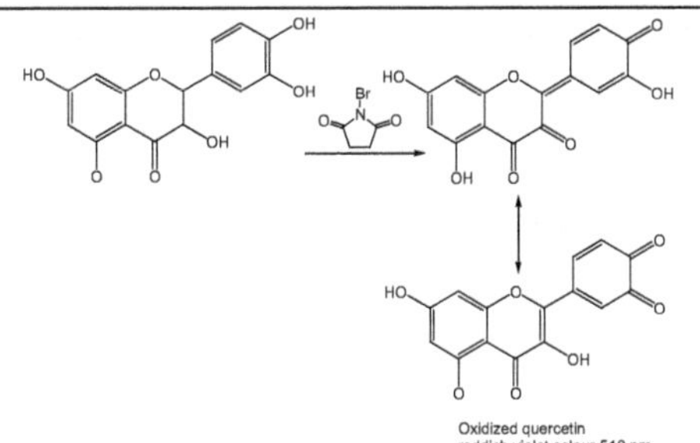

Fig. 9. The mechanism of reduction of quercetin oxidation with *N*-bromosuccinimide [34].

Also, another reagent which is able to oxidize some sulphur atoms present in compounds such as cefotaxime and cefuroxime, is 1-chlorobenzotiazol. As the result of reaction of cephem analogs and 1-chlorobenzotiazol, a product with yellow colour is formed, absorbing radiation at λ_{max} = 298 nm. The suggested possible reaction pathways and absorption spectra are shown in Figure 10 [35].

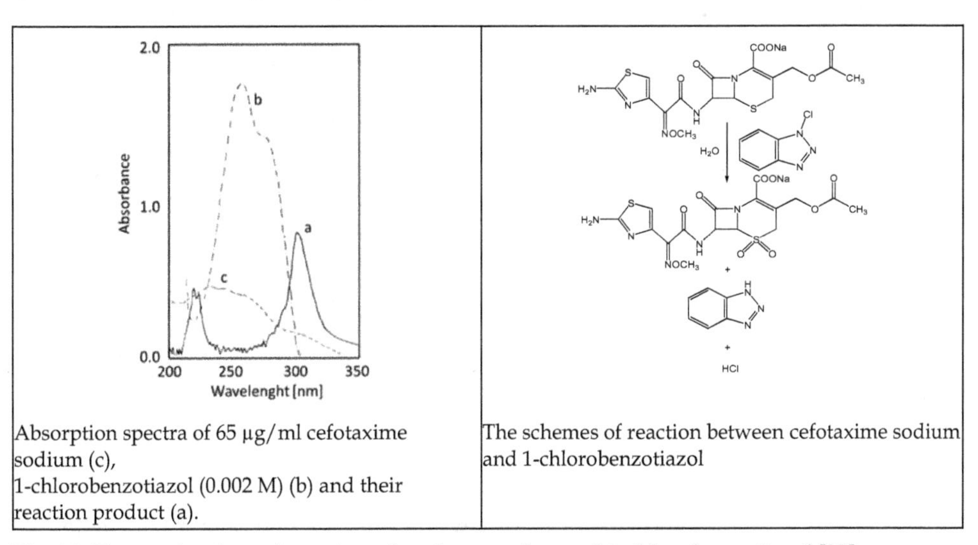

Absorption spectra of 65 µg/ml cefotaxime sodium (c), 1-chlorobenzotiazol (0.002 M) (b) and their reaction product (a).

The schemes of reaction between cefotaxime sodium and 1-chlorobenzotiazol

Fig. 10. The mechanism of reaction of cephem analog and 1-chlorobenzotiazol [35].

In indirect spectrophotometry, redox properties of iodine were used in determination of penam analog (amplicillin, penicillin V, amoxicillin, cloxacillin) and cephem analogs (cefadroxil, ceftezoxime). The method based on formation of hypoiodite, from excess of iodine (which did not react with β-lactam analog) under alkaline conditions. Hypoiodite reduced the intensity of wool fast blue colour (5,9-dianilo-7-phenyl 4,10-disulphbenzo[α]phenazinium hydroxide) by disruption of phenazine chromophore [36].

Especially, for cefadoxil many visible spectrophotometric methods based on redox reactions were proposed (Fig. 11). These methods were based on the reaction with different oxidizing reagents:

- 4-aminoantipyrine in the presence of potassium hexacyanoferrate(III) in alkaline medium (λ_{max} = 505 nm [37]
- 3-methyl-2-benzothiazolinone hydrozone hydrochloride in the presence of ceric ammonium sulphate (λ_{max} = 410 nm) [38]
- 4-aminophenazone in the presence of potassium hexacyanoferrate (III) (λ_{max} = 510 nm)
- 2,6-dichloroquinone-4-chlorimide (Gibb's reagent) (λ_{max} = 620 nm) [39]
- *N*-bromosuccinimide or *N*-chlorosuccinimide in alkali medium (λ_{max} = 395 nm) [40]
- sodium persulfate in alkaline medium (λ_{max} = 350 nm) [41]

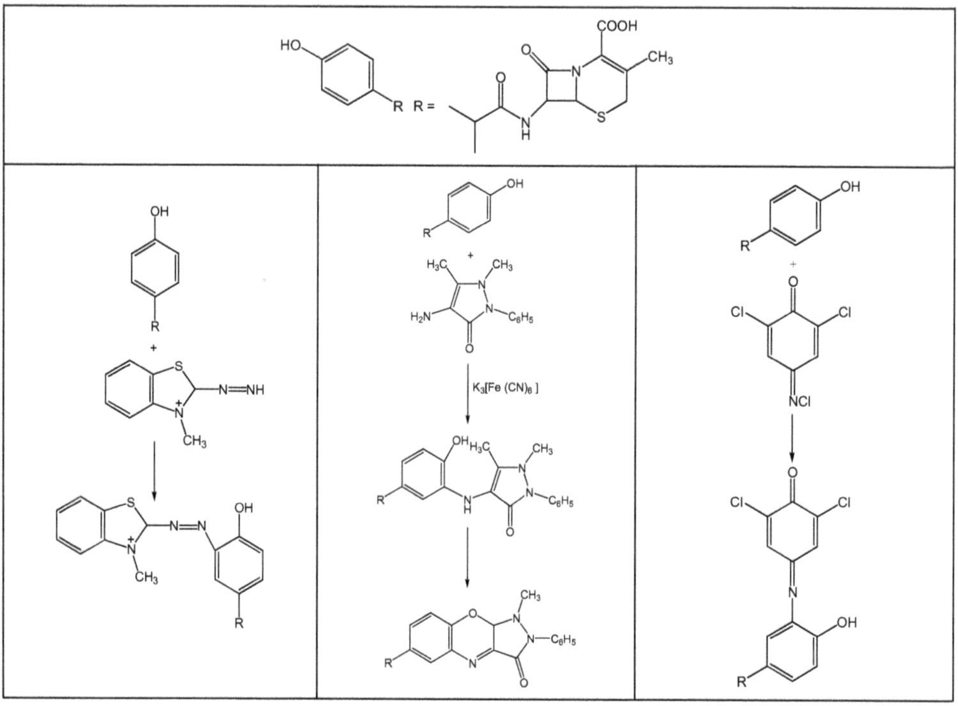

Fig. 11. Schemes of some oxidation reactions of cefadoxil [39].

3.2 Visible spectrophotometric methods based on formation of complex with metal

The formation of colored complexes as a consequence of the interactions between metal ions and analytes (penam and cephem analogs) resulted from:

- a direct reaction between β-lactam analogs and metal ions
- a reaction between the degradation products of β-lactam analogs and metal ions.

Direct complexation of β-lactam analogs and metal ions with formation of yellowish-brown chelate complex was possible due to the presence of suphur atoms in the β-lactam ring and the thiazole ring. Cephem analogs (cefpodoxime, ceftizoxime, ceftazidime, ceftiaxone and cefixime) gave with palladium(II) ions, absorbing complex, in the presence of sodium lauryl sulphate as surfactant, in the range 300–500 nm [42]. B-lactam analogs containing the phenolic ring with free *ortho* position to the hydroxyl group (amoxicillin trihydrate, cefperazone sodium, cefadroxil monohydrate, cefprozil anhydrous) reacted with nitrous acid forming the nitroso derivatives. They were capable of tautometric interconversions to form colored complex in the presence of copper(II) ions (Fig. 12). The stoichiometric ratios (nitroso derivative to copper(II)) were determined by the Job's method 2:1 [43].

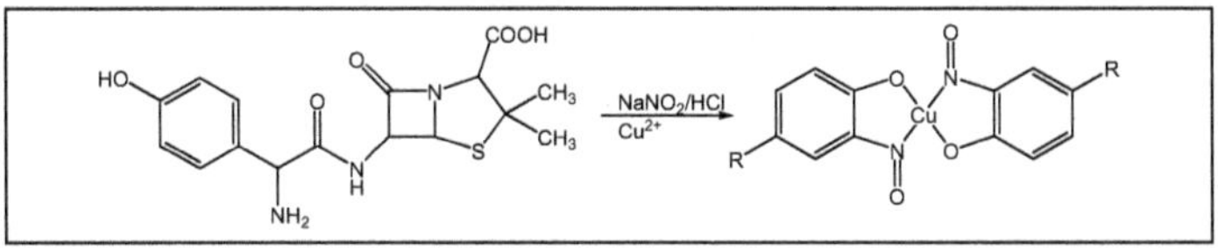

Fig. 12. Scheme of the reaction of nitrous acid, Cu(II) with phenolic β-lactam analogs [43].

The determination of cephalexin, cefixime, ceftriaxone, cefotaxime based on the Bent-French method, in which degradation products of β-lactam analogs with metal ions form the colored complexes, was developed. Hydroxamic acids formed by hydroxiaminolysis of cephem analogs (1:3), formed complexes with iron (II) ions (Fig. 13) [44].

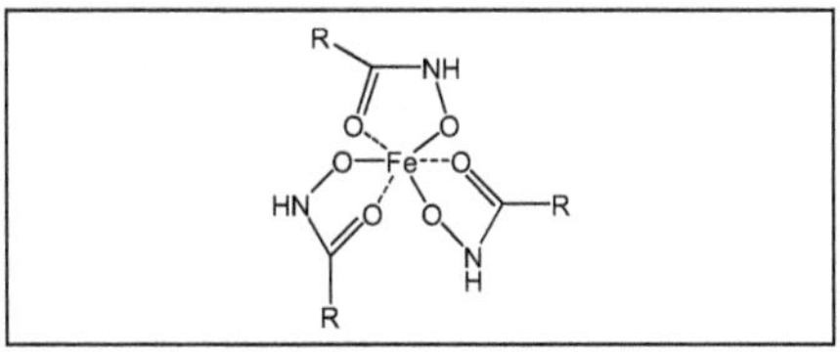

Fig. 13. The scheme of hydroxamic acid-iron(III) [44].

3.3 Visible spectrophotometric methods based on formation of charge-transfer complex

Some drugs, including penam and cephem analogs, are electron donors. Therefore, they form charge-transfer complexes with compounds that are σ- and π-accepptors of electrons. The wavelengths at which the absorption maxima of charge-transfer complexes of β-lactam antibiotics were measured depended on what of reagent was used the acceptor:

- *p*-chloranilic acid gives coloured complex species at 520–529 nm during analysis of cefotaxime sodium, cefuroxime sodium, ceftazidime pentahydrate, cephalexin monohydrate, cefotaxime sodium, cephradine, cephaloridine sodium, cefoperazone sodium (ratio 1:1); cephalotin sodium, cefixime, cefprozil anhydrous, cefazolin sodium, cephapirin (ratio 1:2); cefaclor (ratio 1:4) [45]
- *p*-nitrophenol, 2,4-dinitrophenol, 3,5-dinitrosalycilic acid, picramic acid and picric acid give greenish yellow complexes at 446, 435, 442, 473 and 439 nm, respectively during determination of flucloxacillin (ratio 1:1) [46]
- 7,7,8,8-tetracyanoquinodimethane (TCNQ) gives coloured complex species at 838–843 nm during analysis of cefotaxime sodium, cefuroxime sodium, cephapirin sodium, cefazoline sodium, cephalexin monohydrate, cefadroxil monohydrate, cefoperazone and ceftazidime (ratio 1:1)[47]
- 2,3-dichloro-5,6-dicyano-*p*-benzo-quinone (DDQ) gives coloured complex species at 460 nm during analysis of cephapirin sodium, cefazoline sodium, cephalexin monohydrate, cefadroxil monohydrate, cefoperazone and ceftazidime (ratio 1:1) [47]

- iodine gives coloured complex species at 838–843 nm during analysis of cephapirin sodium, cefazoline sodium, cephalexin monohydrate, cefadroxil monohydrate, cefoperazone and ceftazidime (ratio 1:1) [47].

During formation of charge-transfer complexes between *p*-chloranilic acid and d-donors electrons (A) from group of cephem analogs (D) in polar solvents the radical anion is formed.

$$D + A \rightleftharpoons \underset{\text{complex}}{(D\text{---}A)} \xrightleftharpoons{\text{polar sovent}} \overset{\bullet +}{D} + \overset{\bullet -}{A}$$

Electron transfer from donor to the acceptor moiety occurred with the formation of intensely coloured radical ions with high molar absorptivity value. The formation of charge-transfer complex of some cephem analogs (D) with iodine in 1,2-dichloroethane (J) is observed with the change of colour from violet to lemon yellow:

$$D + J_2 \rightleftharpoons \underset{\text{outer complex}}{D\text{---}\overset{\oplus\ominus}{JJ}} \rightleftharpoons \underset{\text{inner complex}}{[D\text{---}J]^{\oplus}}{}^{\ominus} + J^{\ominus} \xrightleftharpoons{J_2} \underset{\text{tri-iodide ion}}{J_3^{\ominus}}$$

Only in 1,2-dichloroethane, formation of tri-iodide ion pair (inner complex), showing two absorption maxima at 290 nm and 364 nm, was possible. This complex originated from an early intermediate outer complex D---J_2. While the interactions of some cephem analogs with DDQ and TCNQ took place according to the following simple relationship:

$$D + A \rightleftharpoons \underset{\text{complex}}{(D\text{---}A)}$$

3.4 Visible spectrophotometric methods based on formation of ion pair

In visible spetrophotometric analysis of penam and cephem analogs, their ability to form ion-pair was also used. The penam analogs contacting the tertiary amine group (ampicillin, dicloxacillin, flucloxacillin, amoxicillin) and Mo(V)-thiocyanate binary complex in hydrochloric acid give coloured ion-pair formation absorbing at $\lambda = 467$ nm (Fig. 14) [48].

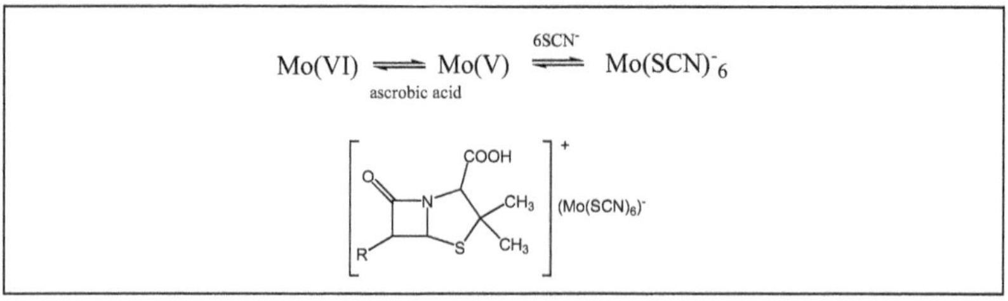

$$Mo(VI) \underset{\text{ascrobic acid}}{\rightleftharpoons} Mo(V) \xrightleftharpoons{6SCN^-} Mo(SCN)^-_6$$

Fig. 14. Scheme of Mo(V)-thiocyanate-β-lactam ion-pair [48].

Some cephem analogs (cephaprin sodium, cefuroxime sodium, cefotaxime sodium, cefoperazone sodium, cefadroxil, ceftazidime, cefazolin sodium and cefaclor) can be determined spectrophotometrically based on formation of ion-pair complex with ammonium reinecekate. In acidic medium at 25 ± 2°C, as the reaction products, the complex, absorbing at 525 nm, was formed according to the scheme [49]:

cephem analog hydrochloride + ammonium reineckate → cephem reineckate + ammonium chloride

3.5 Visible spectrophotometric methods based on coupling with specified reagents

The analysis of phenolic derivatives of penam and cephem analogs were possible by measurement of absorption species formed as a result of reations with specified reagents.

Diazo coupling of β-lactam analogs was conducted with the following compounds:

- benzocaine in thrietylamine medium for determination of cefadroxil and amoxicillin. Stoichiometric ratio of formed species was 1:1 with peaking at λ=455 nm and λ=442 nm, respectively [50] .
- electron-deficient polinitro derivatives for determination of amoxicillin, cefoperazone, cefadroxil, cefprozil. Complexes of the Meisenheimer type were formed [51].

Suggested mechanisms of coupling reaction of phenol derivative of β-lactam analogs were presented in Fig. 15.

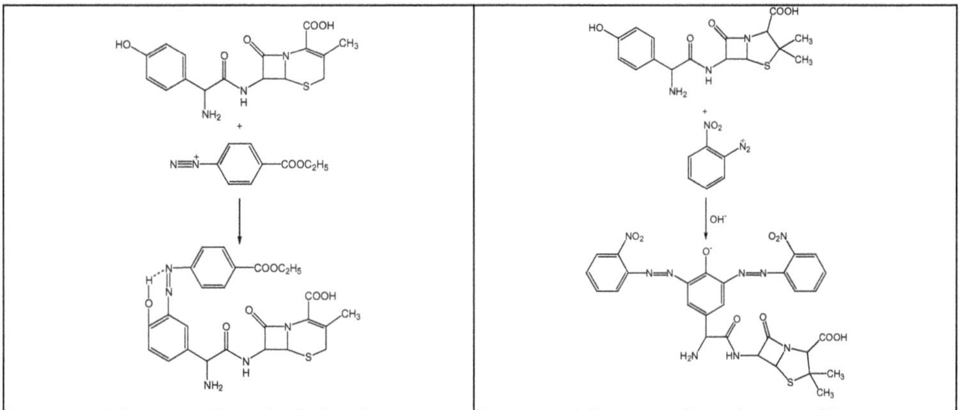

Reaction of diazo coupling of cefadroxil	Reaction of diazo coupling of amoxicillin

Fig. 15. Schemes of reactions of coupling of β-lactam analogs [50-51].

For determination of phenolic derivative of β-lactam analog (cefadroxil) measurement of absorption of formed product in the reaction between it and 4-aminoantipyrine in the presence of alkaline potassium hexacyanoferrate(III) at 510 nm was also proposed (Fig. 16). Potassium hexacyanoferrate(III), being oxidant in this reaction, yielding *N*-substituted quinine imines and in the result was responsible for formation of red-colored antipyrine dye. Additionally, a sequential injection analysis (SIA) spectrophotometric procedure for the determination was reported [52].

1,2-naphthoquinone-4-sulfonic acid is the reagent permitting the nucleophilic substitution reaction in area of amino group of penem (amoxicillin) and cephem (cephalexin) analogs (Fig. 17). The stoichiometric ratio of these species was 1:1 and they absorb at λ = 463 nm [53].

The extension of the methodology for determination of cephalexin by using the H-point standard additions method (HPSAM) and the generalized H-point standard additions methods (GHPSAM) (after solid phase extraction cartridges) permitted also its analysis in urine [53].

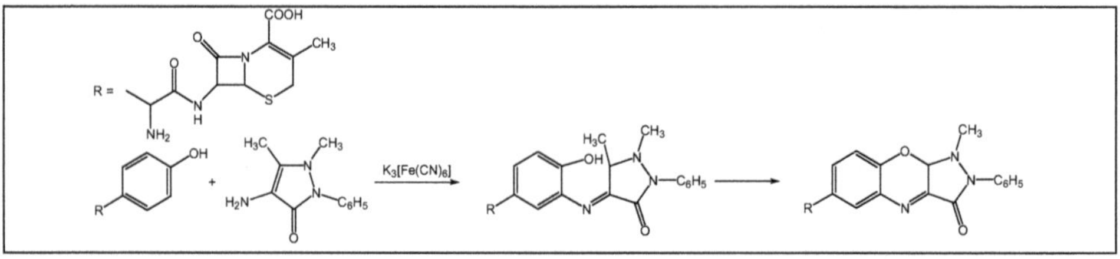

Fig. 16. The reaction mechanism of cefadroxil with 4-aminoantipyrine in the presence of alkaline $[Fe(CN)_6]^{3-}$

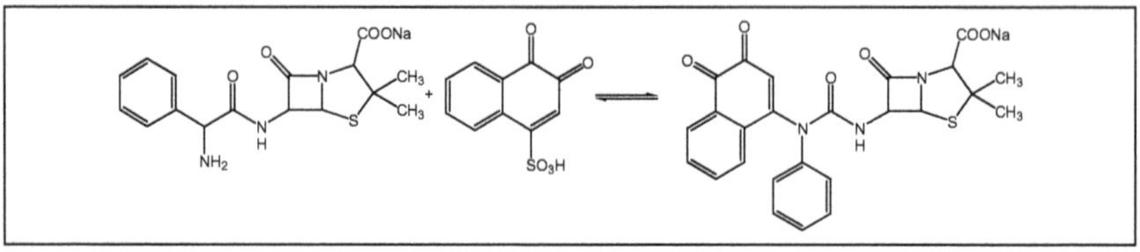

Fig. 17. The reaction mechanism of ampicillin sodium and 1,2-naphthoquinone-4-sulfonic acid [53].

3.6 Visible spectrophotometric methods based on degradation products of β-lactam analogs

The significant instability of β-lactam analogs was also exploited in the spectrophotometrical determination of β-lactam analogs in the visible region.

As the intermediate stages of determination, depending on the affecting factors, the following were present:

- degradation products typical of an acidic environment
- degradation products characteristic of a basic environment.

The formation of degradation products of β-lactam analogs in conditions of a basic hydrolysis was also the first stage during development of visible spectrophotometric methods. As the result of reactivity of degradation products formed in basic medium with some reagents, the determination of the following β-lactam analogs was possible:

- cephem analogs (cefadroxil, cefotaxime), when coupling factor were *N,N*-diethyl-*p*-phenylenediamine sulfate and Fe(III) (λ = 670 nm) or *p*-phenylenediamine dihydrochloride and Fe(III) (λ = 597 nm) [54]
- cephem analogs (cefotaxime, ceftriaxone, cefradine) when reducing factor was potassium iodate (required acidic medium) and a result of the reaction was colour change of leuco crystal violet under the influence of formed iodine (λ = 588 nm) [55]
- cephem analogs (cefotaxime sodium) when coupling factor was 1,10-phenanthroline and ferric chloride (λ = 520 nm) [56]
- penam analogs (amoxicillin, ampicillin) and cephem analogs (cephalexin, cephradine) when reducing factor of formed hydrolyzed products was J_2 (required acidic medium) (λ = 460 nm) [57].

The significant expansion of possibilities for the developed analytical method was the usage of flow injection analysis (FIA).

Similarly, in the case of determination of penam analysis during acidic hydrolysis (1.0 M HCl), formation of non-absorbing degradation products was the intermediate stage of their analysis. Complex which is necessary for achievement of spectrophotometric signals, was formed between penicillamine and palladium(II) chloride, peak at 334 nm (Fig. 18) [58].

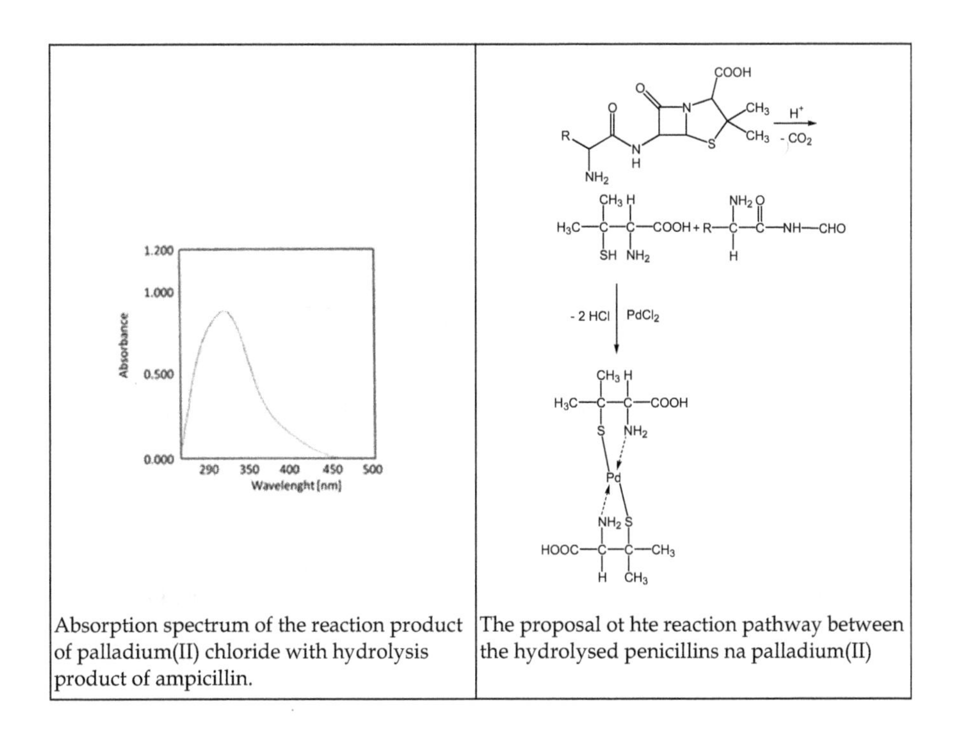

Absorption spectrum of the reaction product of palladium(II) chloride with hydrolysis product of ampicillin.	The proposal ot hte reaction pathway between the hydrolysed penicillins na palladium(II)

Fig. 18. Absorption spectra based on reaction of degradation products of penam analogs and palladium (II) and suggested mechanism of the reaction [58].

In conditions of acidic hydrolysis, determination of cephem analogs took place. Vanadium(IV) after reduction from vanadium(V), reacted with forming degradation products. Colored complexes of some cephem analogs (cephalexin, cephaprine sodium, cefazolin sodium, cefotaxime) were found, absorbing at 515, 512, 518 and 523 nm, respectively (Fig. 19) [59]

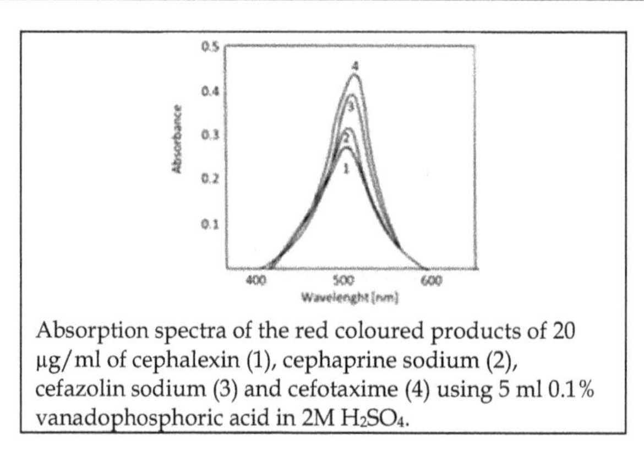

Absorption spectra of the red coloured products of 20 µg/ml of cephalexin (1), cephaprine sodium (2), cefazolin sodium (3) and cefotaxime (4) using 5 ml 0.1% vanadophosphoric acid in 2M H_2SO_4.

Fig. 19. Absorption spectra based on reaction of degradation products of cephem analogs and vanadium(IV) [59].

4. Conclusion

Spectrophotometric determinations of drugs in Vis-UV region are ones from simpler and cheaper methods used in quantitative pharmaceutical analysis. The key problem in application of these methods in analysis of multicomponents mixture is achievement of desired selectivity. Above-mentioned procedures shown that using various methods of chemical derivatization (significant for analysis in Vis region) or ones based on mathematical extends (significant for analysis in UV region) the selective determination of such labile drug as β-lactam antibiotics was possible. Taking into consideration variability of stability of some β-lactam analogs (in the presence of impurities, degradation products, inhibitors of β-lactamases, excipients and other drugs) and pathways of their degradation, confirmation of selectivity of method and values of other validation parameters for each analogs is required. It is important that analytical method used as alternative, referential one should not use spectrophotometric properties of compound as the base of detection. Recently, the development of chemometric procedures, as the complementary method for spectrophotometric ones, is often supported by suitable, high level software and equipment of apparatus allowing to assume that analysis based on calculated algorithms meet all validation criteria and can replace these methods which require pre-treatment of samples. Moreover, these methods permit multiple determination of the same sample and can be used in routine control analysis of intravenous pharmaceutical dosage forms. Additionally, proposed conditions of determination using spectrophotometric methods are in accordance with all criteria identified as the trends of *"green chemistry"*. A enrichment of visible spectrophotometric methods by the application of FIA and SIA permit accept of challenges of modern industrial-scale pharmaceutical analysis [60]. It can be expected that spectrophotometric methods enriched by chemometric procedures will become an important field of pharmaceutical analysis including also labile drugs.

5. Acknowledgment

The authors thank the State Committee for Scientific Research, Poland, for project N N405 683040.

6. References

[1] Cielecka-Piontek J. Michalska K. Zalewski P. Jelińska A. Recent Advances in stability studies of carbapenems. Current Pharmaceutical Analysis 2011;7 213-227.

[2] El-Shaboury S. Saleh G. Mohamed F. Rageh A. Analysis of cephalosporin antibiotics. Journal of Pharmaceutical and Biomedical Analysis.2007;45 1–19.

[3] Cielecka-Piontek J. Michalska K. Zalewski P. Zasada S. Comparative review of analytical techniques for determination of carbapenems. Current Analytical Chemistry 2012; 8 91-115.

[4] ICH. Stability testing of new drug substances and products. In: Proceedings of International conference on Harmonization. Geneva: IFPMA; 2000.

[5] Parisotto G. Ferrao M. Furtado J. Molz R. Determination of amoxicillin content in powdered pharmaceutical formulations using DRIFTS and PLS. Brazilian Journal of Pharmaceutical Sciences 2007;43(1) 89-95.

[6] Tabelbpour Z. Tavallaie R. Agmadi S. Abdollahpour A. Simultanesous determination of penicillin G salts by infrared spectroscopy: Evaluation of combining orthogonal signal correction with radial basis function-partial least squares regression. Spectrochimica Acta Part A 2010;76 452–457.

[7] European Pharmacopoeia 7th ed. 2010

[8] Ivama V. Rodrigues L. Guaratini C. Zanoni M. Spectrophotometric determination of cefaclor in pharmaceutical preparations. Quimica Nova 1999;22(2) 1–6.

[9] Deshpande A. Baheti K. Chatterjee N. Degradation of β-lactam antibiotics. Current Science 2004;78(12) 1684-1695.

[10] Rojanarata T. Opanasopit P. Ngawhirunpat T. Saehuan Ch. Wiyakrutta S. Meevootisom V. A simple sensitive and green bienzymatic UV-spectrophotometric assay for amoxicillin formulations. Enzyme and Microbial Technology 2010;46 292–296.

[11] Campins-Falco P. Sevillano-Cabeza A. Gallo-Martinez L. Bosch-Reig F. Monzo-Mansanet I. Comparative Study on the Determination of Cephalexin in its Dosage Forms by Spectrophotometry and HPLC with UV-vis Detection. Microchimica Acta 1997;126 207–215.

[12] Cantarelli M. Pellerano R. Marchevsky E. Camina J. Simultaneous Determination of Amoxicillin and Diclofenac in Pharmaceutical Formulations Using UV Spectral Data and the PLS Chemometric Method. Analytical Sciences 2011;27 73–78.

[13] Abdel-Hamid M. FSQ spectrophotometric and HPLC analysis of some cephalosporins in the presence of their alkali-induced degradation products. Il Farmaco 1998;53 132–138.

[14] Mahgoub H. Aly F. Uv-spetrophotometric determination of ampicillin sodium and sulbactam sodium in two-component mixtures. Journal of Pharmaceutical and Biomedical Analysis 1998;17 1273–1278.

[15] Murillo J. Lemus J. Garcia L. Simultaneous determination of the binary mixtures of cefsulodin and clavulanic acid by using first-derivative spectrophotometry. Journal of Pharmaceutical and Biomedical Analysis 1995;13(6) 769–776.

[16] Rodenas V. Parra A. Garcia-Villanova J. Gomez M. Simultaneous determination of cefepime and L-arginine in injections by second-derivative spectrophotometry. Journal of Pharmaceutical and Biomedical Analysis 1995;13 1095–1099.

[17] Daabees H. Mahrous M. Abdel-Khalek M. Beltagy Y. Emil K. Spectrophotometric determination of cefprozil in pharmaceutical dosage forms in urine and in the

presence of its alkaline induced degradation products. Analytical Letters 2001;34(10) 1639–1655.
[18] Nuevas L. Gonzalez R. Rodriguez J. Hoogmartens J. Derivative spectrophotometric determination of the thriethylammonium salt of cefotaxime in presence of related compound from the synthesis. Journal of Pharmaceutical and Biomedical Analysis 1998;18 579–583.
[19] Forsyth R. Ip D. Determination of imipenem and cilastatin sodium in Primaxin® by first order derivative ultraviolet spectrophotometry. Journal of Pharmaceutical and Biomedical Analysis 1994;12 1243-1248.
[20] Murillo J. Lemus J. Garcia L. Analysis of binary mixtures of cephalothin and cefoxitin by using first-derivative spectrophotometry. Journal of Pharmaceutical and Biomedical Analysis 1996;14 257-266.
[21] Cielecka-Piontek J. Lunzer A. Jelińska A. Stability-indicating derivative spectrophotometry method for the determination of biapenem in the presence of its degradation products. Central European Journal of Chemistry 2011;9 35–40.
[22] Mohamed A. Salem S. Maher E. Chemometrics-assisted spectrophotometric determination of certain β-lactam antibiotics combinations. Thaiwan Journal Pharmaceutical Sciences 2007;31 9–27.
[23] Lin Z. Liu J. Chen G. A new method of Fourier-transform smoothing with ratio spectra dervative spectrophotometry. Fresenius Journal Analytical Chemistry 2001;370 997–1002.
[24] Zając M. Cielecka-Piontek J. Jelińska A. Development and validation of UV spectrophotometric and RP-HPLC methods for determination of ertapenem during stability studies. Chemia Analityczna 2006;51 761–768.
[25] Cielecka-Piontek J. Jelińska A. The UV-derivative spectrophotometry for the determination of doripenem in the presence of its degradation products. Spectrochimica Acta Part A 2010;77 554–557.
[26] Morelli B. Determination of ternary mixtures of antibiotics by ratio-spectra zero-crossing first- and third-derivative spectrophotometry. Journal of Pharmaceutical and Biomedical Analysis 1995;13(3) 219-227.
[27] Morelli B. Derivative spectrophotometry in the analysis of mixtures of cefotaxime sodium and cefadroxil monohydrate. Journal of Pharmaceutical and Biomedical Analysis 2003;32 257-267.
[28] Murillo J. Lemus J. Garcia L. Application of the ratio spectra derivative spectrophotometry to the analysis of cephradine and clavulanic acid in binary mixtures. Fresenius Journal Analytical Chemistry 1993;347 114–118.
[29] Elragehy N. Abdel-Moety E. Hassan N. Rezk M. Stability-indicating determination of meropenem in presence of its degradation product. Talanta 2008;77 28–36.
[30] Zając M. Cielecka-Piontek J. Jelińska A. Development and validation of UV spectrophotometric and RP.HPLC methods for determination of ertapenem during stability studies. Chemia Analityczna 2006;51 761–768.
[31] Hassan N. Abdel-Moety E. Elragehy N. Rezk M. Selective determination of ertapenem in the presence of its degradation product. Spectrochimica Acta Part A 2009;72 915-921.
[32] Salem H. Saleh G. Selective spectrophotometric determination of phenolic β-lactam antibiotics. Journal of Pharmaceutical and Biomedical Analysis 2002;28 1205-1213.

[33] Al-Momani I. Spectrophotometric determination of selected cephalosproins in drug formulations using flow injection analysis. Journal of Pharmaceutical and Biomedical Analysis 2001;25 751-757.
[34] Saleh G. El-Shaboury S. Mahomed F. Rageh A. Kinetic spectrophotometric determination of certain cephalosporins using oxidized quercetin reagent. Spectrochimica Acta Part A 2009;73 946-954.
[35] Ayad M. Shalaby A. Abdellatef H. Elsaid H. Spectrophotometric determination of certain cephalosporins though oxidation with cerium(IV) and 1-chlorobenzotriazole. Journal of Pharmaceutical and Biomedical Analysis 1999;20 557-564.
[36] Sastry Ch. Rao S. Naidu P. Strinivas K. New spectrophotometric method for the determination of some drugs with iodine and wool fast blue BL. Talanta 1998;45 1227–1234.
[37] Feng S. Jiang J. Fan J. Chen X. Sequential injection analysis with spectrophotometric detection of cefadroxil and amoxicillin in pharmaceuticals. Chemia Analityczna 2007;52(83) 83–92.
[38] Sastry Ch. Rao K. Prasad D. Determination of cefadroxil by three simple spectrophotometric methods using oxidative coupling reactions. Mikrochimica Acta 1997;126 167–172.
[39] Makchit J. Upalee S. Thongpoon Ch. Liawruangrath B. Liawruangrath S. Determination of cefadroxil by sequential injection with spectrophotometric detector. Analytical Sciences 2006;22 591–597.
[40] Salem G. Two selective spectrophotometric methods for the determination of amoxicillin and cefadroxil. Analyst 1996;121 641–645.
[41] Helaleh M. Abu-Named E. Jamhour R. Spectrophotometric determination of selected cephalosporins. Acta Poloniae Pharmaceutica 1998;55(2) 87–91.
[42] Walily A. Gazy A. Belal S. Khamis E. Quantitative determination of some triazole cephalosporins through complexation with palladium (II) chloride. Journal of Pharmaceutical and Biomedical Analysis 2000;22 385-392.
[43] Salem H. Selective spectrophotometric determination of phenolic β-lactam antibiotics in pure forms and in their pharmaceutical formulations. Analytica Chimica Acta2004;515 333–341
[44] Eric A. Karljikovic-Rajic K. Vladimirov S. Zivanov-Stakic D. Spectrophotometric determinations of certain cephalosporins using ferrihydroxamate method. Spectroscopy Letters 1997;30(2) 309–313.
[45] Saleh G. Askal H. Darwish I. El-Shorbagi A. Spectroscopic analytical study for the charge-transfer complexation of certain cephalosporins with chloroanilic acid. Analytical Sciences 2003;19 281–287.
[46] El-Mammli M. Spectrophotometric determination of flucloxacillin in pharmaceutical preparations using some nitrophenols as a complexing agent. Spectrochimica Acta Part A 2003;59 771-776.
[47] Saleh G. Askal H. Radwan M. Omar M. Use of charge-transfer complexation in the spectrophotometric analysis of certain cephalosporins. Talanta 2001;54 1205–1215.
[48] Mohamed G. Spectrophotometric determination of amipicillin dicluxacillin flucloxacillin and amoxicillin antibiotic drugs: ion-pair formation with

molybdenum and thiocyanate. Journal of Pharmaceutical and Biomedical Analysis 2001;24 561-567.

[49] Salem H. Askal H. Colourimetric and AAS determination of cephalosporins using Reineck's salt. Journal of Pharmaceutical and Biomedical Analysis 2002;29 347-354.

[50] El-Ashry S. Belal F. El-Kerdawy M. Wasseef D. Spectrophotometric determination of some phenolic antibiotics in dosage forms Mickrochimica Acta 2000;135 191–196.

[51] Freitas S. Silva V. Araujo A. Canceicao M. Montenegro M. Reis B. Paim P. A multicommuted flow analysis method for the photometric determination of amoxicillin in pharmaceutical formulations using a diazo coupling reaction. Journal of Brazilian Chemistry Society 2011;22(2) 279–285.

[52] Xu L. Wang H. Xiao Y. Spectrophotometric determination of amipicillin sodium in pharmaceutical products using sodium 12 naphtoquinone-4-sulfonic as the chromogentic reagent. Spectrochimica Acta Part A 2004;60 3007-3012.

[53] Gallo-Martinez L. Sevillano-Cabeza A. Campins-Falco P. Bosch-Reig F. A new derivatization procedure for the determination of cephalexin with 12-naphtoquinone 4-sulphonate in pharmaceutical and urine samples using solid-phase extraction cartridges and UV-visible detection. Analytica Chimica Acta 1998;370 115–123.

[54] Metwally F. Alwarthan A. Al-Tamimi S. Flow-injection spectrophotometric determination of certain cephalosporins based on the formation of dyes. Il Farmaco 2001;56 601–607.

[55] Buhl F. Szpilkowska-Sroka B. Spectrophotometric determination of cephalosporins with Leuno crystal violet Chemia Analityczna 2003;48(145) 145–149.

[56] Rao G. Kumar K. Chowdary P. Spectrophotometric methods for the determination of cefotaxime sodium in dosage forms. Indian Journal of Phamraceutical Sciences 2001;63(2) 161–163.

[57] Al-Momani I. Flow-injection spectrophotometric determination of amoxicillin cephalexin ampicillin and cephradine in pharmaceutical formulations. Analytical Letters 2004; 37(10) 2099–2110.

[58] Belal F. El-Kerdawy M. El-Ashry S. El-Wasseef D. Kinetic spectrophotometric determination of ampicillin and amoxicillin in dosage forms. Il Farmaco 2000;55 680–686.

[59] Amin A. Shama S. Vanadophosphoric acid as a modified reagent for the spetrophotometric determination of certain cephalosporins and their dosage forms Monatsefte fur Chemie 2000;131 313–319.

[60] Tzanavaras P. Themelis D. Review of recent applications of flow injections spectrophotometry to pharmaceutical analysis. Analytica Chimica Acta 2007;588 1–9.

8

Identification, Quantitative Determination, and Antioxidant Properties of Polyphenols of Some Malian Medicinal Plant Parts Used in Folk Medicine

Donatien Kone[1,2], Babakar Diop[2], Drissa Diallo[3], Abdelouaheb Djilani[1] and Amadou Dicko[1]
[1]*Université Paul Verlaine-Metz/LCME. 1, Metz*
[2]*Université de Bamako/Faculté des Sciences et Techniques, BP*
[3]*INRSP/Département de Médecine Traditionnelle, Bamako,*
[1]*France*
[2,3]*Mali*

1. Introduction

In biological systems, during the cellular respiration, reactive oxygen species (ROS) like hydroxyl radical (•OH), superoxide anion ($•O_2^-$) and hydrogen peroxide (H_2O_2) are generated, as the natural consequence of oxidation reactions (Tarnawski et al., 2005). Reactive oxygen species (ROS) damage living cells causing lipid, protein, and DNA oxidation (Shukla et al., 2010). They are involved in the development of various diseases such as diabetes, rheumatic disorders (Luximon- Ramma et al., 2002), aging, cancer, cardiovascular or neurodegenerative disorders (Ju et al., 2004; Tarnawski et al., 2005), malaria and gastric ulcer (Gülçin et al., 2006).

The interest in searching natural antioxidants has recently increased. These natural products could be used in food or in medicinal materials to replace synthetic antioxidants which are about to be restricted owing to their side effects such as carcinogenesis (Gülçin et al., 2006). Many medicinal plants contain large amounts of antioxidants, such as polyphenols, which can play an important role in adsorbing and neutralizing free radicals, in quenching singlet and triplet oxygen, or in decomposing peroxides. The compounds that are responsible of antioxidant activity could be used for the prevention and treatment of free radical-related disorders (Gomez-Caravaca et al., 2006). Indeed, the consumption of antioxidants prevents different diseases such as neurological degeneration, inflammatory disorders, coronary diseases, aging and cancers (Djeridane et al., 2006). Hence, the studies on natural antioxidants have gained increasingly greater importance.

A large number of different plants have been studied as new sources of natural antioxidants (Cakir et al., 2003; Lee et al., 2000; Kumaran & Karunakaran, 2007; Muanda et al., 2009). For the first time, we report here the antioxidant properties of extracts from the following six Malian folk medicine plants, which were previously studied for their biological activities:

Anogeissus leiocarpus (DC.) Guill. et Perrot (Combretaceae), *Cissus populnea* Guill. et Perr. (Vitidaceae), *Mitragyna inermis* (Willd.) O. Ktze. (Rubiaceae), *Terminalia macroptera* Guill. et Perrott (Combretaceae), *Vepris heterophylla* R. Let. (Rutaceae) and *Zizyphus mucronata* Willd. (Rhamnaceae). These plants were selected for their traditional used in the treatment of inflammatory diseases such as: malaria, oedema, arthritis, rheumatism, ulcer, gingivitis, conjunctivitis (Burkill, 2000; Malgras, 1992; Arbonnier, 2002; Inngjerdingen et al., 2004). Vonthron-Sénécheau et *al.*, (2003) reported the *in vitro* antiplasmodial activity of the extracts of the leaves of *Anogeissus leiocarpus.* Geidam et al., (2004) reported evidence-proved similar effects of the aqueous stem bark extract from *Cissus populnea* on some serum enzymes in alloxane induced diabetic rats. They have attributed hypoglycaemic properties to these extracts. Aqueous extract from *Mitragyna inermis* has been used by traditional healers for the treatment of various diseases, particularly for hepatic illness, malaria and hypertention. Recently, studies by Ouédraogo et *al.*, (2004) demonstrated the hypotensive, cardiotropic and vasodilatory properties of bark aqueous extract from *Mitragyna inermis.* To identify new antimalarial compounds, Conrad et *al.* (1998) selected *Terminalia macroptera* for an antiplasmodial screening. Moulis et al., (1994) studied the volatile constituents of the leaves of *Vepris heterophylla.* They found that among thirty-three compounds - that were identified by capillary GC - the main constituents were geijerene and pregeijerene. Recently, Mølgaard et *al.,* (2001) reported good activity of the extracts of root bark from *Zyzyphus mucronata* which were tested *in vitro* against tapeworms and schistosomules.

To our knowledge, there are no previous reports concerning *in vitro* antioxidant activities of these plant part extracts. The purposes of this study were to determine the total phenolic and the total flavonoid contents, to evaluate their antioxidant activities using 2, 2'-azino-bis (3-ethylbenzothiazoline)-6-sulfonic acid (ABTS) and 2, 2-diphenyl-1-picrylhydrazyl (DPPH) tests, and finally to identify and to quantify some polyphenolic compounds by using a RP-HPLC coupled to an UV detector.

2. Materials and methods

2.1 Plant material

Professor N'Golo Diarra performed the plant taxonomy in the "Departement of Traditional Medecine" in Bamako, and voucher specimens were deposited in its herbarium as *Anogeissus leiocarpus* (DC.) Guill. et Perrot (Combretaceae) Ref. N° 1559, *Cissus populnea* Guill. et Perr. (Vitidaceae) Ref. N°1368, *Mitragyna inermis* (Willd.) O. Ktze. (Rubiaceae) Ref. N°1394, *Terminalia macroptera* Guill. et Perrott (Combretaceae) Ref. N°1617, *Vepris heterophylla* R. Let. (Rutaceae) Ref. N°2444 and *Zizyphus mucronata* Willd. (Rhamnaceae) Ref. N°2499. The plant material was collected around the district of Bamako in December 2005.

2.2 Apparatus

The HPLC analyses were performed with a Waters 600E coupled to a Waters 486 UV visible tunable detector (SPD-M10Avp) and a Reverse Phases C18 symmetry analytical Alltech Intertsil ODS- 5 µm 4.6mm x 150 mm column. In addition, spectrophotometer analyses were carried out with a Cary 50 scan UV- Visible apparatus (UV Mini 1240).

2.3 Chemical reagents

Standards: catechin and gallic acid, 3,4 dihydroxybenzoic acid (protocatechuic acid), chlorogenic acid, rutin were purchased from Across Organics (France). *p*-coumaric acid, isovitexin and quercetin 3-ß-D-glucoside were obtained from Fluka Chemical Company (France).

Aluminium chloride (AlCl3), ascorbic acid, 2-2'-azino-bis (3-ethylbenzothiazoline-6-sulfonic acid) diammonium salt (ABTS), PBS buffer, 2-2'-azobis (2-methylpropionamidine) dichloride (AAPH), Folin-Ciocalteu's phenol reagent, sodium carbonate (Na_2CO_3), caffeic acid and sodium nitrite ($NaNO_2$), stable free radical DPPH were purchased from Sigma Chemical Company (France). All commercial standards and reagents were of the highest analytical grade.

2.4 Preparation of extracts

Each plant material was dried in a dark ventilated room for 5–7 days. The different parts of the plants (leaves, root barks, and stem barks) were ground to powder and sifted in a sieve (0.750µm).

The extraction of the samples was performed by the ultrasound-assisted method (Kim et al., 2002, 2003). The air-dried plant material (10 g) was extracted using 100 mL of 80% aqueous methanol, the mixture of freeze-dried powder and 80% aqueous methanol was sonicated for 20 min with continual nitrogen gas purging. The mixture was filtered through Whatman N°2 filter paper and rinsing with 50 mL of 100% methanol. The extraction of the residue was repeated using the same conditions and the two filtrates were combined and transferred into a 1 L evaporating flask with an additional 50 mL of 80% aqueous methanol. The solvent was evaporated using a rotary evaporator at 40 °C. The remaining extract concentrate was first dissolved in 50 mL of 100% methanol and diluted to a final volume of 100 mL using distilled deionized water (ddH_2O). The mixture was centrifuged at 1500g for 20 min and stored at -4°C until analyses were performed.

2.5 Determination of the total phenolic and of the total flavonoid contents

The concentration of total phenolics was measured by the method described by (Kim et al., 2003). Briefly, an aliquot (1 mL) of appropriately diluted extracts or standard solutions of gallic acid was added to a 25 mL volumetric flask containing 9 mL of ddH_2O. A reagent blank was prepared using ddH_2O. One mililiter of Folin & Ciocalteu's phenol reagent was added to the mixture and shaken. After 5 min, 10 mL of 7% Na_2CO_3 solution were added and the solution was then immediately diluted to volume (25 mL) with ddH2O and mixed thoroughly. After an incubation of 90 min at 23 °C, the absorbance versus prepared blank was read at 750 nm (Cary 50 Scan UV-Visible apparatus). Total phenolic contents of plant parts are expressed as mg of gallic acid equivalents (GAE) / g dry weight. All samples were analyzed at least in triplicate.

Total flavonoids were measured by a colorimetric assay that was developed by Zhishen et al. (1999). We can add to a 10 mL volumetric flask containing 4 mL ddH_2O either 1 mL of aliquot of appropriately diluted sample or 1 mL of a standard solution of catechin. At zero time, 0.3 mL 5% $NaNO_2$ was added to the flask. After 5 min, 0.3 mL 10% $AlCl_3$ was added.

After 6 min, 2 mL of 1 M NaOH was added to the mixture. Immediately, the reaction flask had to be diluted to volume by the addition of 2.4 mL of ddH2O and thoroughly mixed. Absorbance of the mixture was determined at 510 nm (Cary 50 Scan UV-Visible apparatus) versus prepared water blank. Total flavonoids of plant parts are expressed as mg / g dry weight of catechin equivalents (CE). Samples were analyzed at least in triplicate.

2.6 Determination of total antioxidant activity

Various methods have been introduced for the measurement of the total antioxidant capacity (Delgado-Andrade et al., 2005; Gülçin et al., 2006). In this study antioxidant activity was estimated by the method previously described (Kim et al., 2002, 2003). The *in vitro* antioxidant activities have been determined in two antioxidant tests. Among the different methods permitting to evaluate the antioxidant activities, these two simple stable radical chromogens have a high level of sensitivity and allow the analysis of a large number of samples in a timely fashion (Kim et al., 2002). It was reported that a single method is not enough to evaluate the antioxidant capacity of most of the complex natural products (Ozgen et al., 2006). Antioxidant capacity is expressed as mg of vitamin C equivalent (mg VCE) per g dry weight.

2.7 ABTS radical anion scavenging activity

ABTS radical anions were used according to the method of (Kim et al., 2003). In brief, 1.0 mM of 2, 2′-azobis (2-amidino-propane) dihydrochloride (AAPH), a radical initiator, was mixed with 2.5 mM ABTS in phosphate-buffered saline (pH 7.4) and the mixed solution was heated in a water bath at 68 °C for 13 min. The resulting blue-green ABTS solution was adjusted to the absorbance of 0.650 ± 0.020 at 734 nm with additional phosphate-buffered saline. 20 µl of sample were added to 980 µL of the ABTS radical solution. The mixture incubated in a 37°C water bath under restricted light for 10 min. A control (20 µL 50% methanol and 980 mL of ABTS radical solution) was run with each series of samples. The decrease of the absorbance at 734 nm was measured (Cary 50 Scan UV-Visible apparatus) at an endpoint after 10 min. Total antioxidant capacity of plant parts is expressed as mg / g of dry weight of vitamin C equivalents (VCEAC). The radical stock solution had to be freshly prepared and all measurements of the tested samples were repeated at least three times.

2.8 DPPH radical scavenging activity

The DPPH radical scavenging activity was determined according to the method of Kim et *al.*, (2002). The DPPH radical (100 µM) was dissolved in 80% of aqueous methanol. The plant extract solutions, 0.1 mL, were added to 2.9 mL of the methanolic DPPH solution and the mixture was vigorously shaken and was kept at 23 °C in the dark for 30 min. The decrease of the absorbance of the resulting solution was monitored at 517 nm (Cary 50 Scan UV-Visible apparatus) after 30 min. A control, which consists of 0.1 mL of 50% aqueous methanol and 2.9 mL of DPPH solution, was prepared. The DPPH radical scavenging activity of plant extracts is expressed as mg/g of dry weight of vitamin C equivalents (VCEAC). This measure was taken after 30 min reaction time. The radical stock solution had to be daily prepared and the tests were repeated at least three times.

2.9 HPLC analysis

The HPLC analyses were conducted with a Water 600 Pump apparatus. This apparatus was equipped with a quaternary solvent delivery system, a Rheodyne injector with 20µL sample loop and a UV detector Waters 486 Tunable which was fixed at 280 nm. Throughout this study, Alltech Intertsil ODS-5 C18 reversed phase column (150 mm, 4.6 mm, 5µm particle size) was used. The flow rate of the mobile phase was of 1 mL / min and the gradient elution was adapted from (Nakatani et al., 2000; Bouayed et al., 2007). The solvent composition and the gradient elution program are reported in the table 1.

Times (min)	%A	%B	%C
0	100	0	0
5	65	12	23
11	0	15	85
29	0	22	78
36	0	25	75
42	0	25	75
55	0	35	65
60	0	50	50
70	0	100	0
75	100	0	0

Solvent composition: A =50 mM $NH_4H_2PO_4$ at pH 2.60; B = 80% acetonitrile, 20 %A and C = 200 mM O-phosphoric acid at pH 1.50

Table 1. HPLC solvent gradient elution program

Standards of five phenolic acids and two flavonoids were dissolved in 50% MeOH to make a concentration of 0.5; 0.25; 0.125 and 0.10 mg/mL. The plant part extracts and standards solutions were filtered through 0.45-µm olefin polymer (OP) syringe-tip filters. Then, phenolic acids and flavonoids present in the extracts were identified by matching the retention time against their corresponding standard. In this study, the standards used for comparison were gallic acid, protocatechuic acid, chlorogenic acid, caffeic acid, *p*-coumaric acid, isovitexin and quercetin-3-β-D-glucoside (Figure 1). Quantitative analysis was made according to the linear calibration curves of standards compounds. Three replications were made at least for each standard and plant extract.

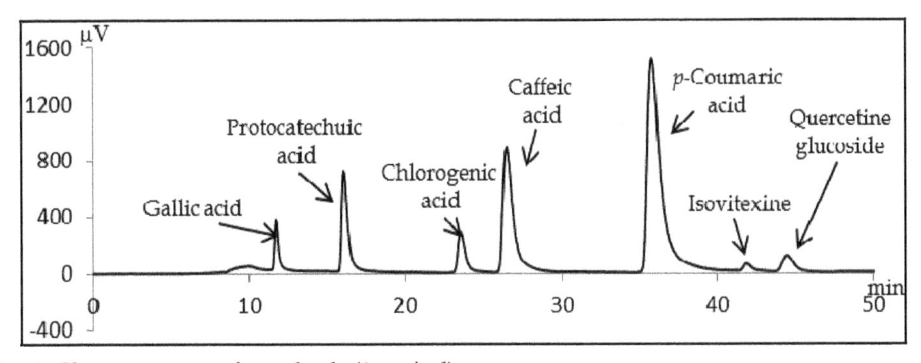

Fig. 1. Chromatogram of standards (1mg/ml)

3. Results

3.1 Total phenolics and total flavonoids

The results of the colorimetric analysis of total phenolics expressed as Gallic Acid Equivalents (GAE) and those of total flavonoids expressed as Catechine Equivalents (CE) are given in the table 2.

Plants name	Plant parts	Total phenolics (mg GAE)	Total Flavonoids (mg CE)
A. leiocarpus	L	223.1 ± 0.2	38.9 ± 1.7
	TB	26.5 ± 0.4	10.3 ± 0.3
C. populnea	RB	76.4 ± 1.1	27.6 ± 1.2
M. inermis	TB	19.5 ± 0.7	11.1 ± 1.3
T. macroptera	TB	48.5 ± 1.3	14.2 ±1.4
	RB	219.6 ± 0.4	33.1 ± 1.3
V. heterophylla	L	51.5 ± 0.5	9.3 ± 0.9
Z. mucronata	L	52.2 ± 0.5	14.4 ± 0.8
	RB	19.3 ± 0.6	9 ± 1.6

L= leaves; TB= trunk barks; RB= root bark.Total phenolics expressed as gallic acid equivalent (GAE), total flavonoids expressed as catechin equivalent (CE).
Values are means of triplicate determination ± standard deviation.
The total phenolic compounds which are present in plant materials were ranged from 19.3 ± 0.6 to 223.1 ± 0.2 mg GAE / g dry weight and the total amount of flavonoids varied from 9 ± 1.6 to 38.9 ± 1.7 mg CE / g dry weight.

Table 2. Total phenolic and total flavonoid contents of the plant parts

3.2 ABTS and DPPH radical-scavenging activity

ABTS and DPPH tests were conducted to evaluate the antioxidant properties of plant part extract on their stable free radicals in comparison to the antioxidant activity of vitamin C, the corresponding results were collected in the figure 2.

The antioxidant activity using ABTS varied from 39 mg to 468 mg VCE per g dry weight. The overall antioxidant capacity of plant parts in VCEAC, which was evaluated by ABTS assay, was in the following order: *T. macroptera* root bark > *A. leiocarpus* leaves > *T. macroptera* trunk bark > *C. populnea* root bark > *A. leiocarpus* trunk bark > *Z. mucronata* leaves > *V. heterophylla* leaves > *M. inermis* trunk bark > *Z. mucronata* root bark.

The antioxidant activity using DPPH ranged from 21 mg to 361 mg VCE per g dry weight. The overall antioxidant capacity of plant parts in VCEAC which was evaluated by DPPH assay decreased in the following order: *T. macroptera* root bark > *A. leiocarpus* leaves >*T. macroptera* trunk bark > *C. populnea* root bark > *Z. mucronata* leaves ≈ *A. leiocarpus* trunk bark > *Z. mucronata* root bark > *V. heterophylla* leaves >*M. inermis* trunk bark.

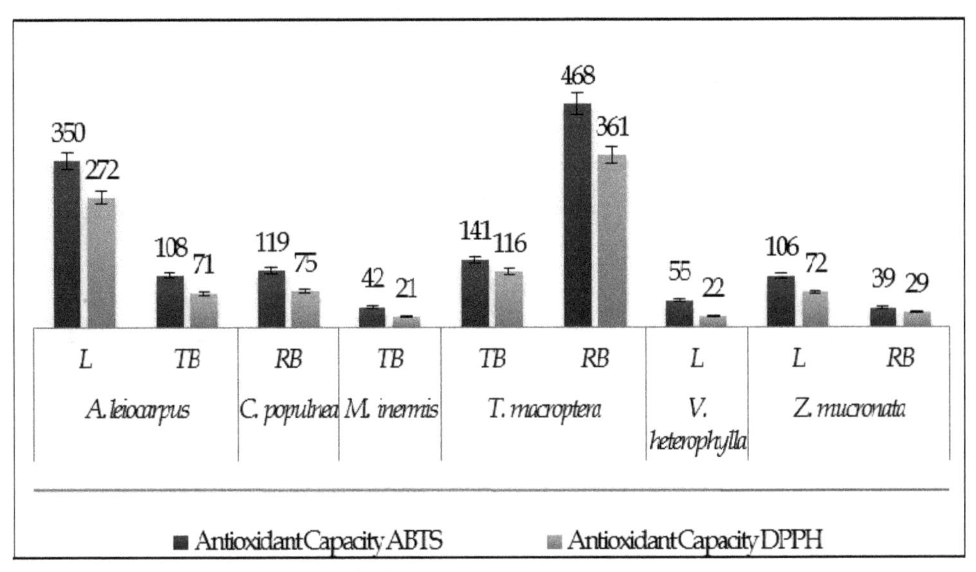

L= leaves; TB= trunk barks; RB= root barks
Values are means of triplicate determination ± standard deviation.

Fig. 2. Vitamin C equivalent antioxidant capacity (VCE mg/g dry weight).

3.3 Analysis of polyphenolic composition in plant part extracts using HPLC

The RP-HPLC results which are summarised in table 3, show that the protocatechuic acid (792.00 - 7.93 mg / 100g dry weight), the *p*-coumaric acid (1833.56 - 4.2 mg / 100g dry weight), the gallic acid (69.00 - 5.14 mg / 100g dry weight) and the chlorogenic acid (2286.08 - 62.09 mg / 100g dry weight) were encountered in high concentration, whereas the caffeic acid (149.86 -43.46 mg / 100g dry weight), the isovitexin (182.22 - 31.46 mg / 100g dry weight) and the quercetin-3-β-D-glucoside (83.53 - 70.89 mg /100g dry weight) were found in low concentration. It should be noted that the trunk and the root barks of *T macroptera* and the leaves of *V. heterophylla* contain the greatest number of compounds similar to the standards, whereas none of the standards was detected for the root barks of *Z. mucronata*.

4. Discussion

The leaves of *A. leiocarpus* had the highest total phenolic contents, which was 4- fold higher than those of the leaves of *V. heterophylla.* The lowest total phenolic and total flavonoid contents were found in the leaves of *V. heterophylla*. In the trunk barks of *T. macroptera*, the total phenolics and the total flavonoids were ranked first, followed by *A. leiocarpus* and then the *M. inermis* ones. The phenolic and flavonoid compounds were found in highest concentration first in the root barks of *T. macroptera* followed by the root barks of *C. populnea.* In contrast, it appeared that the lowest amount of flavonoids was found for the root barks of *Z. mucronata*. The total phenolic and the total flavonoid contents of the root barks of *T. macroptera* were respectively 11-fold and 4-fold greater than those of *Z. mucronata.*

Plant	Plant parts	Gallic Acid	Protocatechuic Acid	Chlorogenic Acid	Caffeic Acid	p-coumaric Acid	Isovitexin	Quercetin
Anogeissus leiocarpus	L	49.1 ± 3.5	67.6 ± 0.3	2286 ± 80	nd	nd	nd	nd
	T B	5.1 ± 0.2	35.5 ± 0.2	nd	nd	221 ± 20	nd	nd
Cissus populnea	R B	nd	26 ± 1.2	62.1 ± 0.5	nd	4.2 ± 0.2	nd	70.9 ± 0.5
Mitragyna inermis	T B	5.9 ± 0.16	53.2 ± 1.2	nd	nd	7.5 ±0.6	nd	nd
Terminalia macroptera	T B	16.4 ± 2.2	7.9 ± 1.3	67 ± 2	nd	59.3 ± 2.2	182 ± 42	nd
	R B	69 ± 1.4	792 ± 7	113.4 ± 1.4	149.9 ± 2.6	1833 ± 4.6	nd	nd
Vepris heterophylla	L	7.5 ± 0.44	10.1 ± 0.7	79 ± 1.7	nd	31.9 ± 2.15	31.5 ± 1.1	nd
Zizyphus mucronata	L	nd	53 ± 0.4	683 ± 20.1	43.5 ±1.4	nd	nd	nd

L= leaves; TB= trunk barks; RB= root barks; nd= not detected.

Table 3. Concentrations of flavonoids and phenolic acids in the medicinal traditional plant parts (mg / 100g of dry material)

The results obtained by ABTS and DPPH tests show that the antioxidant activity order for these different plant parts was approximately similar in both assays. However, the antioxidant capacity using DPPH compared to the one obtained by ABTS essay was underestimated about 33%. Arnao, (2000) and Delgado-Andrade et *al.*, (2005) report the same occurrence and they explain that the DPPH is only dissolved in alcoholic media. In contrast, the ABTS radicals being solubilised in aqueous and in organic media the antioxidant activity measured is due to the hydrophilic and lipophilic nature of the compounds. In addition, at 515 nm near the visible region where the antioxidant activity is measured, interferences occur with the DPPH coloration.

In this study, we found that phenolic compounds are the major contributors to the antioxidant activity, since total phenolics and antioxidant activity showed a good correlation with a correlation coefficient of R^2=0.9208. However, we note that the trunk barks of *A. leiocarpus* exhibit a high antioxidant activity and a low level of total phenol antioxidant. The value of correlation coefficient between total flavonoids and antioxidant activity was R^2 = 0.752 only.

These results showing good antioxidant activity of these plant parts are particularly interesting since the antioxidant agents would induce analgesic, anticarcinogenic, anti-inflammatory, antithrombotic, immune modulating and anti-atherogenic effects (Djeridane et al., 2006).

The results of HPLC analysis were in accordance with those previously reported in the literature. The phytochemical investigations of the different parts of *T. macroptera* led to the isolation of several *C*- and *O*-glycosyl flavones, chlorogenic acid, quercetin, gallic acid (Silva et al., 2000). Chyau et *al.*, (2006) identified 3,4-dihydroxybenzoic acid (protocatechuic acid), *p*-coumaric acid, gallic acid from the leaves of *T. catappa*. Moreover, gallic acid was also present in the trunk barks of *A. latifolia* (Govindarajan et al., 2004). Protocatechuic acid (3,4-dihydroxybenzoic acid) was found in *Mitragyna rotundifolia* (Kang & Hao, 2006). Ojekale et *al.*, (2006) have reported the presence of flavonoids in *C. populnea*. In this study, the presence of isovitexin in the leaves of *V. heterophylla* was identified and quantified.

5. Conclusion

This study permits to evaluate the amount of phenolics, flavonoids and their total antioxidant activity linked to six traditional medicinal plants. Antioxidant activity varied greatly among the different plant parts and was highly correlated with the polyphenolic contents. We take an interest in the leaves of *A.Leocarpus* and in the root barks of *T.Macroptera*, since they exhibited important antioxidant activities and could be attractive sources of natural antioxidants. Moreover, this comparative study permits to identify and determine by RP-HPLC, five individual phenolic acids and two flavonoids that are mainly at the origin of the antioxidant activity in the studied plant parts.

6. Acknowledgement

The authors are thankful to the Service de Cooperation d'Actions Culturelles (SCAC) of the French embassy in Mali for its financial support.

7. References

Arbonnier M. (2002). Arbres, arbustes et lianes des zones sèches d'Afrique de l'Ouest. Paris: 2nd Ed. CIRAD -UICN

Arnao MB. (2000). some methodological problems in the determination of antioxidant activity using chromogen radicals: a practical case. Trends in Food Science & Technology, Vol. 11, pp. 419–421

Bouayed J, Rammal H, Dicko A, Younos C, Soulimani R. (2007). Chlorogenic acid, a polyphenol from *Prunus domestica* (Mirabelle), with coupled anxiolytic and antioxidant effects. Journal of the Neurological Sciences, Vol. 262, pp. 77–84

Burkill HM. (2000).Useful plants of West Tropical Africa, second ed. Royal Botanic Gardens, Kew, London

Cakir A, Mavi A, Yıldırım A, Duru ME, Harmandar M, Kazaz C. (2003).Isolation and characterization of antioxidant phenolic compounds from the aerial parts of Hypericum hyssopifolium L. by activity-guided fractionation. Journal of Ethnopharmacology, Vol. 87, pp. 73–83

Chyau CC, Ko PT, Mau JL. (2006). Antioxidant properties of aqueous extracts from Terminalia catappa leaves. LWT, Vol. 39, pp. 1099-108

Conrad J, Vogler B, Klaiber I, Roos G, Walter U, Kraus W. (1998). Two triterpene esters from Terminalia macroptera bark. Phytochemistry, Vol. 48, pp. 647-650.

Delgado-Andrade C, Rufiaän-Henares JA, Morales FJ. (2005). Assessing the antioxidant activity of melanoidins from coffee brews by different antioxidant methods. J. Agric. Food Chem., Vol. 53, pp. 7832-7836

Djeridane A, Yousfi M, Nadjemi B, Boutassouna D, Stocker P, Vidal N. (2006). Antioxidant activity of some Algerian medicinal plants extracts containing phenolic compounds. Food Chemistry, Vol. 97, pp. 654-660

Geidam MA, Adoga GI, Sanda FA. (2004). Effects of aqueous stem bark extract of Cissus populnea on some serum enzymes in normal and alloxan induced diabetic rats. Pakistan Journal of Biological Sciences, Vol. 7, pp. 1427- 1429

Gómez-Caravaca AM, Gómez-Romero M, Arráez-Román D, Segura-Carretero A, Fernández-Gutiérrez A. (2006). Advances in the analysis of phenolic compounds in products derived from bees. Journal of Pharmaceutical and Biomedical Analysis, Vol. 41, pp. 1220-1234

Govindarajan R, Vijayakumar M, Rao CV, Shirwaikar A, Mehrotra S, Pushpangadan P. (2004). Healing potential of *Anogeissus latifolia* for dermal wounds in rats. *Acta Pharmaceutica*, Vol. 54, pp. 331-338

Gülçin I, Mshvildadze V, Gepdiremen A, Elias R. (2006). Screening of antiradical and antioxidant activity of monodesmosides and crude extract from Leontice smirnowii tuber. Phytomedicine, Vol. 13, pp. 343-351

Inngjerdingen K, Nergard CS, Diallo D, Mounkoro PP, Paulsen BS. (2004). An Ethnopharmacological survey of plants used for wounds healing in dogoland, Mali, West Africa. Journal of Ethnopharmacology, Vol. 92, pp. 233-244

Ju EM, Lee SE, Hwang HJ, Kim JH. (2004). Antioxidant and anticancer activity of extract from Betula platyphylla var. japonica. Life Sciences, Vol. 74, pp. 1013-1026

Kang W, Hao X. (2006). Triterpenoid saponins from *Mitragyna rotundifolia*. *Biochemical Systematics and Ecology*, Vol. 34, pp. 585-587

Kim DO, Lee KW, Lee HJ, Lee CY. Vitamin C Equivalent Antioxidant Capacity (VCEAC) of Phenolic Phytochemicals. (2002). J. Agric. Food Chem., Vol. 50, pp. 3713-3717

Kim DO, Seung WJ, Lee CY. (2003). Antioxidant capacity of phenolic phytochemicals from various cultivars of plums. Food Chemistry, Vol. 81, pp. 321-326

Kumaran A, Karunakaran R. (2007). Activity-guided isolation and identification of free radical-scavenging components from an aqueous extract of Coleus aromaticus. Food Chemistry, Vol. 100, pp. 356-361

Lee KY, Weintraub ST, Yu BP. (2000). Isolation and identification of a phenolic antioxidant from Aloe barbadensis. Free Radical Biology & Medicine, Vol. 28, pp. 261-265

Luximon-Ramma A, Bahorun T, Soobrattee MA, Aruoma OI. (2002). Antioxidant activities of phenolic, proanthocyanidin, and flavonoid components in extracts of Cassia fistula. J. Agric. Food Chem., Vol. 50, pp. 5042-5047

Malgras D. Arbres et arbustes guérisseurs des savanes maliennes. (1992). Ed. Karthala et ACCT, Paris-France

Moulis C, Fouraste I, Keita A, Bessiere JM. (1994). Composition of leaf essential oil from Vepris heterophylla. Flavour and Flagrance Journal, Vol. 9, pp. 35 -37

Mølgaard P, Nielsen SB, Rasmussen DE, Drummond RB, Makaza N, Andreassen J. (2001). Anthelmintic screening of Zimbabwean plants traditionally used against schistosomiasis. Journal of Ethnopharmacology, Vol. 74, pp. 257- 264

Muanda F, Koné D, Dicko A, Soulimani R, Younos C. (2009). Phytochemical composition and antioxidant capacity of three Malian medicinal plant parts. Evidence-based Complementary and Alternative Medcine. (eCAM), Vol. 2011, Article ID 674320. Doi: 10.1093/ecam/nep 109, pp. 1-8

Nakatani N, Kayano S, Kikuzaki H, Sumino K, Katagiri K, Mitani T. (2000). Identification, quantitative determination, and antioxidative activities of chlorogenic acid isomers in prune (Prunus domestica L.). J. Agric. Food Chem., Vol. 48, pp. 5512-5516

Ojekale AB, Lawal OA, Lasisi AK, Adeleke TI. (2006). Phytochemisty and spermatogenic potentials of aqueous extract of *Cissus populnea* (Guill. and Per) stem bark. *The Scientific World Journal,* Vol. 6, pp. 2140–2146

Ouédraogo S, Ranaivo HR, Ndiaye M, Kaboré ZI, Guissou IP, Bucher B, Andriantsitohaina R. (2004). Cardiovascular properties of aqueous extract from Mitragyna inermis. Journal of Ethnopharmacology, Vol. 93, pp. 345–350

Ozgen M, Reese RN, Tulio JR AZ, Scheerens JC, Miller AR. (2006). Modified 2, 2-Azino-bis-3-ethylbenzothiazoline-6-sulfonic Acid (ABTS) Method to measure antioxidant capacity of selected small fruits and comparison to ferric reducing antioxidant power (FRAP) and 2, 2′-Diphenyl-1-picrylhydrazyl (DPPH) methods. J. Agric. Food Chem., Vol. 54, pp. 1151–1157

Silva O, Gomes ET, Wolfender JL, Marston A, Hostettmann K. (2000).Application of High Performance Liquid Chromatography coupled with Ultraviolet spectroscopy and Electrospray Mass Spectrometry to the characterisation of Ellagitannins from Terminalia macroptera Roots. Pharmaceutical Research, Vol. 17, pp. 1396-1401

Shukla KK., Mahdi AA., Ahmad MK., Jaiswar SP., Shankwar SN. & Tiwari SC. (2010). Mucuna pruriens reduces stress and improves the quality of semen in infertile men. Evidence-based Complementary and Alternative Medicine, Vol. 7, No. 1, pp. 137-144

Tarnawski M., Depta K., Grejciun D. & Szelepin B. (2006). HPLC determination of phenolic acids and antioxidant activity in concentrated peat extract - a natural immunomodulator. Journal of Pharmaceutical and Biomedical Analysis, Vol. 41, pp. 182–188.

Vonthron-Sénécheau C, Weniger B, Ouattara M, Tra Bi F, Kamenan A, Lobstein A, Brun R, Anton R. (2003). In vitro antiplasmodial activity and cytotoxicity of ethnobotanically selected Ivorian plants. Journal of Ethnopharmacology, Vol. 87, pp. 221–225

Zhishen J, Mengcheng T, Jianming W. (1999). The determination of flavonoid contents in mulberry and their scavenging effects on superoxide radicals. Food Chemistry, Vol. 64, pp. 555–559

Section 3

FT-IR Spectroscopy

9

Application of Infrared Spectroscopy in Biomedical Polymer Materials

Zhang Li, Wang Minzhu, Zhen Jian and Zhou Jun
State Food and Drug Administration Medical Device Supervising and Testing Center of Hangzhou Zhejiang Insitute for the Control of Medical Device, Hangzhou, China

1. Introduction

Infrared Spectrum (IR) is mainly used to study molecular structure and composition in substances and thus is also called molecular spectrum. When the sample is exposed to infrared light with continuously changing frequency, the molecule absorbs irradiation of certain frequencies and is subject to vibration or rotation, thus to cause the change of dipole moment. The molecule's transition from normal state to excited state weakens the intensity of the corresponding transmitted light in the absorption region, then the infrared software is used to obtain the IR spectrum. Started in 1970, Fourier Transform Infrared Spectroscopy (FTIR) was of high resolution and high scanning speed. It was not only limited to middle infrared (MIR), Spectrum ranges ultraviolet to far infrared section with the assistance of the beam splitter. The main direction of modern analysis, study and development is to combine technology of FTIR with that of the other instrument. For example, FTIR-TGA (Thermogravimetry Analysis) can be used to obtain thermogravimetric curve as well as the IR spectrum of the weight loss material, thus to determine the real composition of vapor generated in the various weight loss stages and the decomposition process.

Ultraviolet and visible absorption spectrum are usually used to study unsaturated organic matter, especially the organic compounds with conjugated system. However, infrared spectroscopy mainly studies chemical compounds with change of dipole moment during vibration. Thus, almost all organic compounds, except single atoms and homonuclear molecule, absorb in the infrared spectrum region. Except for optical isomers, some high polymer of high molecular weight and compounds with slight difference in molecular weight, two compounds of different structure are unlikely to have the same infrared spectrum. Wave number position, number of wave peaks of infrared absorption band and the intensity of absorption band indicate the characteristics of the molecular structure, and thus can be used to identify the structural composure of the unknown objects or its chemical groups. The absorption intensity of the absorption band is related to the contents of the molecular composition or the chemical groups and can be used for quantitative analysis and purity identification. Infrared spectrum analysis has distinctive characteristic and can be used to test gas, liquid and solid samples. Lie other analysis methods, it can be used for qualitative and quantitative analysis and is one of the effective methods for chemical compound identification and molecular structure elucidation.

2. Summary of biomedical polymer materials

Biomedical polymer material is an important component of biological material and is a remarkable functional polymer. It involves in physics, chemistry, biochemistry, medicine, pathology subjects. The synthetic polymer materials and the organisms (natural polymer) have a very similar chemical structure, which determines their similarity in performance and enables high molecular polymer to meet the many complicate and rigorous functional requirements for medical products. Most metal and inorganic materials are incapable in this respect. Presently, high molecular polymers are widely used in medical products.

2.1 Presently there are two types of high molecular polymers: Non-biodegradable and biodegradable polymer materials

Non-biodegradable biomedical polymer material is widely used in manufacturing of adhesives, coating, and artificial lens as well as for repair of the many soft and hard tissues and organs of human body such as ligament, tendon, skin, blood vessel, artificial organs, bone and teeth. Most non-biodegradable biomedical polymer materials have no biological activity and are difficult to bond firmly with tissues and thus may result in toxicity and allergic reaction. Biodegradable biomedical polymer material is mainly used in temporary execution and replacement of tissue and organ functions in clinic or be used as the medicine controlled-release system and delivery carrier, absorbable surgical suture and wound dressings. It is readily biodegradable and the degradation products can be excreted through metabolism. Thus, it has no negative effect to tissue growth. Presently, it has become the key focus in development of biomedical polymer materials.

Non-biodegradable Polymer Materials	Biodegradable Polymer Material
Silicone Rubber	Polyvinyl alcohol
Polyethylene	Modified natural polysaccharides
Polyacrylate	Protein
Polytetrafluoroethene -PTFE, etc.	(Back spinning) synthetic polyester
Dacron	Polypeptide-polyolefine, silk ossein
Carbon-graphite fiber, etc.	PLA

Table 1. Category of Biomedical Polymers

However, biomedical polymer is an interdisciplinary subject and has varying types of classification based on different purposes and practices. For example, it can be categorized based on the source and application purpose of the medical polymer or based on the influence of living tissues to the materials. Presently there is no uniform standard for the classification.

2.2 Biomedical polymer material

Polymer material is generally composed of high polymers and low molecular weight substances. High polymers are divided into homopolymer, copolymer, blends and oligomer; Low molecular weight materials include: 1) additives: regulator, chain transfer agent, terminator and emulsifier; 2) additives: plasticizers, stabilizer, filler and colorant etc.; 3) unreacted monomer, residual catalyst, etc.

Compared with ordinary organic substance, polymer-a long chain connected by multiple monomer units through covalent bonds—features diversified and designable structure, with very high mechanical strength and non-fixed molecular weight. Polymer is partially crystallized or non-crystallized and provides the properties of elastomer and fluid.

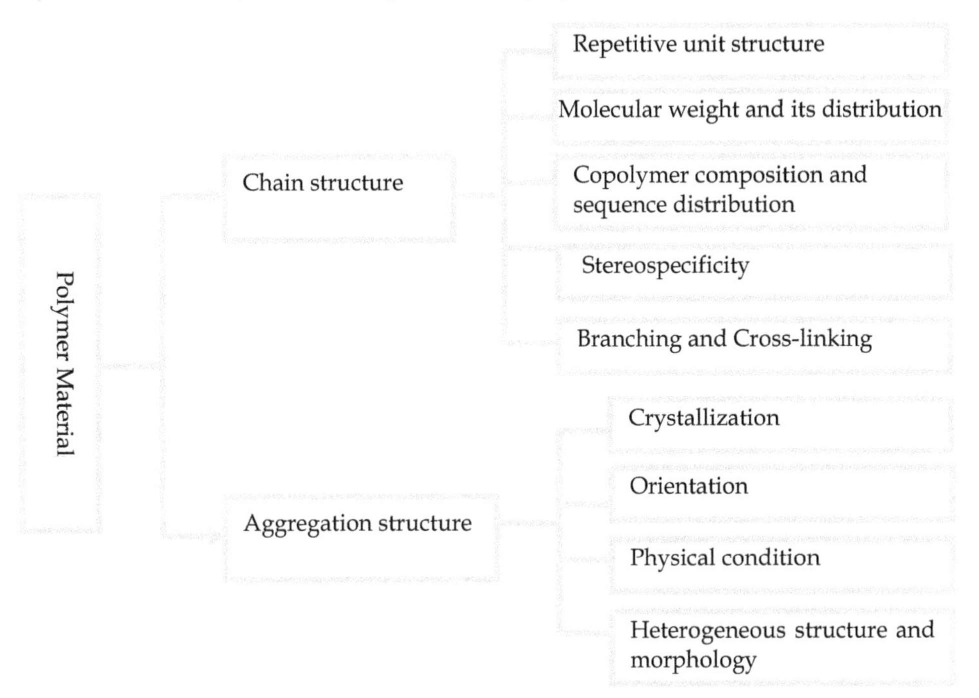

Table 2. Polymer materials

2.3 Basic requirements of biomedical polymer and common materials

Biomedical polymer is directly applied to human body and is closely related to human health. Therefore, the materials used for clinic should be strictly controlled, otherwise, it may cause adverse effect instead of life saving. Below are requirements on properties and performances of the biomedical polymer:

1. Resistance to biological aging: polymer for long-term implantation should have good biological stability.
2. Physical and mechanical stability: different strength, elasticity, size, stability, fatigue resistance and wear resistance for different application.
3. Easy for processing and molding
4. Proper materials with reasonable cost
5. Convenient for sterilization.

Requirements on body effect of biomedical polymer

1. Non-toxic, i.e. chemically inert
2. No pyrogenic reaction
3. Non carcinogenic
4. Non-teratogenic
5. Does not cause allergic reaction or interfere with the body's immune mechanism
6. Does not damage adjacent tissue or cause calcified deposition on material surface.
7. Excellent blood compatibility without causing coagulation when contacting with blood.

Requirements for biomedical polymer production and processing: besides the strict control on biomedical polymer itself, matters harmful to human body shall also be prevented during material production; the purity of the raw materials used in biomedical polymer synthesis shall be strictly controlled, no harmful matter is allowed and the content of heavy mental shall be within the limit; additive processing shall meet medical standard; the production environment should meet proper standard for cleanliness.

The commonly used biomedical polymer materials include Polytetrafluoroethene, polyurethane, polyvinyl chloride, silicone rubber, polypropylene, polysiloxane gel, poly methyl acrylate, chitin derivatives and Polymethylmethacrylate.

2.4 Biomedical polymer material study content

Structure-property relationship: different materials have different properties; the same kinds of materials have different properties; the property of the material not only relevant to its composition but more importantly is relevant to its structure. Moreover, its basic performance and processing performance are sometimes inconsistent with its operational performances.

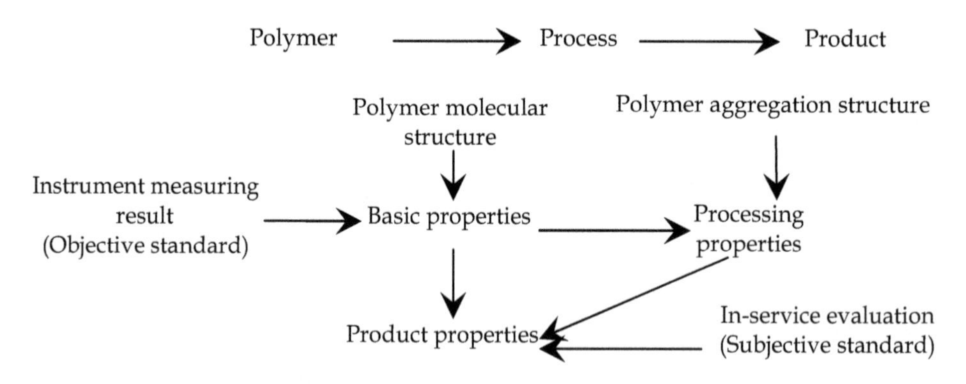

Table 3. Structure-property relationship of Polymer Material

2.4.1 Polymer chain structure study

1. Polymer-based band

Most bands in the spectra are characterized by small molecule band with structure similar to repetitive units, i.e. elemental bands; also there are some unique absorbing bands different from the small molecular organism that belongs to polymer-based bands. Conformation bands; Stereoregularity bands; Conformational regularity band; Crystal band

Such bands are of great significance to the study of polymer.

2. Determination of chemical composition
3. Determination of degree of branching
4. Determination of copolymer composition
5. Determination of the sequence distribution of the copolymer
6. Influence from additives

Adding modifier into certain materials may obtain new material with different performances. But the difference of the infrared spectra maybe small.

2.4.2 Change of polymer materials

Monomer—polymer—Products—Application

2.4.3 Analyses on polymer material

Development of new materials, assimilation of introduced technology and gradual internationalization; discrimination of true and false materials

3. Characteristics of polymer IR spectrum

Infrared spectroscopy is the most effective method to identify the various polymers and additives in polymer material analysis. The main advantages of the infrared spectroscopy include: 1) it does not cause damage to the sample under analysis; 2) it may analyze organic and inorganic compounds of various physical states (gas, liquid and solid) and various exterior forms (elastic, fibrous, thin film, coating and power form); 3) well-developed molecular vibration spectroscopy (the basis of IR spectrum) makes it easy to understand the explanation of the IR spectrum of the compound; 4) a large number of standard infrared spectrogram for various kinds of chemical compounds have already been published in the world and can be referred to in spectra analysis. With application of computer and establishment and improvement of spectral database, the spectral identification will be easier and conclusion will be more reliable. Chemical compounds of different chemical structures have infrared absorption spectroscopy of different characteristics, There is no completely identical spectroscopy except for some isomers. Moreover, each absorption band (band) in the infrared spectrogram represents a certain vibration type of a certain atomic group or radical in the chemical compound. Their vibration frequency (corresponding to the wave numbers of the absorption band on the spectrum) is directly related to the mass and chemical bond strength of the atom in the atomic group or radical. They are consequently subject to changes of proximity structure and different influences of chemical environment. As each molecule of the polymer contains a great number of atoms, it may be considered that the polymer spectrum would be extremely complicate with considerable number of normal vibration. But this is not the case, IR spectra of some polymers are more simple than that of the monomers. This is because polymer chain is made of many repetitive units and each repetitive unit has basically same bond force constant with roughly similar vibration frequency. Moreover, due to limitation of strict selection law, only part of the vibration has infrared activity. The explanation of infrared spectrum of polymer must take into account the molecular chain structure and the aggregate structure of the concerned polymer. Different structural characteristics correspond to different absorption bands: ① absorption bands: reflects the chemical composition of polymer structural unit, connection type of monomers, branching or cross-linking and sequence distribution. ② Conformation bands: such bands are related to the certain conformation of some radical in the molecular chain and have different representations in different morphologies. ③ Stereoregularity bands: these bands are related to the structure of the molecular chain and are therefore identical in the various morphologies of the same high polymer. ④ Conformational regularity bands: these kinds of bands are generated as result of the mutual action of the adjacent radicals in the molecular chain. ⑤ Crystal bands: the bands are formed as a result of the interaction between the adjacent molecular chains in the crystal.

4. Preparation of samples

In order to obtain high quality infrared spectra, a proper method should be selected for sample preparation according to the characteristics of the sample. Close attention should be paid to the following in sample preparation: 1) Purified and separated sample. IR spectrum of mixed sample will be shown in bands of various components and may mislead judgment for the sample. 2) Free of moisture. The infrared spectroscopy is very sensitive to moisture. If there is any trace of moisture in the sample, bands with intensive OH radical characteristic will be appeared near 3400cm^{-1}, affecting the identification of N-H, C-H bonds. 3) Proper concentration and thickness of the sample. In good infrared spectra, most of the absorption bands have transmittance of between 20 to 80%. In especially low concentration or thin thickness, weak absorption bands and band of medium strength will disappear, resulting in inaccurate absorption spectrum. On the contrary, the excessive strong absorption will result in flat peak with no maximum value of the peak or indistinct double peaks.

4.1 Solid sample preparation technology

Common solid samples include polymer and some organic compounds. The following methods are usually adopted for sample preparation:

1. Pressed halide disk method (KBr Pressed Disk Method), mix a quantitative amount of samples (accurately measured for quantitative analysis) And some ground and screened KBr powder in the agate mortar in proper proportion and pulverize them completely until there is no obvious particles in the blends. Please be noted: KBr requires the use of analytical reagent or higher. Qualified KBr should be no absorption in mid-infrared region. Moisture absorption will result in moisture absorption peak. In addition, KBr powder tends to absorb the vapor in the air and should be dried before use. Screen the KBr powder with 200-mesh sieve, then dried under 120°C and put the powder in the dryer for use. The temperature should not be too high(greater than120°C), otherwise it will decompose (see spectra 1).
2. Paste Method

Sample Preparation

Put the fully grounded powder sample in the agate mortar, add 1-2 drops of medium by dropper and blend evenly. Use stainless steel spatula to scoop out the evenly grounded sample, paste it on a window slice, compress with another window slice to a proper thickness and then measure it.

Notes:

All the media used in the method are organic matter and can absorb in certain range. The absorption positions of the media are as shown below:

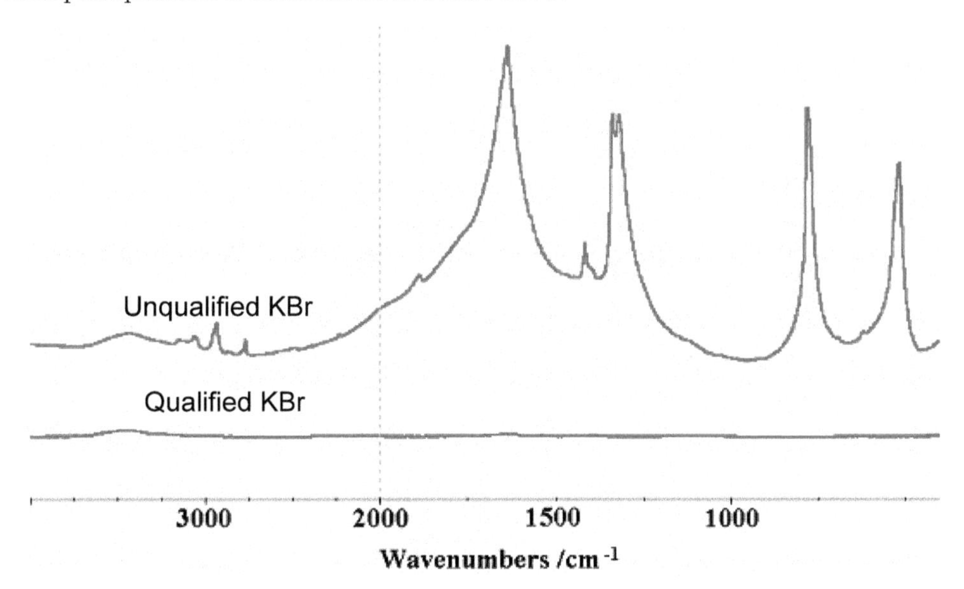

Fig. 1. Absorption requirement of qualified KBr

Name of media	Absorption peak location and assignment
Paraffin oil (long-chain alkane)	3000-2850cm^{-1} ($V_{C\text{-}H2}$, $V_{C\text{-}H3}$) 1468 cm^{-1}, 1397 cm^{-1} ($\delta_{C\text{-}H2}$, $\delta_{C\text{-}H3}$)
Fluorocarbon oil (perfluoroparaffin)	With C-F absorption of different intensity in 1400-1500 cm^{-1}
Hexachlorobutadiene (chloroalkene)	With C=C and C-Cl absorption in 1500-16010 cm^{-1}, 1150-1200 cm^{-1}, 600-1000 cm^{-1}

Table 4. The absorption positions of the media

The IR spectra of above three media have complementary. Complete infrared spectrogram of the sample can be obtained by at least two media. In this consideration, the influence of the media spectrogram should be deducted from spectrum analysis of the sample to baseline.

3. Solution film casting method

Sample preparation method

Dissolve the sample in an appropriate volatile solvent to prepare a solution with concentration of around 2-5%, apply the solution evenly on the watch glass and glass sheet, or drop it directly on the infrared wafers (KBr, NaCl, BaF etc), then solvent volatilize to obtain sample film.

Application Scope: mainly used in sample analysis with both film-forming and solvent dissolving.

Notes:

The selection of solvent and the concentration of the sample prepared play an important role in obtain even and pure film.

Principle for Solvent Selection:

All solvents have infrared spectrum absorption band. Commonly used solvent oil include CCl_4, $CH\ Cl_3$, CS_2 and n-hexane.

Select solvent with low boiling point instead of using solvent with high boiling point and strong polarity, otherwise, the film thickness will be uneven as result of the fast solvent volatilization.

The solubility of the sample in the solvent should be sufficient high. The solution concentration can be regulated.

4. Hot-press film making method

Hot-press device: composed of hydraulic membrane machine, heating die and temperature control device.

Hot-press temperature setup for various common polymers

Sample preparation method

Place a piece of aluminum foil in the core of the die, put the sample on the aluminum foil and then place another piece of aluminum foil on the sample; put on the upper module and then place the die on the tablet press. Control the pressure within 1000-3000Kpa/cm^2. The heating temperature and time for pressing shall be controlled to the extent that there is no thermal decomposition or other chemical changes. After pressing for about 1min, take down the die from the hydraulic press (wear insulating glove to avoid burn), cool to room temperature and then unpack the die to take out the sample tablet. Reduce sample volume if the film is too thick. Reduce temperature if the film turns yellow or has bubbles. Uneven film is caused by low heating temperature, too short heating time or small pressure. Reselect film making condition to remake the film in case of any of the above circumstances.

Application Scope: applicable for non-oxidizing and non-degradable thermoplastic or inorganic polymer materials near the softening point or melting point.

Notes:

Select appropriate hot pressing temperature and pressing time according to the nature of the analyzde sample in order to obtain hot press film in line with infrared transmittance while not destroying structure performance of the sample. Take protective measure against burn when operating in high temperature.

General solid sampling technique can be used in the infrared analysis. However, improvement has been made in actual operation based on the features of the polymer. Mr.

Yang Rui from Tsinghua University has made a detailed research and analysis (in 2011) as below:

Thermoplastic resin: dissolve flow casting film, hot-pressing film or dissolve coated tablet.

Thermosetting resin: such as curing epoxy resin and phenolic resin. Use clean steel file to file sample powder and then use KBr pellet.

Mild cross-link polymer: sample that only swells instead of dissolving in solvent. Grind with KBr in swelling (with solvent) condition, then dry the solvent and pulverize the tablet.

Fiber sample: Filament with diameter of less than 10 microns can be neatly arranged (or be cut into piece) to determine the transmission spectra with KBr pellet. Filament with larger diameter or non-filament fiber should be entangled on the aluminum tablet or be squashed for determination with ART.

Polymer	Temperature /°C	Polymer	Temperature /°C
High-density polyethylene Low density linear polyethylene	170	Nylon 11	220
Low density polyethylene	150	Nylon 66	280
Polypropylene	200	Metaformaldehyde	190
Polystyrene	130	Makrolon	260
PMMA	260	Polybutylene terephthalate	290
PVC	190	Polybutylene terephthalate	250
Nylon 6	250	Teflon	360

Table 5. Reference Temperature for Hot-pressing Film of Common Polymers

Figure 2 shows PMMA infrared spectrum from different sample preparation methods. The PMMA infrared spectrum from powder pulverized tablet and hot-pressing film are nearly same on wavenunbers, and slightly different on wave intensity. But there is another additional infrared eigen wave (at 760 cm^{-1}) for PMMA infrared spectrum from solution film casting method (chloroform resolving).

4.2 Liquid sample preparation technology

4.2.1 Material of window slice used for organic liquid testing

KBr and NaCl are the most commonly used window slice materials used for determination of IR spectrum of the organic solution. But KBr is widely used than NaCl. KBr is the most suitable window slice material for testing organic liquid in middle infrared.

Rinse the KBr wafer immediately after testing. As KBr is non-dissolvable with in soluble, remove the organic solvent from the surface of the wafer by anhydrous ethanol instead of clean water, then dry with crystal-tipped tissue.

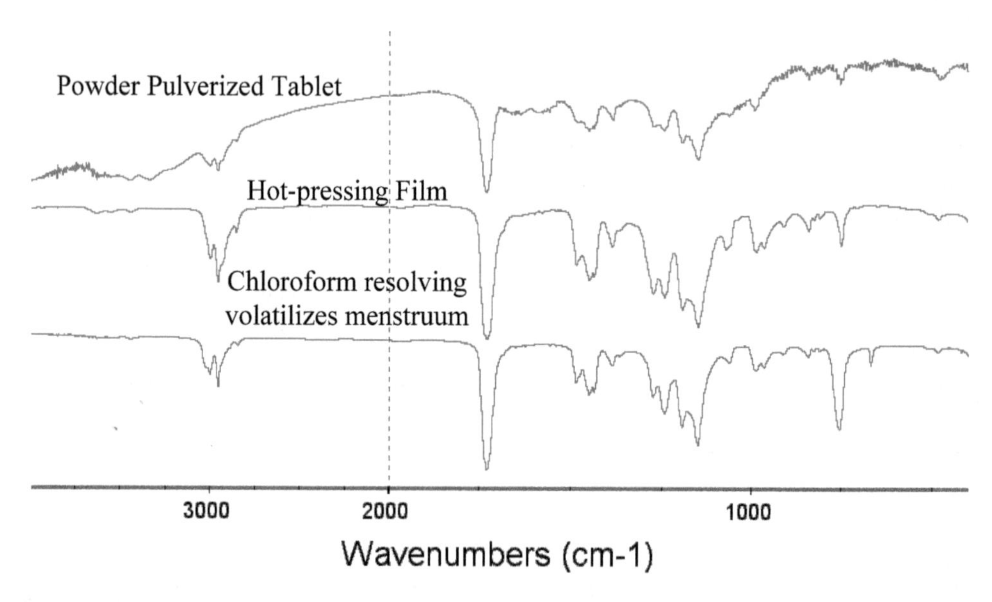

Fig. 2. PMMA Infrared Spectrum from Different Sample Preparation Methods

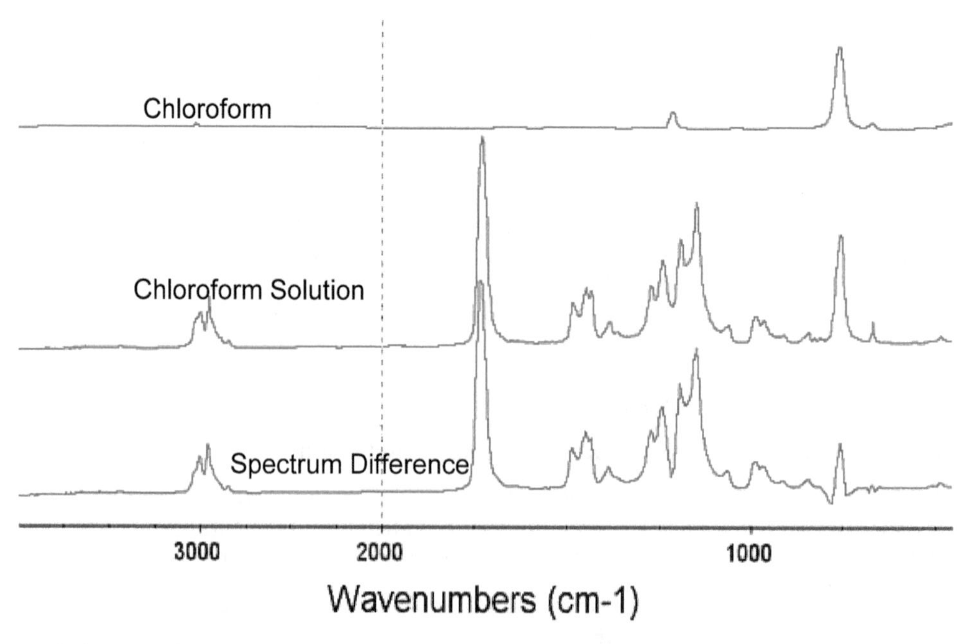

Fig. 3. PMMA Chloroform Solution and Difference Spectrum of Chloroform

4.2.2 Window slice for testing aqueous solutions

The most commonly used Window slice materials for IR spectroscopy of aqueous solution samples is BaF2 wafer, followed by CaF2 wafer. Though BaF_2 wafer is non-dissolvable in water, it can be dissolved in acid and ammonium chloride and can react with phosphate and sulfate to generate barium phosphate or barium sulfate respectively, thus erode the surface of the wafer. When testing metal salt solution, the metal ion will exchange with barium ion and thus erode wafer surface.

4.2.3 Preparation technology for liquid samples with different boiling points

1. Liquid with low boiling points

As the sample has low boiling point, a sealed tank should be used to prevent evaporation of sample. The thickness of the liquid membrane is decided by the nature of the sample. The stronger the polarity, the thinner the tanker is.

2. Liquid with high boiling point and low viscosity

Liquid sample with high boiling point, good flow property and low viscosity may be dropped between the two window slices of the removable liquid tank, then compress the automatically formed even liquid film for measurement.

3. Sample with high boiling point and high viscosity

For qualitative analysis of plasticizer and pyrolysates used in the samples with high boilling point and high viscosity, such as grease, polymers, apply a small amount of sample on KBr window slice by stainless spatula, and scrape evenly, then put it on the sample rack for measurement.

5. Analysis of infrared spectroscopy in biomedical polymer material

The key difference between medical polymer material and other polymer materials is that the former has both medical functionality and biocompatibility and resorts to chemical or physical means to achieve polymeric modification of polymer materials. Fourier Transform Infrared Spectroscopy (FT-IR) is an effective method to analyze polymer materials and its modification.

5.1 The transformation degree of dental composite resin after polymerization may directly affect the biological, chemical and mechanical strength. The most urgent problem in the application and development of the composite resin material is to remove various unfavorable factors that may affect solidification of the composite resin to maximize its transformation degree. FTIR spectroscopic technology may comprehensively study the polymerization of chemical curing and visible-light curing composite resin and the influences of various factors to the degree of polymerization as well as the relationship between transformation degree of the composite resin and the various indirect indicators. The existing study on FTIR indicates: Of different brands of dental composite resin available in market, the transformation degree of visible-light curing resin is superior to that of the chemical-curing resin; the double bond transformation degree and the mechanical properties of the resin have positive correlation with the contents of the catalyst and the reductive and have negative correlation with the contents of the inhibitors.

5.2 As the polymer surface properties affect the cohesiveness, wettability and biocompatibility of the polymer and its actual application, the study focusing on improving polymer properties through polymer surface modification. The quantitative analysis of the polymer surface composition is the important basis for the property study. Attenuated total reflectance infrared spectroscopy (ART-FTIR) technology is one of the most effective methods to test the information of the material construction of the polymer surface without interference of the test samples. Just as transmission infrared spectrum, ATR-FTIR provides information [1-5] about chemical structure, three-dimensional structure, molecular orientation and hydrogen bond of the material. The existing study[6] uses attenuated total reflectance Fourier transform infrared spectroscopy (FTIR-ATR) to test the surface composition of polyethylene glycol/ polyethylene blends (PEG/PE) film. With the corresponding characteristic peak intensity ratio as the basis for the quantitative determination, ATR correction procedures, and NMR-FTIR correction equation[6~7] : Y=0.346 48-1.336 55X+1.268 37X2,, are used for quantitative analysis of the relative composition of the polyethylene glycol chain and the polyethylene chain on the blend surface to obtain a better result of the reproducibility and comparability. The quantitative analysis of the relative composition of polyethylene glycol chain and the polyethylene chain on the surface layer of the blend film is achieved through working curve method. See figure as below:

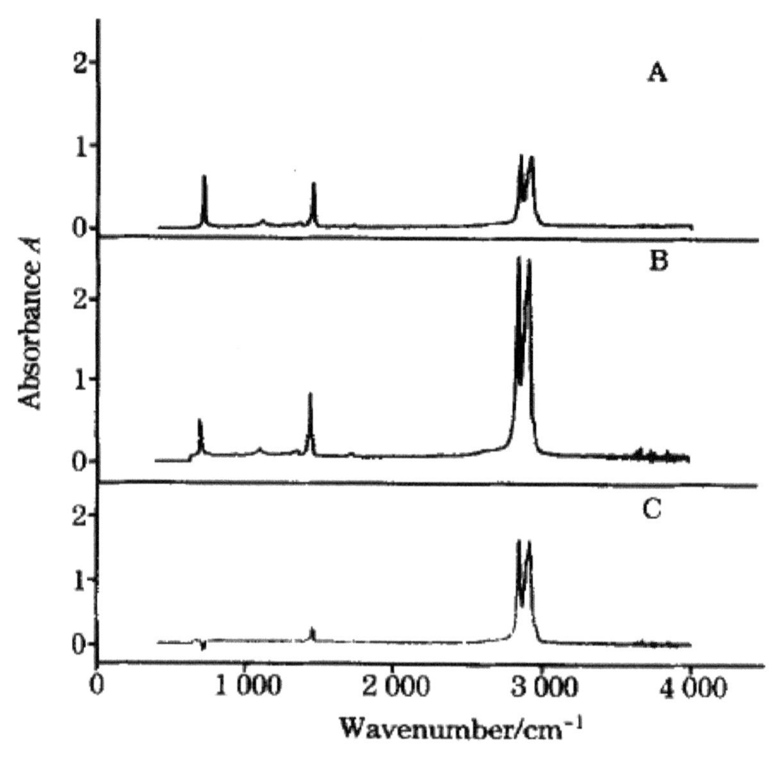

A. ATR spectrumwithout calibration; B. Corrected ATR spectrum;
C. The difference spectrum between spectra A and B

Fig. 4. ATR spectra of blend of PEG (2000) and LLDPE

5.3 Another key focus of study is to use combined infrared spectroscopy and computer technology to make quantitative analysis of the chemical structure of the auxiliary materials added to the medical polymer, such as additive, adhesives and plasticizer. Spectrum subtraction technology can be used to identify the additives in the high polymer products. Medical infusion devices are made of conventional polymer material polyvinyl chloride (PVC) and 2-ethylhexyl phthalate (DEHP) is added to plasticize rigid polyvinyl chloride (PVC), with additive dosage of 40-60%. Study has verified that DEHP can enter human body through venous transfusion, respiratory tract and skin and bring damage to human health. This has become focus of academic research and disputes and has attracted attention from media. Though DEHP's toxicity and carcinogenicity has been fully confirmed in experimental animals, its adverse effect in human body is still controversial. Using infrared spectroscopy subtract technology to analyze PVC infrared spectrogram of PVC and the infrared spectrogram of plasticized PVC may determine the kernel of material construction of the plasticizer. FTIR spectrum subtraction may also be used in polymer end-group analysis, polymer oxidation and degradation reaction analysis and inter-molecular analysis.

Below is the infrared spectra of traditional bis (2-ethylhexyl) phthalate (DEHP) plasticized rigid PVC and PVC materials used in medical infusion equipment.

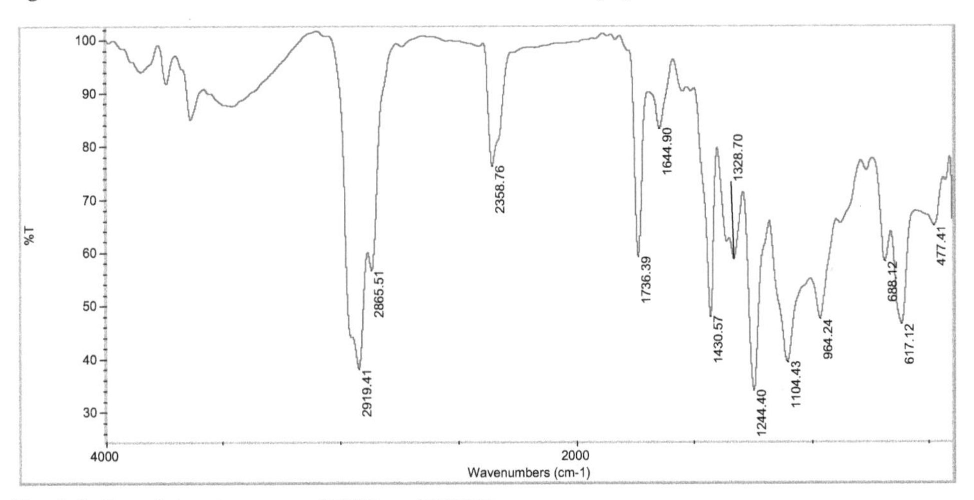

Fig. 5. Infrared spectrogram of PVC and DEHP

5.4 Polymer materials in ophthalmology

Contact lenses are the most common polymer product in ophthalmology. The basic requirements for this type of materials are: ①excellent optical properties with a refractive index similar to cornea; ② good wettability and oxygen permeability; ③ biologically inert, degradation resistant and not chemically reactive to the transfer area; ④ with certain mechanical strength for intensive processing and stain and precipitation prevention. The common used contact lens material includes poly-β-hydroxy ethyl methacrylate, poly-β-hydroxy ethyl methacrylate-N-vinyl pyrrolidone, poly-β-hydroxy ethyl methacrylate, Poly-β-hydroxy ethyl methacrylate - methyl amyl acrylate and polymethyl methacrylate ester-N-vinyl pyrrolidone, etc. The artificial cornea can be prepared by silicon rubber, poly methyl acrylate, poly-casein or other thin films. The main body of the artificial lens can be made of polymethacrylate. The researchers have attached increasingly great importance to the qualitative analysis of polymer materials for this kind of medical equipments. As explicitly specified in YY0477-2004 "Orthokeratology Using Gas Permeable Rigid Contact Lens", infrared spectrum analysis is adopted to determine the components of the lens materials. The following drawing is the infrared spectrum analysis of this material: point 2961cm-1 is the methyl characteristic peak of methyl acrylate; points 1104 cm-1 and 1046 cm-1 are characteristic absorption peaks of siloxane, points 1730 cm-1,1227 cm-1 and 1199 cm-1 are ester peaks of methyl acrylate; points 893 cm-1 and 7556 cm-1 are the structure characteristic peaks of polymethacrylate; also it is worth knowing that carbonyl peak on point 1769 cm-1 is the characteristic peak of fluoro-alkylated methyl acrylate.

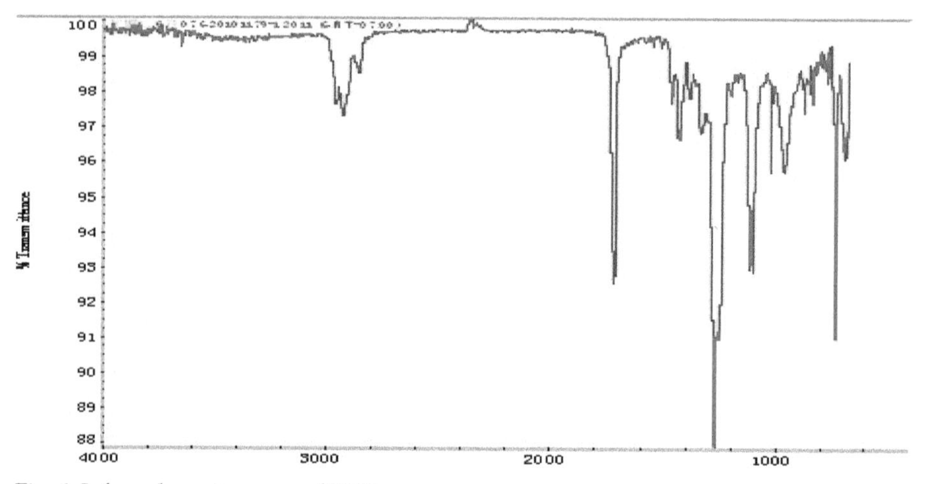

Fig. 6. Infrared spectrogram of PVC

5.5 The innovation of instrument performance and the development of computer application technology provide possibility for combined use of the previous stand-alone analysis instruments. The combined use of various types of instruments has greatly improved the accuracy and reliability of the analysis and testing results. With combination use of thermogravimetric analysis and IR spectrum and other analytical methods in recent years, thermogravimetric analysis is increasingly playing an important role in study of thermal behavior in chemical materials. In comparison with traditional thermogravimetric analysis method, TGA-IR spectrum combined analysis can directly and accurately determine the various physical-chemical change during the heating proves and identify the chemical composition of decomposition or degradation products during the during the weight loss process. Thus, it has been a key experimental method in studying thermostability and thermal decomposition (degradation) process of various inorganic, organic and polymer materials and proves a promising prospect in respect of thermal performance analysis of the materials. As Fourier transform infrared spectrometer is of high noise-signal ration and high precision, it may detect the slight intensity change and frequency shift of the infrared bands in the sample before and after the heat treatment. Thus to provide structural differences of polymer film for three different thermal stages of high molecular polymer, from high-elastic state slow cooling, high-elastic state quenching to heat treatment below temperature Tg. The study shows the conformation changes of PVC films with different thermal histories in heating process. Meanwhile, in FTIR measurement, the sample subjected to heat treatment below Tg temperature occurred sudden change of conformation in the temperature range corresponding to enthalpy absorption peak of differential scanning calorimetry (DSC).

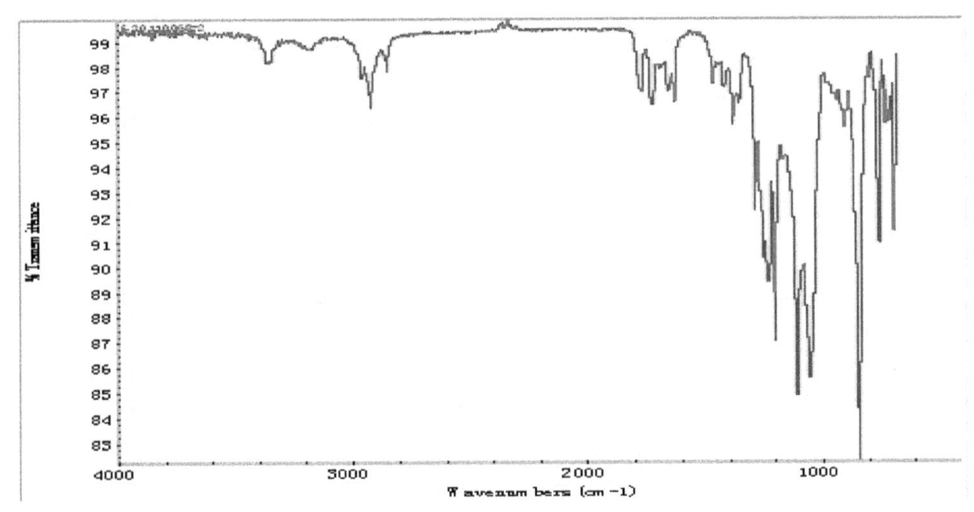

Fig. 7. Spectrum of Orthokeratology Using Gas Permeable Rigid Contact Lens

6. Prospect

FTIR is becoming widely used in the field of medical polymer materials, especially for quantitative analysis of the material properties. At present, the study on application of infrared spectroscopic technology or combined application of IR spectrum with other technologies in polymer material, to be still growing. With rapid development of scientific technology, the research in IR spectrum is further deepened. IR spectrum is not only widely used in polymer materials but is also widely used in pharmaceuticals, foods and environmental science. The gradually improved infrared detection method may be used on vivo analysis of pathological tissue and greatly contribute to the rapid and accurate diagnosis of the diseases. Meanwhile, the disease mechanism and progression maybe clarified through analysis of infrared spectrogram of tissues or cells.

7. References

[1] Yang Qun, Wang Yi-lin, Yao Jie, et al. Spectroscopy and Spectral Analysis, 2006, 26(12): 2219.

[2] Jiang Zhi, Yuan Kai-jun, Li Shu-fen, et al. Spectroscopy and Spectral Analysis, 2006, 26(4): 624.

[3] Bergberiter D E, Srinivas B. Macromolecules, 1992, 25: 636.

[4] Lee K W, Kowalczyk S P, Shaw J M. Macromolecules, 1990, 23: 2097.

[5] Francis M, Mirabella J R. Appl. Spectrosc. Rev., 1985, 21: 45.

[6] Qian Hao, Zhu Ya-fei, XU Jia-rui Spectroscopy and Spectral Analysis, 2003, 23 (4): 708-713
[7] Chen Jia-xing, Joseph A. Gardella Jr. Appl. Spectroscopy, 1998, 52(3): 361.

10

Organic Compounds FT-IR Spectroscopy

Adina Elena Segneanu, Ioan Gozescu*,
Anamaria Dabici, Paula Sfirloaga and Zoltan Szabadai
National Institute for Research and Development in Electrochemistry and Condensed Matter, Timisoara (INCEMC-Timisoara)
Romania

1. Introduction

General spectral range of electromagnetic radiation with a wavelength greater than 750 nm (i.e. with the number of wavelength below 13000 cm^{-1}) bears the name of the domain infrared (IR). In this field samples absorb electromagnetic radiation due to transitions of vibration of the structure of molecules, molecular transitions in vibrations crystalline network (if the sample is in the solid state of aggregation) or due to transitions of molecular rotation. Subdomain of spectral wavelengths between 2500 - 50000 nm (respectively the wave numbers 4000 to 200 cm^{-1}) bears the name of the *middle infrared domain.*

From the point of view of analytical control of medicinal products, this domain is the most used. At the base of absorption is being generated electromagnetic radiation in this area spectral transitions are the vibrations of individual molecules or of crystalline network (if the sample examined is solid). Show effects such transitions caused by the vibrations of individual molecules provides information about molecular structure of the sample examined, and show effects such crystalline network to identify a particular forms of crystallization of the substance of interest.

The most frequent use of the absorption spectrophototometry in the middle infrared field lies in the identification of substances through molecular vibration. The wavelength (i.e, the wave numbers) of the of the absorption band are characteristic chemical identity of the substance in question. The intensity of the absorption bands allows quantitative analysis of the samples but, unlike in the ultraviolet and visible, in the infrared field diffuse radiation is much more refreshing, and for this reason quantitative determination infrared, are affected by notable errors.

From the standpoint of analytical use, the spectra of molecular vibration is enjoying increased popularity in comparison to the study of the crystal latice's vibrations. A molecule may be considered to be a vibrator with more than one degree of freedom, able to execute more modes of vibration. In each mode of vibration every atom in the molecule oscillates about their own position of equilibrium. Such oscillations have different amplitudes for

* Corresponding Author

different atoms of the molecule, but at a certain mode of vibration, each atom in the molecule oscillates with the same frequency. In other words, the frequency of oscillations of the atoms in molecule is characteristic of a particular mode of oscillation of the molecule.

A molecule composed of N atoms has several possible modes of oscillation. In each mode of oscillation (in principle) all the atoms of molecule perform periodic shifts around level position with a frequency of oscillation mode which is a feature of the assembly. Because each of the N atoms can run periodic shifts in 3 perpendicular directions each other, the assembly of N atoms can have 3N ways of motion. But, those displacements that correspond to moving molecule as a whole (not deform the geometry of the molecule) and movements, which correspond to entire molecules rotation about an axis (also without deforming the molecule's geometry), do not represent actual oscillation (associated with actual deformation of the molecule).

These displacements (3 in number) and rotations around the three orthogonal axis (also 3 in number) are eliminated of the total number of atomic movements possible. Therefore, a molecule is, in general, (3N - 6) distinct modes of oscillation and in each of these (3N - 6) modes of oscillation each atom oscillates with frequencies characteristic individual modes of vibration. A special case represents molecules whose structures are linear, because in these cases the inertia of the molecule, in relation to the axis flush by molecule, it is practically zero. For this reason, in the case of a linear molecules consisting of N atoms, the number of modes of vibration is 3N - 5.

2. The vibration of a diatomic molecule

For an understanding of the vibrations of a polyatomic molecule, should be first a preliminary analysis of the oscillations of a molecule composed of two atoms linked by covalent binding. Such a molecule, with N= 2 atoms, shows N = 3x2 - 5 = 1 modes of vibration. The steering as defined by the covalent binding of the two atoms is the only special steering, it is ordinary to accept that atoms will move (in a periodic motion) after direction of the covalent connection. Assembly oscillation may be considered in relation to several systems of reference. It may choose as origin of the system of reference the center of gravity of the diatomic assembly.

In this case the both atoms perform periodic shifts in relation to this reference point. The mathematical analysis of oscillations is advantageous to place the reference origin in one of atoms. In this case, however, in the place of mass m_A and m_B of the atoms of the molecule A-B are used *reduced mass* (noted with μ) of the assembly dimolecule. Reduced mass is calculated from the **mA** and **mB** of atoms of the dimolecule assembly in accordance with following relationship.

$$\frac{1}{\mu} = \frac{1}{m_A} + \frac{1}{m_B} \quad (1)$$

Rigorous deduction of the relationship (1) can be found in literature on the subject. During the oscillation, the kinetic energy, $\mathbf{E_c}$, and potential energy $\mathbf{E_p}$ of the assembly are varying, periodically. If the system does not radiate energy to environment, or do not accept energy from the environment, then the amount of E_c and E_p remains constant during oscillation. Potential energy is dependent on the single variable of the diatomic system (namely, the deviation of the **Δr** inter-atomic distance to **r**0) which is variable in time. Potential energy dependence of the **Δr** (i.e. lengthening the deformation of the diatomic molecule) is expressed, in the harmonic approximation, of the relationship (2).

$$E_p = \frac{1}{2} \cdot k \cdot \Delta r^2 = \frac{1}{2} \cdot k \cdot (r - r_0)^2 \quad (2)$$

In the relationship (2) the coefficient 'k' is *constant of force*, size that characterises the strength of inter-atomic connection in the molecule. On the basis of expression (2) the potential energy of the diatomic assembly, using the mechanics in this quantum mechanics, may deduct quantified values ('allowed') of diatomic oscillator.

These values of energy 'allowed' shall be calculated on the basis of the expression (3) by substituting for the number of quantum vibration (**n**vib) integers numbers (0, 1, 2, . .

$$E_{vib}(n_{vib}) = E_c + E_p = h \cdot \nu_0 \cdot \left(n_{vib} + \frac{1}{2} \right) \quad (3)$$

The expression (3) shows that the energy $\mathbf{E}_{vib}$ (the sume of the kinetic energy $\mathbf{E}_c$ and potential energy $\mathbf{E}_p$) has a state of vibration allowed to diatomic system depends on the number of vibration quantum $\mathbf{n}_{vib}$.

The lower value of energy (in the fundamental vibration's state diatomic system) is obtained by replacing **n**vib= 0 in the relationship (3). In the relationship (3) h is the size Planck constant. (6,626075 x 10^{-34} Js). If diatomic molecule fundamental changes from the vibration (**n**vib = 0) in the state of vibration excited immediately above (**n**vib = 1), then change of energy **ΔE**vib**(0→1)** is expressed by the relationship (4).

$$\Delta E_{vib}(0 \rightarrow 1) = h.\nu 0 \quad (4)$$

This value to change the vibration energy determines how often (or the number of wavelength) at which diatomic molecule shows preferential absorption of radiation.

In principle, diatomic molecule can pass from the fundamental ($\mathbf{n}_{vib}$ = 0) in a excited state (for example, corresponding $\mathbf{n}_{vib}$ = 2) but, those quantum transitions in which the number is changing more than one establishment are prohibited by the rules of selection.

Rigorous justification of the rules of selection is treated in detail in literature on the subject.

Preferred frequency (ν_0), the favorite number of wave **n**vib = 0 to which a small diatomic molecule absorbs radiation (hence to which generates a strip of absorption) as the transition (0→1), is expressed quantitatively the relationship (5)

$$\nu_0 = \frac{1}{2 \cdot \pi} \cdot \sqrt{\frac{k}{\mu}} \quad ; \quad \tilde{\nu}_0 = \frac{1}{2 \cdot \pi \cdot c} \cdot \sqrt{\frac{k}{\mu}} \quad (5)$$

In the relationship (5) 'C' is the speed of propagation of electromagnetic radiation in a vacuum. At the harmonic approximation, the dependency **Δr** is sinusoidal. But in the case of molecules, the potential energy is dependent on the momentary deflection **Δr** of the system in a manner more complicated, so the approximation describes successfully the harmonic oscillations limited to a diatomic molecules. As a result of difficulties with mathematical order but the description of molecular oscillations, especially in the case poliatomice molecules, it accepts harmonic approximation."

$$E_p = De \cdot \left[1 - e^{-\beta \cdot (r - r_0)} \right]^2 \quad (6)$$

Figure 1 represents the dependency of potential energy E_p of a diatomic molecules to the momentary distance (r) in a approximately more faithful than the harmonic (based on parabolic dependence). In a more or less accurate in the description diatomic vibration of molecules, energy dependence potential (E_p) by the distance inter atomic (r) is described by a function of type Morse (6) in place of a relationship of type (2).

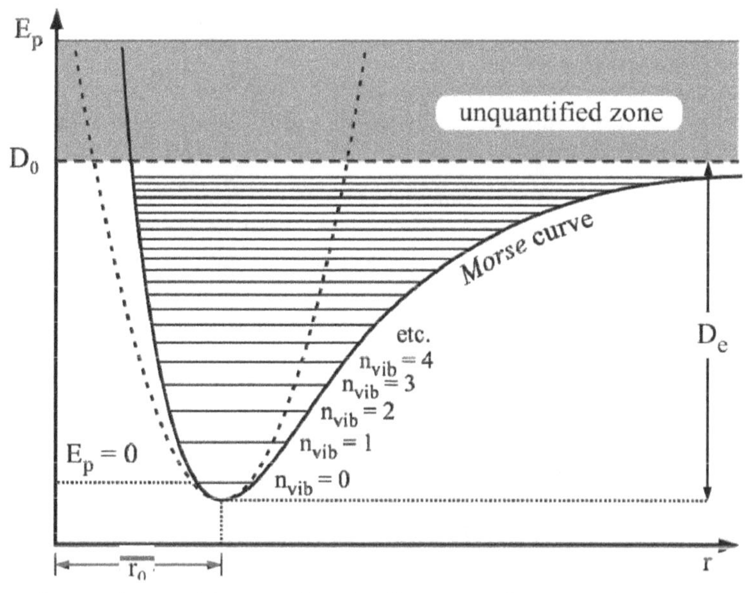

Fig. 1. Dependency of potential energy Ep of a diatomic molecule to the momentary distance

In the function (6.) the coefficient **β** depends on the mass reduced (μ) of the assembly diatomic, in accordance with relationship (7).

$$\beta = v_0 \cdot \sqrt{2 \cdot \pi^2 \cdot \mu \cdot De} \quad (7)$$

Continuous curve in figure 1 graphically represents the function (6). Morse function curve is compared with the curve corresponding harmonic approximation (parable with the interrupted curve).

The horizontal lines, arranged on the inside of the cavity Morse curve, shows the values allowed (quantifiable) of the energy of vibration of the assembly diatomic. Advanced to a deformation of the length of connection inter atomic, the energy potential of deformation tends toward a limit value (**D_0**) over which the energy of deformation of the assembly shall cease to be quantified.

In an approximation more accurate, 'anharmonic', in the phenomenon of vibration, the amounts permitted of the energy of oscillation are expressed a relationship similar to (3), with the difference that the anarmonic approximation. Status of vibration energy depends on the binomial $\mathbf{n}_{vib}$ quantum number after an expression of the degree 2 in relation:.

$$E_{vib}(n_{vib}) = E_c + E_p = h \cdot \nu_0 \cdot \left(n_{vib} + \frac{1}{2}\right) - h \cdot \nu_0 \cdot x \cdot \left(n_{vib} + \frac{1}{2}\right)^2$$
$$\nu_0 = \frac{1}{2 \cdot \pi} \cdot \sqrt{\frac{k}{\mu}} \qquad (8)$$

The coefficient 'x' in the relationship (8) characterized quantitatively anarmonicity of molecule diatomic vibration, i.e. the drift behavior system from the model of harmonic vibration.

3. Potential energy dependence of the inter atomic distance of a diatomic molecule in Morse potential energy approximation

In inharmonic approximation of the vibration of diatomic molecules of the selection rule, relating to the variation in $\mathbf{n}_{vib}$ allowed for the quantum number, it is not so strict as in the case described harmonics. The model does not exclude the possibility inharmonic transitions between the status of vibration to which variation **nvib** quantum number to be 2,3 , etc. , in practice IR spectrophotometry.

Transitions associated with variations in higher than the unit are called harmonics of the upper fundamental transition (i.e., the transition that starts at the same lower status, but for which $\mathbf{\Delta n}_{vib} = 1$).

The appearance of the absorption bands assigned to upper harmonics inherent in spectra are observed frequently in IR (especially in the case polyatomic molecules), but as a rule occur with intensities that are smaller than corresponding fundamental bands.

Strips of the upper harmonics associated with fundamental tape appear at frequencies (or wave numbers) which are approximately multiples whole frequency (or the wave-number) fundamental.

Another practical consequence of the inharmonicity of vibration of molecules is the rise of the inter-combination bands in the IR absorption spectra.

These bands of absorptions are observed at frequencies equal to the sum or the difference between two frequencies or fundamental frequency of a fundamental and a harmonic one. By cause of bands of combination appear various normal modes of oscillation of the molecule. The high harmonics and the bands of combination in IR absorption spectra cause considerably complications in their interpretation.

4. Vibration of polyatomic molecules

In the case the vibration diatomic molecules atoms can oscillate just in the direction of connection covalent) binding atoms. In the case of molecules consisting of several atoms (N atoms) the description of the assembly oscillations, even in harmonic approximation, is significantly more complicated. In principle, each atom in the structure molecule can execute, independently of the other atoms of the same molecule, the three-way oscillation linear independent (after three axes orthogonal coordinate attached each of atoms).

Therefore, the N atoms can run periodic shifts after 3N directions. In other words, a polyatomic molecule, consisting of N atoms, has 3N degrees of freedom of movement of constituent atoms. From this number, not all directions of movement of individual atoms correspond to deformations of real three-dimensional structure of the molecule, as oscillations in which each constituent atom is moving at the same time in the same direction, with the same amplitude and phase. They are equivalent to move whole molecules (translation molecule) without deforming them.

Also, movements that the N atoms can be synchronized in such a way that the assembly rotation movements correspond to atomic molecule as a whole, without causing structural deformation. Whereas entire molecule can be translated into 3 directions linear independent of each other and can rotate around a three axis-oriented perpendicular to each other. Of the total number of 3N directions of movement are deleted atomic 6 shifts (that is 6 degrees of freedom) in order to obtain the number (3N - 6) for detailed rules for the movement of atoms, therefore the same number of modes of oscillation actual molecular structure. .

It must be remembered, however that if molecule is not free, but is linked to a structure with crystalline comparable forces with those which act between atoms of molecule, then moving molecule as a whole from its position of equilibrium means a deformation, But, in this case is not of the molecule, instead of the structure supra molecular (crystalline lattice) in which molecule is a constituent.

In the particular case of a molecules polyatomic (with N atoms) having structure linear (all the atoms constituents are willing co-linear), the number of actual oscillation of the molecule is equal to (3N - 5). In this case is deleted of the total number of 3N possible directions of displacement 3. Degrees of freedom correlated with translation no deformation of the entire molecule and only 2 degrees of freedom corresponding to rotation molecule around two mutually orthogonal axis and perpendicular to the longitudinal axis of the molecule.

The explanation lies in the fact that in this case of the inertia of the molecule to the third axis of rotation (the flush to the longitudinal axis of the molecule) is vertical, so that rotation around this axis is virtually builds up kinetic energy.

At each of the (3N - 6) (i.e. (3N - 5)) possible ways of real oscillation, in principle each of constituents of atoms vibrating molecule running around their own positions of equilibrium. But in a particular mode of oscillation, individual atoms oscillate after other directions and with other amplitudes, but with the same frequency (feature mode of vibration in question). Directions and the amplitudes individual travel, and the frequency (common) of oscillation are characteristic mode of vibration.

The vibrating atoms in a molecule polyatomic can be described as a function of internal coordinates instead of cartesian coordinates. Such are eliminated those movements which correspond translations to atoms and rotation molecule without deformation. Internal co-ordinates of a molecules polyatomic can be defined in different ways. How to define the most frequently involves the covalent connection between pairs of atoms connect (l_{ab} for atoms a and b), Angles between connections covalent) binding centered on an atom common (αabc for atoms a and c bound by common atom (b) and diedral angles θabcd (the angle of the plans P and R containing three connections between four **atoms covalent) binding) (Figure 2).**

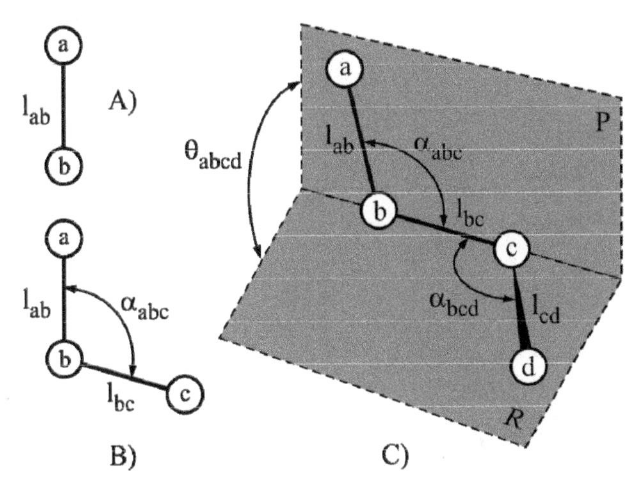

Fig. 2. Define Internal coordinates l , α and θ

Each vibration mode of the molecule can be described as a time-dependent periodic variation of all the internal coordinates. For a specific molecular structure, consisting of N atoms, the whole structure runs ((3N-6) or (3N-5)) modes of preferential oscillations, involving in (3N-6) (or (3N-5)) ways the internal coordinates of the molecule. These "preferential" types of oscillations represent the normal oscillations of the molecule. In each normal type of oscillation, all the internal coordinates of the molecule oscillate at a common frequency (in principle), specific for that type of oscillation.

In each normal mode of oscillation, the internal coordinates are involved in varying degrees. For a normal oscillation is characteristic of the internal coordinate of the molecule (e.g., a covalent bond length) is more involved than the others, then it may be said (with some tolerance) that normally oscillation that the respective oscillation type is characteristic for the respective internal coordinate (e.g., the length of the covalent bond).

Figure 3 represents the fine vibration structure of the fundamental electronic state and of the first excited electronic states, for the case of a hypothetical triatomic molecule. This type of molecule has 3·3-6=3 normal vibration modes. Oscillation modes are also represented in Figure 3, indicating the direction and the direction of the relative shift of the individual atoms in one of the half period of oscillation. Each electronic state consists in a number of vibration states characterized by the vibration quantum numbers $\mathbf{n}_{vib}$. For each normal way of vibration (indicated by "1", "2" and "3" in Figure 3), the vibrations in the electronic states may be differently arranged. In other words, for different normal types of vibrations, the fundamental electronic state is differently "split" in substrates of vibration. In practice, the most probable vibrational transition occurs between the vibration states corresponding to the quantum numbers n_{vib}=0 and n_{vib}=1 for each normal vibration (transitions represented in Figure 3 by bold arrows). There is only a diminished probability to also occur, for every type of normal vibration, transitions between the vibration states corresponding to the quantum numbers n_{vib}=0 and n_{vib}=2 (transitions depicted in Figure 3 by dashed arrows). This kind of transition at a normal vibration type generates its first superior harmonic.

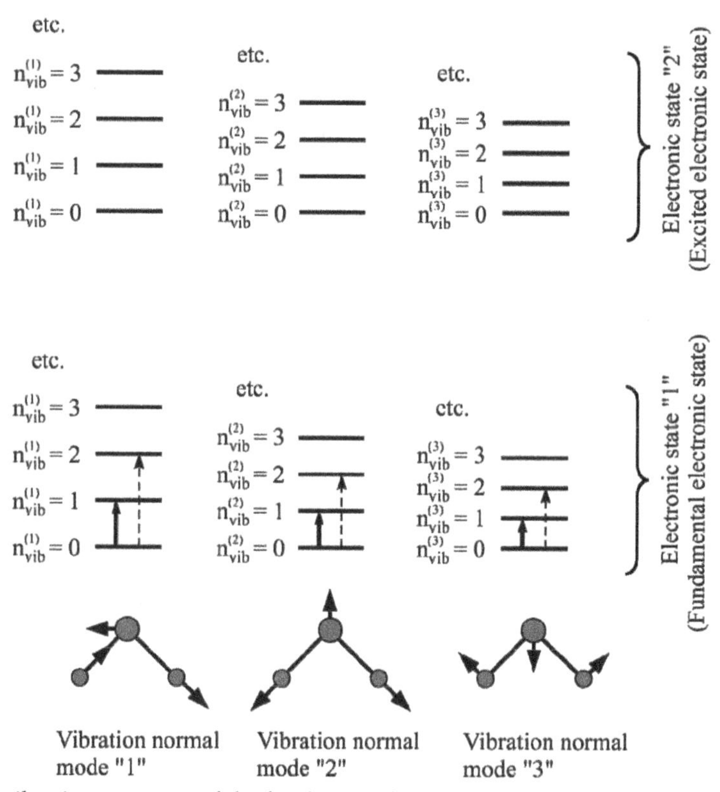

Fig. 3. Fine vibration structure of the fundamental and first excited electronic states

The transition between the vibrational states is initiated by the absorption of radiation with proper frequency (or wave number). The transition is visualized, in principle, by the appearance of the absorption band in the absorption spectrum of the analyzed substance. The incident radiation on the sample. Probability that is able produce transitions between the vibrational states of the fundamental electronic state), is located in the infrared (IR) domain. In principle, the IR absorption spectrum of a molecule contains a number of absorption bands that is equal to the normal modes of vibration of the molecule in question. Thus, in the case illustrated in Figure 3, the molecule would present the first absorption band corresponding to the normal mode of vibration no. "1" at the highest value of the wave number (the energy gap is maximum), the second absorption band corresponding to the normal mode of vibration no. "2", at an intermediate wave number (the energy gap is intermediate) and the third absorption band corresponding to the normal mode of vibration no."3", at the lowest value number of the wave number (the energy gap is the smallest). The superior indices of the vibration quantum numbers that are present in Figure 3 $\left(n_{vib}^{1}, n_{vib}^{2}, n_{vib}^{3}\right)$ refers to the serial number of the respective normal vibration mode. The above made statement made, that each normal vibration type generates an absorption band in the IR spectrum of the analyzed substance, is true only in principle. It is common that not all the normal vibration types of a molecule (3N-5 or 3N-6, for an N atoms molecule) to present absorbtion bands.

If the covalent bonds in a molecule (or a molecular fragment) have comparable force constants or the masses of the atoms involved in the covalent bonds are close, then those atoms are involved, to some extent, in all possible normal modes of vibration of the molecule (or molecular fragment). In these situations, the individual internal coordinates (bond lengths or individual angles) do not independently vibrate; their vibrations are coupled, generating the appropriate number (N) of normal vibrations.

If one of the atoms linked by a covalent bond has a much smaller mass than the other atom (for example, the C-H, N-H, O-H, S, H, P-H bonds), the reduced mass of the ensemble of bound atoms is almost equal to the mass of the lighter atom and in this case the oscillation of the bond length is quasi-independent from the oscillation of the rest of the molecule. It may be said, in this case, that the absorption band associated with the normal mode of vibration affects the covalent bond, and it is characteristic of the presence of that chemical bond in the molecule structure.

On the other hand, if two atoms in a molecule are connected by a significantly stronger covalent bond then the other covalent bonds of the molecule (e.g., the isolated double or triple bond, as in one of the cases: >C=C<, >C=O, >C=N-, -N=N-, -C-C-), than the vibration of the respective bond can be also considered quasi-independent from the oscillation of the rest of the molecule. In this case, we can say that in one of the normal modes of vibration of the molecule practically occurs only the elongation oscillation of that bond, so the optical absorption band, generated in the IR spectrum by the normal mode of vibration, is characteristic for the presence of that covalent bond

Often, a group of atoms (or even a functional group) have their "own" normal ways of vibrations, quasi-independent of the rest of the molecule vibrations. In these situations, the normal "own" modes of the group of atoms is manifested in the IR absorption spectrum as a corresponding number of "group characteristic bands". An example is the characteristic band of the amidic functional group.

5. Practical aspects of infrared spectrophotometric analysis

The IR absorption spectrum is the graphical representation of a measure of energy depending on a measure of wave of the involved radiation. IR practice has established the use of the wave number (reciprocal of the wave length and proportional to the frequency of the radiation) and of the percentage transmission (T%) or absorbance (A), as related to the energy of the radiation.

There are significant differences between the UV-Vis and IR absorption spectra. IR spectra, even those of samples in condensed states, are generally characterized by a large number of well defined, sharp bands, with easily localizable positions. Therefore, IR spectra are useful for the fast, non-destructive identification of the chemical substances, and it is extremely unlikely that two substances that are chemically different to have, accidentally, identical IR spectra. Instead, the quantitative determinations in the IR spectral range are more difficult because diffuse radiation in the IR spectrophotometers is greater than in the UV-Vis spectrophotometer, so the error sources affect the results of quantitative determinations more than in the UV or Vis range.

Another difference between the two spectral areas consists in the transparency (and usability) of auxiliary materials (glass, optics, etc.). Spectra of liquid samples are recorded using similar cells to those used in UV-Vis spectrophotometry, except that glass walls are made of specific transparent materials (NaCl, KBr, KCl, ZnSe, As_2S_3, KRS-5, and others). The thickness of the sample in IR spectrophotometry are usually much smaller (0.05 mm - 1 mm) than those found in the UV-Vis absorption spectrophotometry (1 - 10 mm).

The IR absorption spectra can be recorded for solid, liquid or gaseous samples. The most common presentation state of samples in drug control is solid state. Commonly practiced method for obtaining IR spectra of substances or mixtures of pharmaceutical interest in solid form, consists in incorporate them into a solid, microcrystalline medium (for exemple potassium bromide). This method of sample preparation is called "inclusion in the tablet (or pill) of potassium bromide." To achieve such a compressed is triturated a small amount medium (1-2 mg) of solid interest with 200-250 mg of potassium bromide microcrystalline. Potassium bromide used for this purpose must be high purity (purity "for spectroscopy") and dried before use for several hours at 180 ° C. The triturating of solid mixture containing the substance of interest and potassium bromide, is running in agate mortar medium (glass or porcelain mortar is not appropriate). After the grain sufficiently fine, solid mixture is placed in a special mold will compress high pressure medium (about 10 ton-force) with a hydraulic press. Before applying pressure, air is removed from the stencil to prevent inclusion of air microbubbles in solid mass during pressing, that may produce microfissure in mass tablet at the end pressure . During pressing, the potassium bromide microcrystalline are sinterising forming a solid transparent, optically homogeneous. A compressed pellet carried out in ideal conditions is transparent without opaque area. In the spectral range 4000 - 300 cm^{-1} potassium bromide shows very good transparency, which is why this mode is used preferentially for sample preparation. Whereas, the included technique of sample in a potassium bromide matrix keeps crystallization form of the sample solid, the IR spectrum obtained by compression in potassium bromide is dependent from the crystallization of the sample. For substances that shows polymorphic, the IR absorption spectrum of solid samples , included in compressed potassium br An essential difference between the UV-Vis spectra registration procedure and the IR field is that in the UV-Vis domain, the solvents absorbtion is insignificant, so their absorption can be compensate, in the case of IR domain all used solvents presents its own band absorption, sometimes even more powerful, that in these areas of the spectrum of energy received by the detector is too small to differentiate the strong absorption of the solvent from the sample absorption that exceeding only in small extent the solvent absorbtion. For this reason, in the IR absorption spectra of the solutions are frequently areas where the IR radiation detector is inactive, and the signal recorded in these areas is irrelevant. To view the sample spectrum in these areas, is repeat the recording spectrum in another solvent that is transparent.omide, allows the identify the crystallization form of the interest substance. Another technical detail is related by the fact that in IR absorption domain the solvents presents absorption bands, sometimes quite strong. Some IR spectrophotometers operating in double beam mode (similar to the double beam spectrophotometers used in the spectral UV-Vis). In the case of UV-Vis spectrophotometers is introduced, in the right reference optical path, a vat filled with pure solvent, and the right of the second optical path is introduced a vat of the same thickness, filled with solution (solvent and solute). The electronics parts of the spectrophotometer compare the absorbances of vats located in the two optical paths and subtract the absorbance of the solvent, located in the reference route, from solution absorbances located in the route of sample. Because the interest substances absorbance is marked in UV-Vis, and the absorbance of solvent is insignificant, the difference between the absorbances associated with two optical route is almost always positive.

In IR domain, where the absorbances of dissolved substances are comparable with those of solvents. It is easily understood that if a certain place in the spectrum (the number of wavelengths $\tilde{\upsilon}$) the solvent has a strong absorption band (molar absorptivity large solvent $\boldsymbol{\varepsilon_0(\tilde{v})}$ and the solute does not absorb significantly at that wavenumbers ($\boldsymbol{\varepsilon(\tilde{v})}$ small), then, the same layer thickness "d", the absorbance values measured in the two opticalroute ($\boldsymbol{A_{ref}(\tilde{v})}$ for route reference and $\boldsymbol{A(\tilde{v})}$ for solution route), are expressed by the relations (9).

$$\begin{aligned} A_{ref}(\tilde{v}) &= d\cdot\varepsilon_0(\tilde{v})\cdot C_0 \\ A(\tilde{v}) &= d\cdot\varepsilon_0(\tilde{v})\cdot C_0^* + d\cdot\varepsilon_0(\tilde{v})\cdot C \end{aligned} \tag{9}$$

In the (9) relation, $\mathbf{C_0}$ şi $\mathbf{C^*_0}$ represents molar concentration of pure solvent in the two cells (placed in reference route and in the sample route respectively), and C is the molar concentration of solute in the cells placed in the route sample. It is obvious that $\mathbf{C^*_0 < C_0}$ because in the sample cells is in addition to solvent, a quantity of solution. The signal recorded by spectrometry, $\boldsymbol{\Delta A(\tilde{v})}$, to the number of wave $\boldsymbol{(\tilde{v})}$, is the difference between the both absorbance of relationship (9)

$$\Delta A(\tilde{v}) = A(\tilde{v}) - A_{ref}(\tilde{v}) = d\cdot[\cdot\varepsilon 0(\tilde{v})\cdot C^*_0 - C_0) + \varepsilon(\tilde{v})\cdot C] \tag{10}$$

The first term in the right side of the parenthesis right above relationship is negative, because $\mathbf{C^*_0 < C_0}$. If $\varepsilon_0(\tilde{v})$.is significantly higher than $\boldsymbol{\varepsilon(\tilde{v})}$, then it can happen that $\boldsymbol{\Delta A(\tilde{v})}$ to have a negative value for the number of wave $\tilde{v}$. Obviously, such a negative value of absorbance is an artifact, without spectrophotometric real significance. To overcome this problem, manifested in solutions recording spectra, are used in the reference route a cell with variable thickness. The solution is placed in a cell of fixed thickness"**d***", lower than thickness"**d**", in the reference route. The absorbances $\boldsymbol{A^*_{ref}(\tilde{v})}$ and $\boldsymbol{A(\tilde{v})}$, and the difference of absorbances, $\boldsymbol{\Delta A^*(\tilde{v})}$, associated with the two optical route, the new working conditions, are given by the relations (11).

$$\begin{aligned} A^*_{ref}(\tilde{v}) &= d^*\cdot\varepsilon_0(\tilde{v})\cdot C_0 \\ A(\tilde{v}) &= d\cdot\varepsilon_0(\tilde{v})\cdot C_0^* + d\cdot\varepsilon(\tilde{v})\cdot C \end{aligned} \tag{11}$$

If we choose suitable variable thickness of the cell in the reference route, the expression $\Delta A^{*}(\tilde{\nu}) = A(\tilde{\nu}) - A^{*}ref(\tilde{\nu}\sim) = \varepsilon_0(\tilde{\nu})\cdot(d\cdot C^{*}0 - d^{*}\cdot C0) + d\cdot\varepsilon(\tilde{\nu})\cdot C$

If we choose suitable variable thickness of the cell in the reference route, the expression **($d.C^{*}_0$ - $d^{*}.C_0$)** is null for any value of wavenumber (as the expression does not depend on the number of wavelengths), so the choice of suitable thickness **d*** resolve the problem reported for the entire spectrum. In this case the absorbance of the solute depends on its concentration, according to the relation BLB.

$$\Delta A^{*}(\tilde{\nu}) = A(\tilde{\nu}) - A_{ref}(\tilde{\nu}) = d\cdot\varepsilon(\tilde{\nu})\cdot C \tag{12}$$

Because IR absorption spectra are generated by transitions between vibrational states of sample molecules and the frequency of normal modes vibration depends (in addition to the force constants associated with deformations of the molecule) of the masses of atoms, it is expected that the replacement of atoms in molecular structure sample with different isotopes of the respective atoms (isotopic marking of the molecule) to induce dramatic changes of the IR absorption spectrum of the sample.

By isotopic marking in the known positions of molecule and by confronting these changes with changes in IR absorption band positions, significant conclusions can be drawn on whether the different atoms are involved in the normal modes of vibration of the molecule. If an atom of molecule is replaced by its heavier isotope, then IR absorption bands is moving to lower wavenumbers. The most significant movement is found in these absorption bands corresponding to normal vibration modes which involves mostly the replaced atom with heavier isotope.

The biggest relative change in mass of an atom by isotopic substitution is made for replacement of the hydrogen atom (isotope 1H) with deuterium (2H isotope). It follows that by sample deuterating, the IR absorption bands associated with chemical bonds in which one of the atom is hydrogen, suffer very significant movement toward smaller wave numbers.

Isotopic displacement of absorption bands is useful and for choice of suitable solvent in those cases where the IR absorption spectrum should be recorded in solution. Because of own absorption, some solvent (eg chloroform, $HCCl_3$) can not be used except in those domain where this is sufficiently transparent. The spectral regions in which the chosen solvent substantially absorbed are not used. But if using deuterated solvent (e.g. deuterocloroform, $DCCl_3$), this have unusable areas at other wavenumbers. Original solvent (undeuterated) and deuterated solvent presents identical dissolution properties, but are complementary with respect to transparency in the IR spectral range.

6. Aspects of construction and specific features of operating mode for Fourier transform spectrophotometers (FTIR)

Old design spectrophotometers work similar with those for UV-Vis domain, i.e. are composed of radiation source, monochromator designed to select a desired wavelength radiation, the sample chamber and the radiation detector. In IR domain, diffuse radiation presents more serious problems then in ultraviolet and visible domain. Thus, in IR domain, the ratio of useful signal and noise is more disadvantageous. The new concept of Fourier transform IR spectrophotometers meant an important step in achieving spectra with high quality even for difficult samples where the spectrophotometers with traditional construction proved to be powerless.

Construction scheme and specific features of operating mode for a Fourier Transform Infrared Spectrophotometer (FTIR - "*Fourier* **T**ransform **I**nfra**R**ed") are presented in Figure 4. It is noted that the optical assembly has no monochromator, which is replaced by a Michelson interferometer type. The polychromatic radiation from **LS** (light source) source is transmitted through concave mirror $\mathbf{M_1}$ (mirror 1) and radiation divider $\mathbf{BS_2}$ (beam splitter 2) to sample **S** (sample). After crossing the sample, the radiation reach the radiation divider $\mathbf{BS_1}$ (beam splitter 1) that divides the flow of radiation in two tracks: one for the mirrors $\mathbf{M_2}$ (mirror 2) and another for the mirrors $\mathbf{M_3}$ (mirror 3). Mirrors $\mathbf{M_2}$ and $\mathbf{M_3}$ turn back the radiation to the radiation divider $\mathbf{BS_1}$. Reaching its surface, radiations which had different routes, merge and produce a interference phenomena. The only constructive element moving while recording the spectrum is the set of $\mathbf{M_3}$ mirrors. If the mirrors $\mathbf{M_3}$ are in position **A**, the optical path difference, corresponding to the two optical paths is null, thus the radiations which turn back on the surface of the radiation divider produces an interference maximum. By translation of $\mathbf{M_3}$ mirrors, optical path difference δ between the two routes is changed progressive.

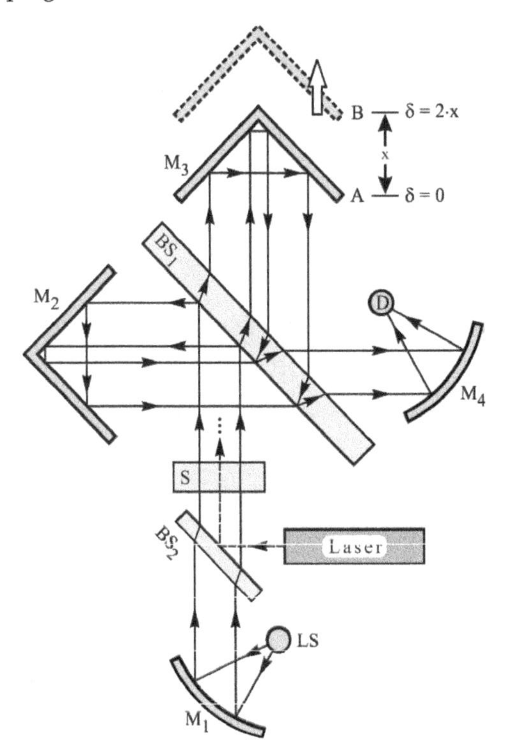

Fig. 4 Construction and operating scheme for Fourier Transform Infrared Spectrophotometer

It can be shown that the detector **D**, which measures the intensity of interference as a function of $\mathbf{M_3}$ mirrors position (so depending on the optical path difference δ between the two routes) records an interferogram which depends on inverse Fourier transform of emission spectrum of the source **LS** and on inverse Fourier transform of transparency (transmission) spectrum of the sample **S** (sample). After Fourier transform of detector **D** signal and some additional mathematical operations on detector signal the transmission (or optional absorption spectrum) spectrum of the sample **S** in known form is obtained.

Figure 4 shows the typical interferogram recorded by the detector (representing the light flux, which reached the detector, as a function of the optical path difference δ associated with $\mathbf{BS_1}$ - $\mathbf{M_2}$ - $\mathbf{BS_1}$ and $\mathbf{BS_1}$ - $\mathbf{M_2}$ - $\mathbf{BS_1}$).

To know the exact positions of absorption maxima in the IR spectrum of the sample, the position of mirrors $\mathbf{M_3}$ must be known exactly commensurable with the radiation wavelength in each moment of this whole movement. Therefore, together with the radiation of source **LS**, it is sent another radiation, this time the radiation is monochromatic, coming from a laser emitting in the visible (usually red radiation) or near infrared (often with a wavelength of 1064 nm) range. The interferogram produced by monochromatic laser radiation is practically a sinusoidal function. This sinusoidal signal, also noted by the detector **D**, is superimposed on the signal generated by the sample. By tracking the interference maximum and minimum (sinusoidal type) of the laser radiation, it can be indicated the current location of the mirrors $\mathbf{M_3}$, in each moment of the recording operation, with an accuracy comparable to the wavelength of laser radiation.

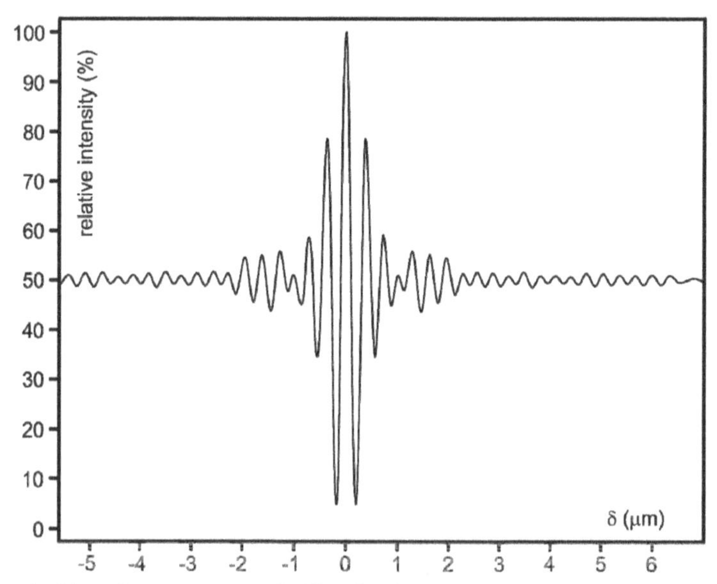

Fig. 5. The typical interferogram recorded by the detector

Recording of an IR spectrum of a sample based on Fourier Transform method has many advantages:

a. using a spectral interferometer ensures the achievement of a resolution much higher than that offered by spectrophotometers with dispersion or a high signal / noise for a given resolution;
b. lack of slits in the optical assembly of the Fourier transform spectrophotometers removes a series of disadvantages related to the fact that for spectrophotometers with dispersion the optical image of the input slit is deformed due to dispersive optical element (prism or diffractive optical network);
c. the signal / noise ratio achieved in Fourier transform spectrophotometers it is more advantageous with several magnitude orders compared with dispersion spectrophotometers;
d. because of the signal / noise ratio advantage, recording of an interferogram (a single displacement of mirrors $\mathbf{M_3}$ from position corresponding to $\boldsymbol{\delta} = 0$ to a position with a extreme $\boldsymbol{\delta}$ value) can be achieved in a very short time (less one second) reason why, within a reasonable time, interferogram recording operation can be repeated for several times, followed by the mediation of the obtained signals.

The last aspect is particularly important for difficult samples, which absorb infrared radiation in a very advanced position. For these samples, a single interferogram recording, with all the inherent advantages of Fourier multiplexing technique, signal / noise ratio is often unsatisfactory. In these cases, overlapping a larger number of records (the number can be **N**), followed by calculating the arithmetic average of the records, significantly improves the signal / noise ratio. In theory errors can be demonstrated that the overlap (acquisition) of N records, followed by mediation of the obtained interferogram, the signal / noise ratio is improved by a factor equal to $\sqrt{N}$ in comparison with a single record case. Thus, and in IR spectra (obtained by the Fourier transform of the interferogram) the signal / noise ratio is improved by acquisition of spectra.

The electrical signal of the detector is digitized with an appropriate electronic interface and data (pairs of wavenumber values vs. absorbance or wavenumbers vs. transmission percentage) are stored in a file created by a computer. The stored data can be later processed in different ways. Thus, you can add or subtract algebraic different spectra, can make corrections of baseline, can reduce noise by techniques different than acquisition (e.g. "signal smoothing") or it can be presented the absorbance derived as a function of scanned size (wavenumber). Derivation of the original spectrum as a function of wavenumer often surprises some details of the IR spectrum which are harder to observe in its original form (absorbance vs. wavenumber).

7. Organic compounds

Elucidation of the molecular structure is especially important in organic chemistry. An analytical method for the identification of functional groups from organic compounds uses one of the most physical properties of a chemical compound: the infrared absorption spectrum. Compared with other physical properties: melting point, refractive index, or specific gravity which offer only a single point of comparison with other substances, the IR spectrum of a specific compound, gives a multitude of important information (position of bands, band intensity). The intensity is indicative of the number of a particular group contributing to absorption.

It is well known that molecules absorb a unique set of IR light frequencies, because the frequency of vibration involved depends on the masses of atoms involved, the nature of the bonds and the geometry of the molecule. A molecule absorbs only those frequencies of IR light that match vibrations that cause a change in the dipole moment of the molecule. Each organic molecule, with the exception of enantiomers, has a unique infrared spectrum. This is because symmetric structures and identical groups at each end of one bond will not absorb in the IR range. The spectrum has two regions. The *fingerprint* region is unique for a molecule and the *functional group* region is similar for molecules with the same functional groups.

The entire spectral pattern is unique for a given compound. The bands that appear depend on the types of bonds and the structure of the molecule.

In a complicated molecule many fundamental vibrations are possible, but not all are observed movements which do not change the dipole moment for the molecule or the those which are so much alike that they coalesce into one band.

IR is usually preferred when a combination of qualitative and quantitative analysis is required. It is often used to follow the course of organic reactions allowing the researcher to characterize the products as the reaction proceeds.

For the analysis, the samples can be liquids, solids, or gases. The only molecules transparent to IR radiation under ordinary conditions are monatomic and homonuclear molecules such as Ne, He, O_2, N_2, and H_2. One limitation of IR spectroscopy is that the solvent water is a very strong absorber and attacks NaCl sample cells.

Computerized spectra data bases and digitized spectra are widely used today especially in research, chemistry, medicine, criminology, etc

8. Interpretation of spectra

Identification of a molecular structure from the IR spectrum can be realized using information from correlation tables and absorbances from the functional group region of the spectrum and comparison of the obtained spectrum with those of known compounds or obtain a known sample of a suspected material.

A preliminary examination of a spectrum use requires the examination of two important spectrum areas: functional group region (**4000-1300 cm^{-1}**) and the **909-650** cm^{-1} region. The characteristic stretching frequencies for important functional groups such as OH, NH, and C=O occur in this portion of the spectrum. The absence of absorption in the assigned ranges for the various functional groups can usually be used as evidence for the absence of such groups from the molecule. The absence of absorption in the **1850-1540** cm^{-1} region excludes a structure containing a carbonyl group.

Strong skeletal bands for aromatics and heteroaromatics fall in the **1600-1300** cm^{-1} region of the spectrum. These skeletal bands arise from the stretching of the carbon-carbon bonds in the ring structure. The lack of strong absorption bands in the **909-650** cm^{-1} region generally indicates a aliphatic structure. Aromatic and heteroaromatic compounds display strong out-of-plane C-H bending and ring bending absorption bands in this region. The intermediate portion of the spectrum, **1300-909** cm^{-1} is usually correspond to the fingerprint region. The absorption pattern in this region is complex, with bands originating in interacting vibrational modes. Absorption in this intermediate region is probably unique for every molecular species. For example, in the cases of hydrocarbons, organic compounds classified as saturated or unsaturated based on the absence or presence of multiple bonds, the energy of the infrared light absorbed by a C-H bond depends on the hybridization of the hybrid orbital, in the order of $sp^3>sp^2>sp$. The sp^3-hybridized C-H bonds in saturated hydrocarbons absorb in the 2850-3000 cm^{-1} region. The sp^2-hybridized C-H bonds from alkenes absorbs at **3080** cm^{-1}. A sp-hybridized C-H bond in a molecule, alkyne absorbs infrared at 3320 cm^{-1}. Another classification of hydrocarbons can be made based on absorptions due to the carbon-carbon bond. Carbon-carbon bond strength increases in the order of single>double>triple. Saturated hydrocarbons all contain carbon-carbon single bonds that absorb in the 800-1000 cm^{-1} region. But, unsaturated hydrocarbons also contain carbon-carbon single bonds that absorb in this same region. So, this interval can not be considered as fingerprint region because most organic compounds have carbon-carbon single bonds.

The alkanes give an IR spectrum with relatively few bands because there are only CH bonds that can stretch or bend.

The next table present the characteristic group frequencies of organic molecules.

Class	**Group**	**Wavenumber (cm^{-1})**
Hydrocarbons		
Alkane	C-H	2850-3000
	C-C	800-1000
Aromatic	C-H	3000-3100
	C=C	1450-1600
Alkene	C-H	3080-3140
	C=C	1630-1670
Alkyne	C-H	3300-3320
	C-C	2100-2140
Oxygen Compounds		
Alcohol	O-H	3300-3600
	C-O	1050-1200
Ether	C-O	1070-1150
Aldehyde	C=O	1720-1740
	C-H	2700 -2900
Carboxylic Acids	C=O	1700-1725
	O-H	2500-3300
	C-O	1100-1300
Ester	C=O	1735-1750
	C-O	1000-1300 (2 bands)
Ketone	C=O	1700-1725
Acyl halides	C=O	1785-1815
Anhydrides	C=O	1750;1820 (2 bands)
	O-C	1040-1100
Amides	C=O	1630-1695

Class	Group	Wavenumber (cm^{-1})
	N-H	1500-1560
Isocyanates,Isothiocyanates, Diimides, Azides, Ketenes	-N=C=O, -N=C=S -N=C=N-, $-N_3$, C=C=O	2100-2270
Nitrogen compounds		
Amines	N-H	3300-3500
	C-N	1000-1250
	NH_2	1550-1650
	NH_2 & N-H	660-900
Nitriles	C≡N	2240-2260
Oxidized Nitrogen Functions		
Oxime **(=NOH)**	O-H	3550-3600
	C=N	1665± 15
	N-O	945± 15
Amine oxide (N-O)	aliphatic	960± 20
	aromatic	1250± 50
N=O	nitroso	1550± 50
	nitro	1530± 20;1350± 30
Alkyl bromide	C-H	667
Sulfur compounds		
Thiols	S-H	2550-2600
Esters	S-OR	700-900
Disulfide	S-S	500-540
Thiocarbonyl	C=S	1050-1200
Sulfoxide	S=O	1030-1060
Sulfone	S=O	1325± 25; 1140± 20
Sulfonic acid	S=O	1345
Sulfonyl chloride	S=O	1365± 5;1180± 10
Sulfate	S=O	1350-1450
Phosphorous compunds		
Phosphine	P-H	2280-2440 950-1250
Phosphonic acid	(O=)PO-H	2550-2700
Esters	P-OR	900-1050
Phosphine oxide	P=O	1100-1200
Phosphonate	P=O	1230-1260
Phosphate	P=O	1100-1200
Phosphoramide	P=O	1200-1275
Silicon compounds		
Silane	Si-H	2100-2360
	Si-OR	1000-1110
	$Si-CH_3$	1250± 10

Table 1. Schematic representation of the Infrared Group Frequencies classification

9. References

E.O. Brigham: "The Fast Fourier Transform", Prentice-Hall, Inc., 1974

P.L. Polavarapu: "Vibrational Spectra: Principles and Applications with Emphasis on Optical Activity", Elsevier Science B.V. 1998;

Y. Wang, R. Tsenkova, M. Amari, F. Terada, T. Hayashi, A. Abe, Y. Ozaki: 'Potential of Two-Dimensional Correlation Spectroscopy in Analysis of NIR Spectra of Biological Fluids. I. Two-Dimensional Correlation Analysis of Protein and Fat Concentration-Dependent Spectral Variations of Milk", Analusis Magazine, 1998:26, M64-M69

R.N. Bracewell: "The Fourier Transform and Its Applications", McGraw Hill, 2000

D. Baurecht, U.P. Fringeli: "Quantitative Modulated Excitation Fourier Transform Infrared Spectroscopy", Review of Scientific Instruments, 2001:72, 3782-3792

M-J. Paquet, M. Laviolette, M. Pézolet, M. Auger: "Two-Dimensional Infrared Correlation Spectroscopy Study of the Aggregation of Cytochrome C in the Presence of Dimyristoilphosphatidylglycerol", Biophysical Journal, 2001:81, 305-312

Y. Kauppinen, J. Partanen: "Fourier Transforms in Spectroscopy", Wiley-VCH Verlag Gmbh, 2001

J.M. Chalmers, P.R. Griffiths (Editors): "Handbook of Vibrational Spectroscopy. Theory and Instrumentation", John Wiley & Sons Ltd., 2002

M. Cho: "Two-Dimensional Optical Spectroscopy", CRC Press, 2009

P.W. Atkins, Physical Chemistry. 2nd Ed. San Francisco: W.H. Freeman and Company, 1982. Discussion of vibrational spectra from a quantum mechanical view.

B.W. Cook, K. Jones. A Programmed Introduction to Infrared Spectroscopy. New York: Heyden & Son Inc., 1972. Excellent resource for the beginning spectroscopist.

R. T. Morrison, R. N. Boyd. Organic Chemistry. 5th Ed. Boston: Allyn and Bacon, Inc., 1987. Provides a brief description of spectroscopy. Includes relevant IR spectra for each family of organic compounds.

R. L., Shriner, R. C., Fuson, D. Y., Curtin, T. C. Morrill. The Systematic Identification of Organic Compounds. 6th Ed. New York: Wiley, 1980. Contains a brief section on IR spectroscopy. Mainly a text for identification of compounds by chemical tests.

R. M., Silverstein, G. C., Bassler, T. C. Morrill. Spectrometric Identification of Organic Compounds. 4th Ed. New York: Wiley, 1981. Description of mass spectrometry, IR spectrometry, 1H NMR spectrometry, 13C spectrometry, and UV spectrometry.

A. Lee Smith,. Applied Infrared Spectroscopy: Fundamentals, Techniques, and Analytical Problem-Solving. New York: Wiley, 1979. Comprehensive treatment of IR spectroscopy. Includes history, instrumentation, sampling techniques, qualitative and quantitative applications.

G. Socrates, Infrared Characteristic Group Frequencies. 2nd Ed. Chichester: Wiley, 1994. A comprehensive reference of correlation tables.

A.Streitweiser, Jr., C. H. Heathcock. Introduction to Organic Chemistry. 2nd Ed. New York: Macmillan Publishing Co., Inc., 1981. Introductory organic text with a section on IR spectroscopy. Includes spectroscopic information as each family is presented.

Section 4

Fluorescence Spectroscopy

11

Current Achievement and Future Potential of Fluorescence Spectroscopy

Nathir A. F. Al-Rawashdeh
United Arab Emirates University, Department of Chemistry, UAE
Jordan University of Science and Technology, Department of Chemistry, Jordan

1. Introduction

Spectrofluorometric methods of analysis are the most commonly analytical techniques and continue to enjoy wide popularity. The wide availability of the instrumentation, the simplicity of procedure, sensitivity, selectivity, precision, accuracy, and speed of the technique still make the spectrofluorometric methods attractive. These features make fluorescence spectroscopy an attractive technique as compared to other forms of optical spectroscopy or other analytical techniques such as chromatography and electrophoresis. Fluorescence spectroscopy has been used widely as a tool for quantitative analysis, characterization, and quality control in the pharmaceutical, environmental, agricultural, nanotechnology and biomedical fields.

The emission of light from an excited electronic state of a molecular species is called luminescence. The discovery and characterization of luminescence begun from the 15th century. In 1506 Nicolas Monardes was the first to describe the bluish opalescence of the water infusion from the wood of a small Mexican tree. In 1612 Galileo described the emission of light (phosphorescence) from the famous Bolognian stone, which discovered by Vincenzo Casciarolo, a Bolognian shoemaker. Galileo wrote: "It must be explained how it happens that light is conceived into the stone, and is given back after some time, as is childbirth". Even though, some of the first scientific reports of luminescence appeared in the middle of the 18th century. In 1845 Sir J.F.W. Herschel reports on an experiment he did twenty years earlier. Herschel made the first observation of fluorescence from quinine sulfate (quinine: (*R*)-(6-methoxyquinolin-4-yl)-((2*S*, 4S, 8R)- 8-vinylquinuclidin-2-yl)methanol, $C_{20}H_{24}N_2O_2$, quinine absorbs in the UV region), he observed that an otherwise colorless solution of quinine in water emitted a blue color under certain circumstances. Herschel concludes that a species in the solution, "exert its peculiar power on the incident light" and disperses the blue light. The experiment can be repeated simply by observing glass of tonic water exposed to sunlight. Often a blue glow is visible at the surface (Rendell, 1987).

The phenomenon of fluorescence was known by the middle of the nineteenth century. British scientist Sir George G. Stokes first made the observation that the mineral fluorspar exhibits fluorescence when illuminated with ultraviolet light, and he coined the word "fluorescence". In 1852, Sir G.G. Stokes studied the same compound (quinine) that has been used by Herschel and found that the fluorescing (*emitted*) light has longer wavelengths than the excitation (*absorbed*) light, a phenomenon that has become to be known as the Stokes shift. Stokes' paper demonstrated the fundamental property of fluorescence, which simply can be summarized as a photon of ultraviolet radiation collides with an electron in a simple atom, exciting and elevating the electron to a higher energy level. Subsequently, the excited electron relaxes to a lower level and emits light in the form of a lower-energy photon (*higher wavelength*) in the visible light region. In 1871 Adolph Von Baeyer, a German chemist, synthesized the fluorescent dye, fluorescein. In 1880 Edmund Bequerel showed that certain metal ion complexes emit radiation with a very long decay time (Rendell, 1987).

Jabłoński and others developed a modern theoretical understanding of Stokes observation some 70 years later. In the 1920s and 1930s Jabłoński investigated polarized light and fluorescence and was able to show that the transition moments in absorption and emission are two different things. Thus the foundation for the concept of anisotropy was laid. For that and other accomplishments Jabłoński has been referred to as the father of fluorescence and his work has had a major impact on the theoretical understanding of photophysics (Rendell, 1987; Lakowicz, 2006).

How and why do certain molecules known as fluorophores or fluorescent molecules (such as: dyes, polyaromatic hydrocarbon, or heterocyclic,...etc.) emit different colors of light?. Briefly the answer for this question is that some molecules are capable of being excited, via absorption of light energy, to a higher energy state, also called an excited state. The energy of the excited state, which cannot be sustained for long, "decays" or decreases, resulting in the emission of light energy. This process is called *fluorescence*. To "fluoresce" means to emit light via this process. A fluorophore is a molecule that is capable of fluorescing due to the presence of certain chromophores within a molecule. In its ground state, the fluorophore molecule is in a relatively low-energy, stable configuration, and it does not fluoresce. When light from an external source hits a fluorophore molecule, the molecule can absorb the light energy. If the energy absorbed is sufficient, there are multiple excited states or energy levels that the fluorophore can attain, depending on the wavelength and energy of the external light source. Since the fluorophore is unstable at high-energy configurations, it eventually adopts the lowest-energy excited state, which is semi-stable. The length of time that the fluorophore is in excited states is called the *excited lifetime*, and it lasts for a very short time, ranging from 10^{-15} to 10^{-9} seconds. Next, the fluorophore rearranges from the semi-stable excited state back to the ground state, and the excess energy is released and emitted as light. The emitted light is of lower energy, and thus longer wavelength, than the absorbed light. This means that the color of the light that is emitted is different from the color of the light that has been absorbed. Emission of light returns the fluorophore to its ground state. The fluorophore can absorb light energy again and go through the entire process repeatedly (Lakowicz, 2006).

The cyclical fluorescence process, shown in Figure 1, can be summarized as: 1. Excitation of a fluorophore through the absorption of light energy, the excitation wavelength is usually the same as the absorption wavelength of the fluorophore; 2. A transient excited lifetime with some loss of energy, during this period, the fluorophore undergoes conformational changes and is also subject to possible interactions with its molecular environment, with two important consequences: first, the energy of higher excited state is partially dissipated as a heat, yielding a relaxed lowest singlet excited state from which fluorescence emission originates, second, not all the molecules initially excited by absorption (stage 1) return to the ground state by fluorescence emission, as other processes such as collisional quenching, fluorescence energy transfer, and intersystem crossing may also depopulate the first excited state ; and 3. Return of the fluorophore to its ground state, accompanied by the emission of light. The light energy emitted is always of a longer wavelength (*lower energy*) than the light energy absorbed, due to the energy dissipation during the transient excited lifetime, as shown in Step 2. Consequently, the ratio of the number of fluorescence photons emitted (stage 3) to the number of photons absorbed (stage 1) represents the *fluorescence quantum yield* (Lakowicz, 2006).

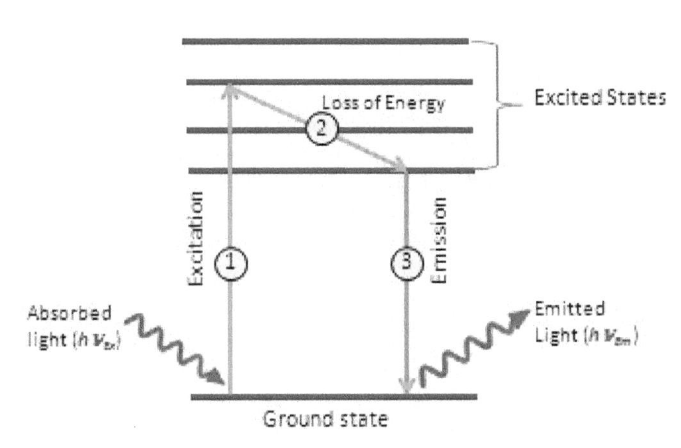

Fig. 1. The Jablonski diagram illustrates the three stages involved in the creation of an excited electronic singlet state by optical absorption and subsequent emission of fluorescence

A fluorophore can repeatedly undergo the fluorescence process—in theory, indefinitely. This is extremely useful, because it means that one fluorophore molecule can generate a signal multiple times. This property makes fluorescence a very sensitive technique for visualizing microscopic samples—even a small amount of the stain can be detected. In reality, however, the fluorophore's structural instability during the excited lifetime makes it susceptible to degradation. High-intensity illumination can cause the fluorophore to change its structure so that it can no longer fluoresce—this is called photobleaching. When a fluorescent sample, such as a slide with mounted tissue, is photobleached, the fluorophores are no longer promoted to an excited state, even when the required light energy is supplied (Lakowicz, 2006).

Now that we've introduced the general process of fluorescence, let's take a look at the basic properties of the light spectrum and its importance in fluorescence. The visible spectrum (Figure 2) is composed of light with wavelengths ranging from approximately 380 nanometers to 750 nanometers.

Light waves with shorter wavelengths have higher frequency and higher energy. Light waves with longer wavelengths have lower frequency and lower energy. As we stated before, an excited fluorophore emits lower-energy light than the light it absorbed. Therefore, there is always a shift along the spectrum between the color of the light absorbed by the fluorophore during excitation, and the color emitted.

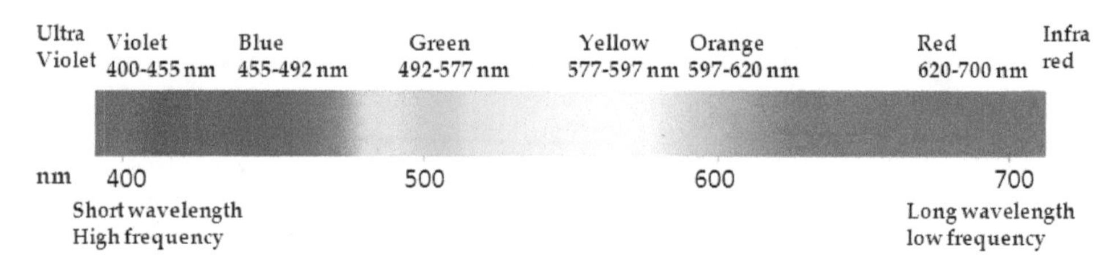

Fig. 2. Visible light spectrum

For example, let's say that we have a tube that contains a particular fluorescent dye. If we shine 480 nanometer light at the dye solution, some of the fluorophore molecules will become excited. However, the majority of the molecules are not excited by this wavelength of light. As we increase the excitation wavelength, say to 520 nanometers, more molecules are excited. However, this is still not the wavelength at which the proportion of excited molecules is maximal. For this particular dye, 550 nanometers is the wavelength that excites more fluorophores than any other wavelength of light. At wavelengths longer than 550 nanometers, the fluorophore molecules still absorb energy and fluoresce, but again in smaller proportions. The range of excitation wavelengths can be represented in the form of a fluorescence excitation spectrum, which looks like the spectrum shown in Figure 3.

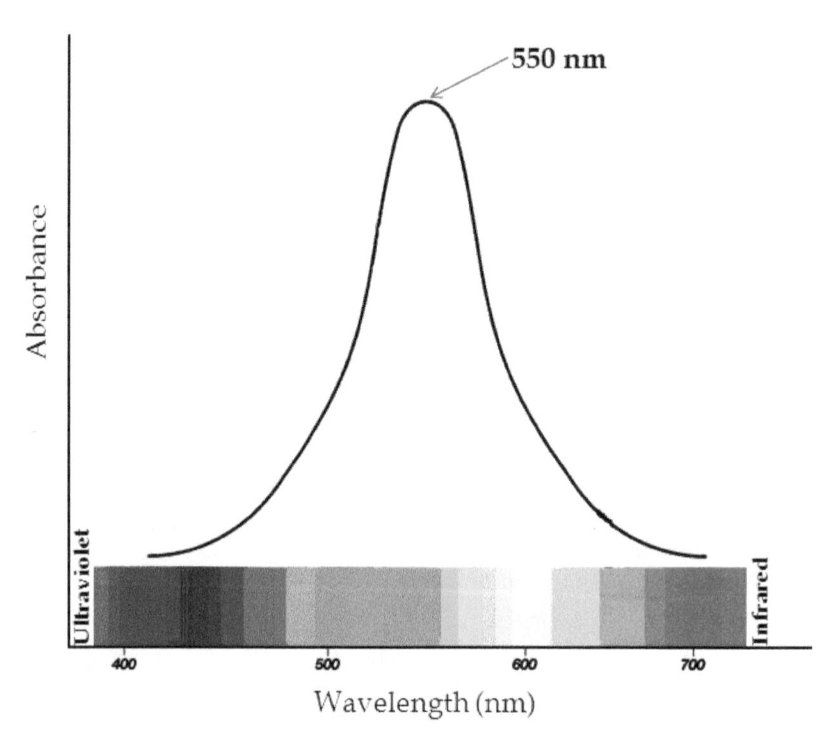

Fig. 3. Excitation spectrum of dye solution recorded at λ_{em} = 570 nm, the excitation wavelength maximum at 550 nm.

A fluorescent molecule absorbs light over a range of wavelengths—and every chemical molecule has a characteristic excitation range. However, some wavelengths within that range are more effective for excitation than other wavelengths. This range of wavelengths reflects the range of possible excited states that the fluorophore can achieve. Thus for each fluorescent molecule, there is a specific wavelength—the excitation maximum—that most effectively induces fluorescence. Now let's look at the light that is emitted by the fluorophore molecules when they are excited at the optimal excitation wavelength. Just as fluorophore molecules absorb a range of wavelengths, they also emit a range of wavelengths. There is a spectrum of energy changes associated with these emission events. When we excite the previously described dye solution at its excitation maximum, 550 nanometers, light is emitted over a range of wavelengths. A molecule may emit at a different wavelength with each excitation event because of changes that can occur during the excited lifetime, but each emission will be within the range. Although fluorophore molecules emit same intensity of light, the wavelengths and therefore the colors of the emitted light are not homogeneous. However, a larger population of molecules has most intensely fluorescence at 570 nanometers (Lakowicz, 2006).

Based on this distribution of emission wavelengths, we say that the emission maximum of this fluorophore is 570 nanometers. The range of wavelengths is represented by the Fluorescence Emission Spectrum, Figure 4.

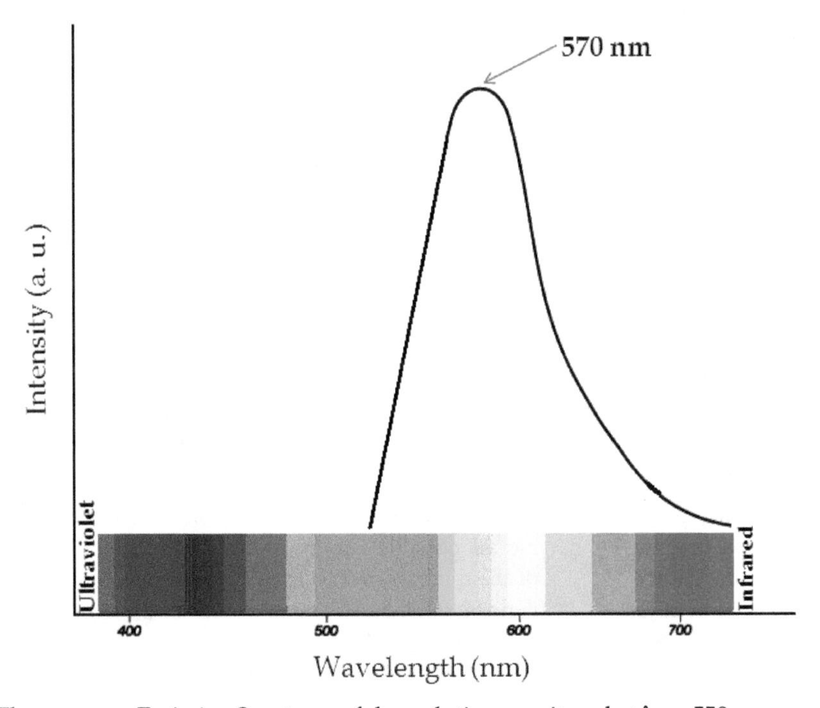

Fig. 4. Fluorescence Emission Spectrum of dye solution monitored at $\lambda_{ex} = 550$ nm

The emission intensity is proportional to the amplitude of the fluorescence excitation spectrum at the excitation wavelength. Fluorescence emission intensity depends on the same parameters as absorbance—defined by the Beer-Lambert law as the product of the molar extinction coefficient, optical path-length, and concentration—as well as on the fluorescence quantum yield, the intensity of the excitation source, and the efficiency of the instrument and, in dilute solutions, is linearly proportional to these parameters.

The summary points of this introduction to fluorescence are: 1. Fluorophores are molecules that, upon absorbing light energy, can reach an excited state, and then emit light energy. 2. The three-stage process of excitation, excited lifetime, and emission is called fluorescence. 3. Fluorophores absorb a range of wavelengths of light energy, and also emit a range of wavelengths. Within these ranges are the excitation maximum and the emission maximum. Because the excitation and emission wavelengths are different, the absorbed and emitted lights are detectable as different colors or areas on the visible spectrum.

The purpose of this chapter is to review the articles on the interior cited aspects published since 2000 about various aspects of application of fluorescence spectrophotometry in chemical analysis.

2. Theoretical and instrumental aspects

2.1 Basic theory of fluorescence

This section provides a basic tutorial on specific topic in luminescence, namely fluorescence, and fluorescent instrumentations. To be able to understand the basic theoretical principles of luminescence spectroscopy, which include the electronic transitions, one should have a basic background on quantum mechanics and atomic orbitals, which was developed by Schrödinger in 1926. A tutorial review of Schrödinger's wave equation is out of the scope of this chapter, but briefly the most fruitful outcome of solving Schrödinger's wave equation is a set of wave functions (*called orbitals*), and their corresponding energies. An orbital is described by a set of three quantum numbers (Principal (n), Angular momentum (l), and magnetic (m_l) quantum numbers). Later a fourth quantum number, that describes the magnetic field generated due to the spinning of an electron on its own axis, was discovered and named as spin magnetic quantum number (m_s). The spin quantum number has only two allowed values: +1/2 and −1/2. According to Pauli Exclusion Principle, no two electrons in the same atom can have identical sets of four quantum numbers, n, l, m_l, *and* m_s. Thus any two electrons in same orbital (n, l, m_l) must have different spins either m_s=+1/2 or m_s= -1/2. The total spin is defined by $S = \Sigma m_s$ and the Multiplicity (M) is defined as $M = 2S+1$(M=1, 2, 3… Singlet, Doublet, and Triplet, respectively,..) as shown in Figure 5.

The photophysical processes that occur from absorption to emission are often shown in a so-called Jabłoński diagram. Of course all possible energy routes cannot be encompassed in single figure, and different forms of the diagram can be found in different contexts. Figure 7 is a simple version of Jabłoński diagram, where absorption ($S_0 \rightarrow S_1$ or $S_0 \rightarrow S_2$), fluorescence (emission not involving spin change, $S_1 \rightarrow S_0$), phosphorescence (emission involving spin change, $T_1 \rightarrow S_0$), intersystem crossing (non-radiative transition with spin change, as an example from singlet to triplet states; $S_1 \rightarrow T_1$), internal conversion (non-radiative transition either to lower state when vibrational energy levels "match" or to lower state by collisional deactivation, $S_2 \rightarrow S_1$)), vibration relaxation (within the vibrational levels in any excited electronic state) as well as intermolecular processes (radiationless relaxations) are shown.

Other intermolecular processes (e.g. quenching, energy transfer, solvent interaction etc.) are omitted (Rendell, 1987; Lakowicz, 2006).

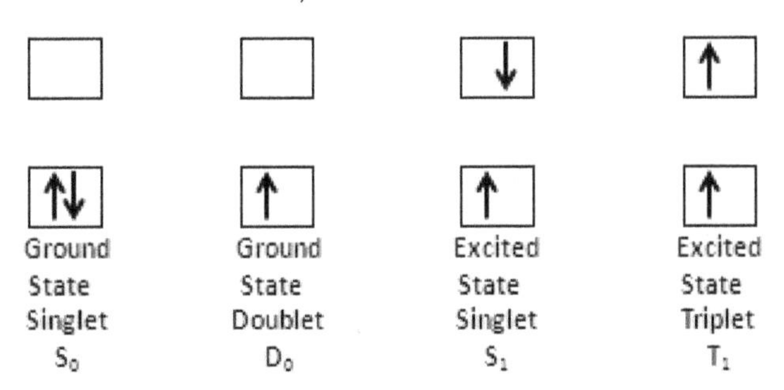

Fig. 5. Possible energy states according to their spin multiplicity

Once a molecule is excited by absorption of light it can return to the ground state with emission of fluorescence, but many other pathways for de-excitation are also possible, these are summarized in Figure 6.

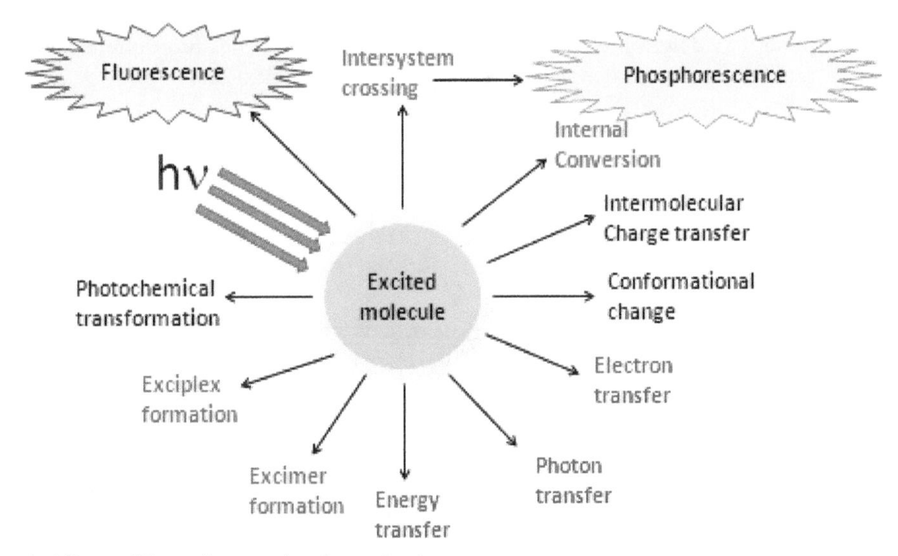

Fig. 6. All possible pathways for de-excitation processes

Jablonski diagram (Fig. 7) explains the mechanism of light emission in most organic and inorganic luminophores. Upon absorption of the light by a molecule, the electron promoted from ground electronic state (S_0) to an excited state that should possess the same spin multiplicity (such as, S_1, S_2,….) this process usually occurs within ~10^{-15} s. This excludes the triplet excited state as the final state of electronic absorption because transitions between states of different multiplicities are improbable "*forbidden*" (e.g. T→S or S→T). According to the quantum mechanical selection rules for electronic transitions, spin state should be maintained upon excitation because it is harder for an electron to go from a singlet state to triplet state since the spin has to be flipped (i.e. *change in spin during the electronic transitions is not allowed*). Therefore, to go from a singlet to a triplet state ($\Delta M = 1$) is so-called forbidden transition and occurs with a small rate constant and typically too weak to be of much importance. At room temperature the higher vibrational energy levels are in general not populated (less than 1% according to Boltzmann statistics). The magnitude of the absorbed energy decides which vibrational level of S_1 (or S_2) becomes populated. This process is very fast and happens within 10^{-15} s. In the next 10^{-12} s the molecule relaxes to the lowest vibrational level of S_1, a process called internal conversion. Since emission typically occurs after 10^{-9} s the molecule is fully relaxed at the time of emission, hence, as a rule, emission occurs from the lowest vibrational level of S_1 (Kasha.s rule) and the fluorescence spectrum is generally independent of the excitation wavelength. After emission the molecule returns to the ground state, possibly after vibrational relaxation. This completes the simplest case of fluorescence: excitation, internal conversion, emission and relaxation. The energy lost to the surroundings, due to vibrational relaxation and internal conversion is the reason why a Stokes' shift (*defined as the energy difference between the emission and absorption peak maxima for the same electronic transition*) is observed (Figure 8). The Stokes shift is due to the fact that some of the energy of the excited fluorophore is lost through molecular vibrations that occur during the brief lifetime of the molecule's excited state. This energy is dissipated as heat to surrounding solvent molecules as they collide with the excited fluorophore. The magnitude of the Stokes shift is determined by the electronic structure of the fluorophore, and is a characteristic of the fluorophore molecule (Rendell, 1987; Lakowicz, 2006).

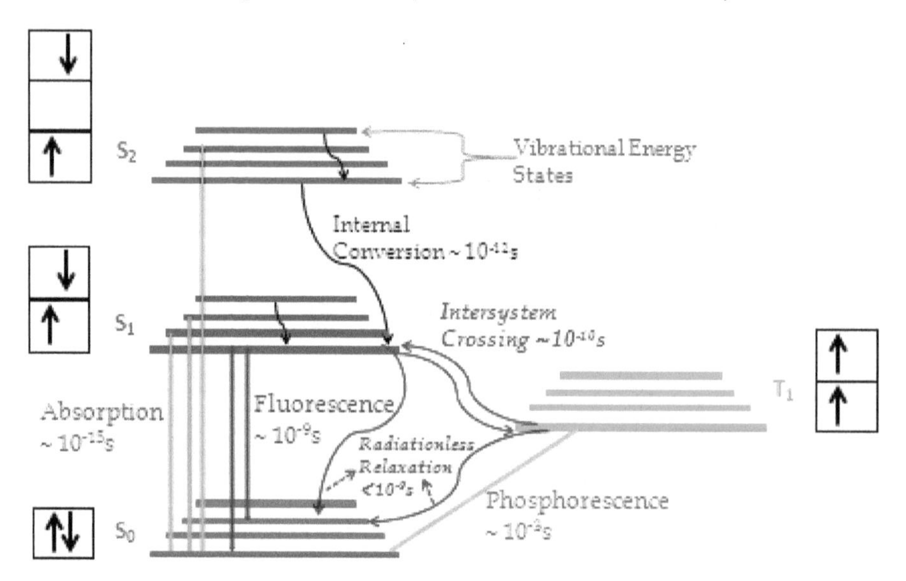

Fig. 7. The Jablonski diagram. Four electronic levels are depicted along with four vibrational energy levels.

Since the energy spacing between the vibrational levels in S_0 or S_1 is of the same size, there often exist mirror image symmetry between the emission spectrum and the $S_0 \rightarrow S_1$ absorption spectrum (not the $S_0 \rightarrow S_2$ absorption) (Fig. 8), needless to say there are plenty of exceptions to the rule. The transitions that will predominate can be justified by Franck-Condon principle. The principle states that since the electronic absorption of light occurs in extremely short time (~10^{-15}s), thus during the time scale of absorption the nuclei are assumed to be frozen, that is that the transitions between various electronic levels are vertical (Rendell, 1987; Lakowicz, 2006).

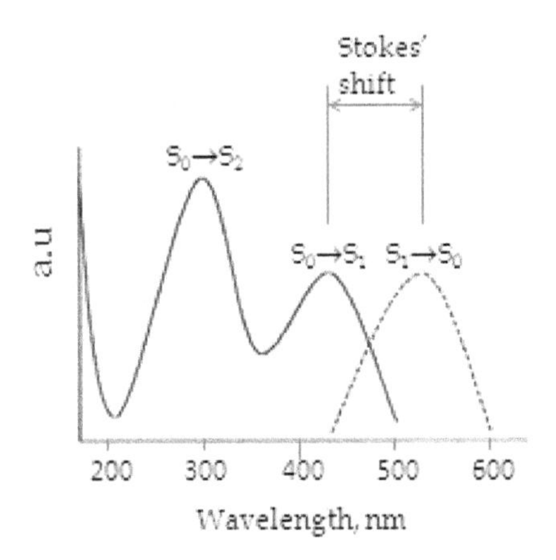

Fig. 8. Typical absorption and emission spectra.

The typical molecular photoluminescence relaxation processes that illustrated by Jablonski diagram (Fig. 7) can be classified to two main type of transition, these are radiative and nonradiative transitions. The non-radiative relaxation processes are vibrational relaxation (a rapid relaxation of excited molecules from higher to lowest vibrational level, occur within ~10^{-14}-10^{-12} s), internal conversion (a rapid relaxation of excited molecules to the lowest energy singlet excited state (S_1) from higher energy excited singlet state (such as S_2 in Fig. 7), occur within a time scale of 10^{-12} s), and intersystem crossing (relaxation between excited states of different spin multiplicity, such as $S_1 \rightarrow T_1$ in Fig. 7, occur within a time scale of 10^{-8} s). Intersystem crossing occurs more slowly than internal conversion since it's a less probable process than internal conversion due to spin multiplicity is not conserved.

The radiative processes are fluorescence and phosphorescence (Figure 9) (Lakowicz, 2006). The fluorescence refers to the emission of light associated with a radiative transition from an excited electronic state that has the same spin multiplicity as the ground electronic state ($S_1 \rightarrow S_0$, Fig. 7). Fluorescence transitions are spin allowed, thus they occur very rapidly and the average lifetimes of the excited states responsible for fluorescence are typically 10^{-9}-10^{-5} s. Phosphorescence refers to the emission of light associated with a radiative transition from an excited electronic state that has a different spin multiplicity from that of the ground electronic state ($T_1 \rightarrow S_0$, Fig. 7). Phosphorescence transitions are spin forbidden, thus they are less probable than spin allowed transitions and the average lifetimes of the excited states responsible for phosphorescence are typically 10^{-3} s. However, spin forbidden transitions become more probable when a significant interaction between the spin angular moment and the orbital angular momentum is observed (spin-orbit coupling increases), this can be observed in the presence of heavy atoms. Furthermore, in solutions, the presences of paramagnetic species such as molecular oxygen increase the probability of intersystem crossing and consequently make the spin forbidden transitions more probable.

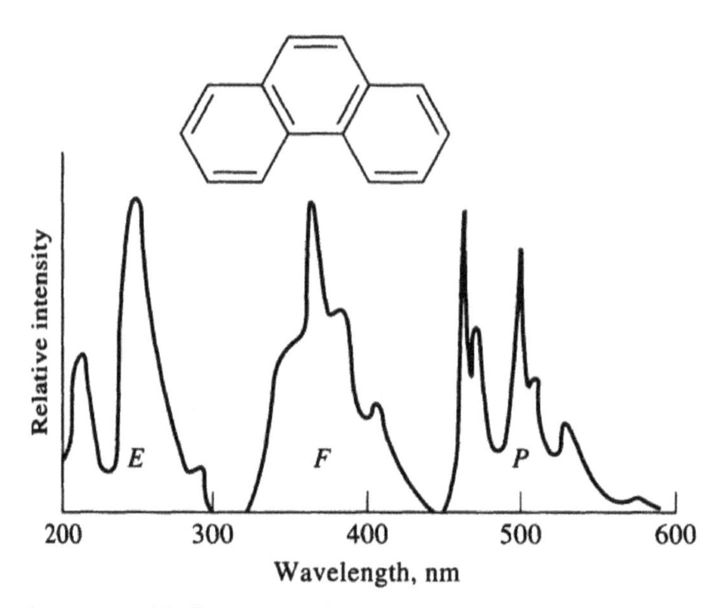

Fig. 9. Typical excitation (E), fluorescence (F), and phosphorescence (P) spectra of phenanthrene (Lakowicz, 2006).

The relative intensity of fluorescence peak is controlled by the Frank-Condon principle, but also the total fluorescence peak intensity (I) is related to the fluorescence quantum yield (Φ_F), which defined as the ratio of number of photons emitted to number of photons absorbed. Furthermore, the fluorescence quantum yield (Φ_F) can be expressed as the rate of photons emitted divided by the total rate of depopulation of the excited state (Equation 1) (Rendell, 1987; Lakowicz, 2006).

$$\Phi_F = \frac{k_F}{k_F + \Sigma k_{nr}} \tag{1}$$

Where k_F and k_{nr} are the rate constant of fluorescence and non-radiative processes, respectively. The fluorescence quantum yield (Φ_F) value in the range of 0.0 to 1.0. If the non-radiative relaxation is fast compared to fluorescence ($k_{nr} > k_r$), Φ will be small, that is the compound will fluoresce very little or not at all. Often different non-radiative events are limited in the solid phase, and long-lived luminescence (e.g. phosphorescence) is often studied in frozen solution or other solid phases. Quenchers make non-radiative relaxation routes more favorable and often there is a simple relation between Φ and the quencher concentration. The best-known quencher is probably O_2, which quench almost all fluorophores; other quenchers only quench a limited range of fluorophores. If a molecule is subject to intramolecular quenching, Φ may yield information about the structure.

The factors that affect the fluorescence quantum yield (Φ_F) are: (1) the excitation wavelength (λ_{ex}), the short wavelengths' break bonds and increase the rate constant of dissociation processes, the most common excitation wavelength are involving the $n \rightarrow \pi^*$ and $\pi \rightarrow \pi^*$

transitions; (2) lifetime of the excited state, the transition probability measured by the molar absorbitivity (ε), large ε implies short lifetime, thus largest fluorescence are observed from short lifetime and high ε state, as an example $\pi^* \rightarrow \pi > \pi^* \rightarrow n$ (10^{-9}-10^{-7} s > 10^{-7}-10^{-5} s); (3) structure of the molecule, emission of light is favored in aromatic molecules (conjugated systems) involving $n \rightarrow \pi^*$ and $\pi \rightarrow \pi^*$ transitions, and fluorescence increased by number of fused rings and substitution on or in the ring, such as pyridine, pyrrole, quinoline and indole; (4) rigidity of the structure, fluorescence quantum yield increases with increasing the rigidity of the molecules specially with chelation, such as fluorine and biphenyl; and (5) the fluorescence quantum yield is highly dependence on the temperature, pH and solvent (Rendell, 1987; Lakowicz, 2006).

The total fluorescence intensity (*I*) is given by Equation 2:

$$I_F = 2.303 I_o \Phi \varepsilon c b \tag{2}$$

Where I_o is the incident radiant power, the term εcb is deduced from the well known Beers' law expression for the absorption (ε is the molar absorptivity, c is the molar concentration of the fluorescent substance, and b is the path length of the cell). The Beers' law is valid only for diluted solutions ($\varepsilon bc < 0.05$) (Guilbault, 1990)

When an emission spectrum is obtained, data are typically collected for more than 0.1 sec. at each wavelength increment (typically 1nm), but since fluorescence lifetimes typically is measured in nanoseconds, it follows that the obtained spectrum is a time-average of a many events. The time averaging loses much information, and time-resolved experiments are often the more interesting when a system is investigated. The fluorescent lifetime of the excited state, τ_F, is the average time a molecule stays in the excited state before returning to ground state. Thus τ_F can be expressed as the inverse of the total depopulation rate as in Equation 3 (Rendell, 1987; Lakowicz, 2006).

$$\tau_F = \frac{1}{k_F + k_{nr}} \tag{3}$$

Where k_F and k_{nr} are the rate constant of fluorescence and non-radiative processes, respectively.

Typically fluorescence lifetime values are in the 5-15 ns range. The expression in Eq. 3 is related to the expression for Φ_F, in that way that they have a common denominator. Actually an approximation of τ_F can be obtained by measuring Φ_F in aired and degassed solutions.

In the absence of non-radiative relaxation (k_{nr}= 0), the lifetime becomes the inverse of k_r and is often called the natural lifetime, denoted τ_N. For many compounds the natural lifetime can be calculated from the measured lifetime τ and the measured quantum yield Φ_F, Equation (4) (Rendell, 1987; Lakowicz, 2006).

$$\tau_N = \frac{\tau_F}{\Phi_F} \tag{4}$$

It is important to notice that the fluorescent lifetime is what is experimentally obtained, and the natural lifetime can be calculated.

The fluorescence lifetime, τ_F, is determined by observing the decay in fluorescence intensity (decay profile) of a fluorophore after excitation. Immediately after a molecule is excited the fluorescence intensity will be at a maximum and then decrease exponentially according to Equation 5 (Rendell, 1987; Lakowicz, 2006).

$$I(t) = I_o e^{-t/\tau_F} \tag{5}$$

Thus after a period of τ_F the intensity has dropped to 37% of I_0, that is 63% of the molecules return to the ground state before τ_F. In many cases the above expression needs to be modified into more complex expressions. First of all it is assumed that the instrument yields an infinite (or very) short light pulse at time zero. In cases where τ_F is small I_0 must be replaced by a function, which describes the lamp profile of the instrument. Also, more than one lifetime parameter is often needed to describe the decay profile, which is I(t) must be expressed as a sum of exponentials. Finally the concept of anisotropy should be mentioned. Anisotropy is based on selectively exciting molecules with their absorption transition moments aligned parallel to the electric vector of polarized light. By looking at the polarization of the emission the orientation of the fluorophore can be measured. The anisotropy of the system is defined as (Equation 6) (Rendell, 1987; Lakowicz, 2006):

$$r = \frac{I_{\parallel} - I_{\perp}}{I_{\parallel} + 2I_{\perp}} = \frac{3\langle \cos^2(\theta) \rangle - 1}{2} \tag{6}$$

The oval in Figure 10 symbolized the absorption transition moment. Vertical polarized excitation light enters along the x-axis and $I_{\perp}$ and $I_{\parallel}$ are measured along the y-axis, setting the emission polarizer perpendicular and parallel to the excitation polarizer respectively. Θ is the angle of the emission to the z-axis (see Figure 10), the squared brackets indicates that it is the average value. If all absorption transition moments are aligned along the z-axis then $I_{\perp} = 0$ and $\theta = 0$, leading to $r = 1$, the maximum anisotropy.

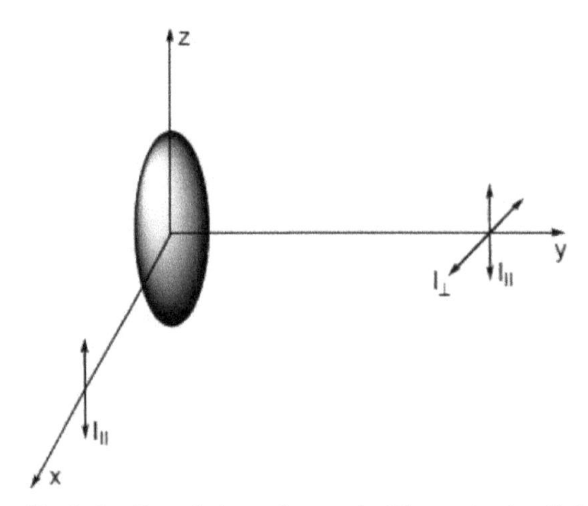

Fig. 10. The absorption dipole is aligned along the z-axis. The excitation light is vertically aligned and follows the x-axis. Emission is measured along the y-axis.

By combining anisotropy with time-resolved measurements it is possible to measure the mobility of a fluorophore. Immediately after excitation all excited molecules will be oriented along a common axis. In the solid phase the system will retain its anisotropy until emission. However, if the fluorophores are free to move, the anisotropy of the system will decrease before emission.

2.2 Instrumentation

The principal sketch of a typical fluorescence spectrophotometer is shown in Figure 11. It consists of a light source, an excitation and emission monochromator (*grooves/mm*), polarizers (*prisms*), sample chamber and a detector (*such as photomultiplier tube*). For steady state measurements the light source usually consists of a 450W xenon arc lamp, and for time resolved measurements it is equipped with nanosecond flash lamp. Most simple spectrometers have a similar geometry, but often extra detectors and/or light sources are fitted resulting in a T- or X-geometry.

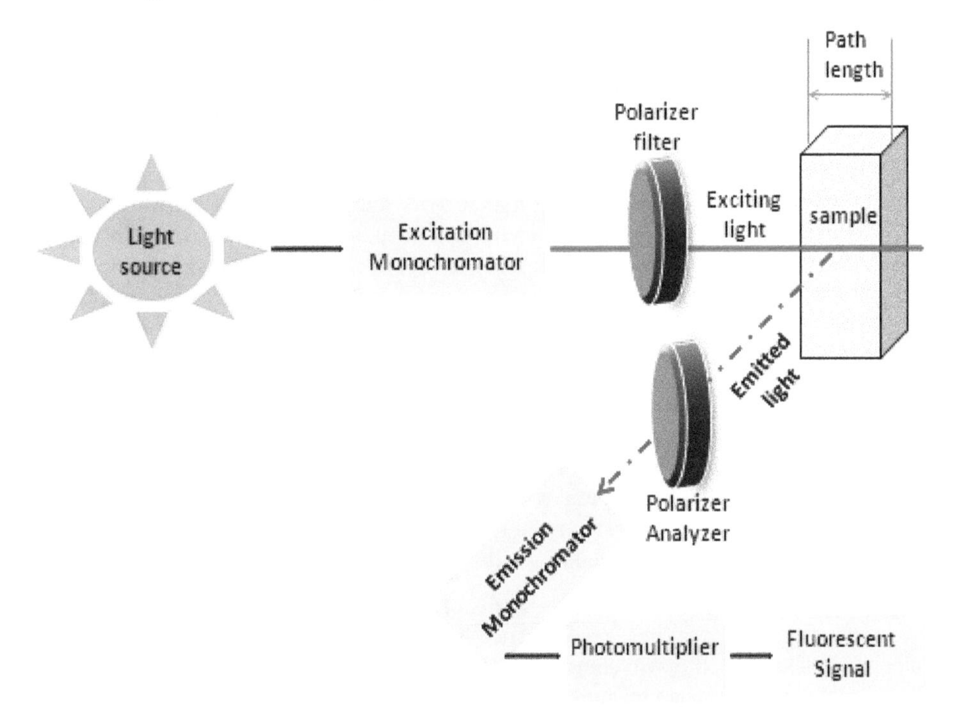

Fig. 11. Schematic representation of a fluorescence spectrophotometer.

The light source produces light photons over a broad energy spectrum, typically ranging from 200 to 900 nm. Photons impinge on the excitation monochromator, which selectively transmits light in a narrow range centered about the specified excitation wavelength. The transmitted light passes through adjustable slits that control magnitude and resolution by further limiting the range of transmitted light. The filtered light passes into the sample cell causing fluorescent emission by fluorphors within the sample. Emitted light enters the emission monochromator, which is positioned at a 90° angle from the excitation light path to eliminate background signal and minimize noise due to stray light. Again, emitted light is transmitted in a narrow range centered about the specified emission wavelength and exits through adjustable slits, finally entering the photomultiplier tube (PMT). The signal is amplified and creates a voltage that is proportional to the measured emitted intensity. Noise in the counting process arises primarily in the PMT. Therefore, spectral resolution and signal to noise is directly related to the selected slit widths (Guilbault, 1990; Rendell, 1987; Lakowicz, 2006).

Not all fluorimeters are configured as described above. Some instruments employ sets of fixed band pass filters rather than variable monochromators. Each filter can transmit only a select range of wavelengths. Units are usually limited to 5 to 8 filters and are therefore less flexible. Fiber optics are also employed for "surface readers", to transmit light from the excitation monochrometers to the sample surface and then transport emitted light to the emission monochrometers. This setup has the advantage of speed, but has the disadvantages of increased signal to noise, due to the inline geometry, and smaller path length which increase the probability of quenching.

Fluorescence requires a source of excitation energy. There are several main types of light sources that are used to excite fluorescent dyes. This section introduces the types of commonly used excitation sources and presents some of the ways that filters can be used to optimize your experimental result. The most popular sources used for exciting fluorescent dyes are broadband sources such as the mercury-arc and tungsten-halogen lamps. These produce white light that has peaks of varying intensity across the spectrum. In contrast, laser excitation sources, which will be described later, offer one or a few well-defined peaks, allowing more selective illumination of your sample. More recently, high-output light emitting diodes, or LEDs, have gained popularity due to their selective wavelengths, low cost and energy consumption, and long lifetime.

When using broadband white light sources it is necessary to filter the desired wavelengths needed for excitation; this is most often done using optical filters. Optical filters can range from simple colored glass to highly engineered interference filters that selectively allow light of certain wavelengths to pass while blocking out undesirable wavelengths. For selective excitation, a filter that transmits a narrow range of wavelengths is typically used. Such a filter is called a band pass excitation filter.

The high intensities and selective wavelengths of lasers make them convenient excitation sources for many dyes. The best performance is achieved when the dye's peak excitation wavelength is close to the wavelength of the laser. Compact violet 405 nm lasers are replacing expensive UV lasers for most biological work. The most commonly used lasers are the 488 nm blue-green argon laser, 543 nm helium-neon green laser and 633 nm helium-neon red laser. Mixed-gas lasers such as the krypton-argon laser can output multiple laser lines and therefore may still require optical filters to achieve selective excitation. While a given dye's excitation maximum may not exactly match the laser's peak wavelength, the high power of the laser can still produce significant fluorescence from the dye when exciting at a suboptimal wavelength. Filters are important for selecting excitation wavelengths. They are also important for isolating the fluorescence emission emanating from the dye of interest. Detecting the fluorescence emission of a sample is complicated by the presence of stray light arising from sources other than the emitting fluorophore—for example, from the excitation source. This stray light must be kept from reaching the light-sensitive detector in order to insure that what the instrument "sees" is due only to the fluorescence of the sample itself. When a single dye is used, a filter that blocks out the excitation light to reduce background noise, but transmits everything else is often a good choice to maximize the signal collected. Such a filter is called a long pass emission filter (Guilbault, 1990; Rendell, 1987; Lakowicz, 2006).

If multiple dyes are used in the sample, a band pass emission filter can be used to isolate the emission from each dye. Careful filter selection helps to ensure that the detector registers only the light you are interested in—the fluorescence emitted from the sample.

LEDs are relatively new light sources for fluorescence excitation. Single-color LEDs are ideal for low-cost instrumentation, where they can be combined with simple long pass filters that block the LED excitation and allow the transmission of the dye signal. However, the range of wavelengths emitted from each LED is still relatively broad. Currently their application may also require the use of a filter to narrow the bandwidth.

There are many options for light sources for fluorescence. Selecting the appropriate light source, and filters for both excitation and emission, can increase the sensitivity of signal detection to an astounding degree. Making fluorescence labeling one of the most sensitive detection technologies available.

With recent advances in sensitive array detectors, fiber optic wave guides, high speed electronics and powerful software, many new generations of spectrofluorometers have been developed. These new spectrofluorometers use charged couple devices (CCDs) or photodiode arrays to replace the photomultipliers and avalanche photodiodes used in conventional spectrometers. They offer excellent performance for a wide range of spectroscopic applications from the UV to the near IR. Because of their unique combination of outstanding sensitivity, high speed, low noise, compactness, instantaneous capture of full spectra, low cost and robustness, these detectors have revolutionized spectroscopic detection. A quick glance at today's instrumentation market indicates the popularity of the CCD as the detector of choice. The overwhelming benefits of array detectors are simultaneous and multi-wavelength data acquisition. On the other hand, the use of fiber optics as light guidance allows a great modularity and flexibility in setting up an optical measurement system. Recent applications and a critical comparison between simple luminescence detectors using a light-emitting diode or a Xe lamp, optical fiber and charge-coupled device, or photomultiplier for determining proteins in capillary electrophoresis are presented by Casado-Terrones et al. (Casado-Terrones, 2007).

In summary, from the preceding discussion, four essential elements of fluorescence detection systems can be identified: (1) an excitation source, (2) a fluorophore, (3) wavelength filters to isolate emission photons from excitation photons and (4) a detector that registers emission photons and produces a recordable output, usually as an electrical signal or a photographic image. Regardless of the application, compatibility of these four elements is essential for optimizing fluorescence detection.

For the sample holders, the majority of fluorescence assays are carried out in solution, the final measurement being made upon the sample contained in a cuvette or in a flowcell.

Cuvettes may be circular, square or rectangular (the latter being uncommon), and must be constructed of a material that will transmit both the incident and emitted light. Square cuvettes or cells will be found to be most precise since the parameters of path length and parallelism are easier to maintain during manufacture. However, round cuvettes are suitable for many more routine applications and have the advantage of being less expensive. The cuvette is placed normal to the incident beam. The resulting fluorescence is given off equally in all directions, and may be collected from either the front surface of the cell, at right angles to the incident beam, or in line with the incident beam. Some instruments will provide the option of choosing which collecting method should be employed, a choice based upon the characteristics of the sample. A very dilute solution will produce fluorescence equally from any point along the path taken by the incident beam through the sample. Under these conditions, the right-angled collection method should be used since it has the benefit of minimizing the effect of light scattering by the solution and cell. This is the usual measuring condition in analytical procedures.

2.3 Sample preparation

Fluorescence is a very sensitive technique. This is the one criterion that makes it a viable replacement to many radioisotope-labeling procedures. However, it is extremely susceptible to interference by contamination of trace levels of organic chemicals. Potential sources of contamination are ubiquitous since any aromatic organic compound can be a possible source of fluorescence signal. For example, the researcher is a possible source of this type of contamination since oils secreted by the skin are fluorescent. Good laboratory procedure is essential in preventing solvents and chemicals from becoming contaminated with high background fluorescence that could hinder low-level measurements. Solvents should be of the highest level purity obtainable commercially. In addition, care must be taken to eliminate all forms of solid interference (suspended particulates such as dust and fibers). These will float in and out of the sampling area of the cuvette via convection currents, and cause false signals due to light scattering while they remain in the instrument's beam.

Fluorescence spectra and quantum yields are generally more dependent on the environment than absorption spectra and extinction coefficients. For example, coupling a single fluorescein label to a protein reduces fluorescein's quantum yields ~60% but only decreases its molar extinction coefficient by ~10%. Interactions either between two adjacent fluorophores or between a fluorophore and other species in the surrounding environment can produce environment-sensitive fluorescence.

Many environmental factors exert influences on fluorescence properties. All fluorophores are subject to intensity variations as a function of temperature, pH of the aqueous medium, and solvents polarity. In general fluorescence intensity decreases with increasing temperature due to increased molecular collisions that occur more frequently at higher temperatures. These collisions bleed energy from the excited state that produces fluorescence. The degree of response of an individual compound to temperature variations is unique to each compound. While many commercially available dyes are selected for their temperature stability, care should be taken to eliminate exposure of samples to drastic temperature changes during measurement. If possible, the temperature of instrument's sample compartment should be regulated via a circulating water bath. At lower assay temperatures, higher fluorescence signal will be generated. It has been found that a 50%

decrease in the fluorescence signal of yellow-green microspheres when exposed to 160°C for 15 minutes (Guilbault, 1990).

Fluorescence variations due to pH changes are caused by the different ionizable chemical species formed by these changes. The results from these pH variations can be quite drastic since new ionization forms of the compound are produced. Fluorescence spectra may be strongly dependent on solvent. This characteristic is most usually observed with fluorophores that have large excited-state dipole moments, resulting in fluorescence spectral shifts to longer wavelengths in polar solvents. As the polarity of environment decreases, the fluorophore shows a shift to longer wavelength with an increase in fluorescence quantum. Also, in polar environments the fluorescence quantum yield decreases with increasing temperature, while in nonpolar environment very little change in the fluorescence quantum yield was observed. Never the less, the environmental sensitivity of a fluorophore can be transformed by structural modifications to achieve desired probe specificity (Guilbault, 1990; Rendell, 1987; Lakowicz, 2006).

In summary, all fluorescence data required for any research project will fall into one of the following categories: (1) the fluorescence emission spectrum, (2) the excitation spectrum of the fluorescence, (3) the quantum yield, and (4) the fluorescence lifetime. In a typical emission spectrum, the excitation wavelength is fixed and the fluorescence intensity versus wavelength is obtained. Early examination of a large number of emission spectra resulted in the formulation of certain general rules: (1) in a pure substance existing in solution in a unique form, the fluorescence spectrum is invariant, remaining the same independent of the excitation wavelength (known as Kasha's rule), (2) the fluorescence spectrum lies at longer wavelengths than the absorption, (3) the fluorescence spectrum is, to a good approximation, a mirror image of the absorption band of least frequency. These general observations follow from consideration of the Jablonski diagram shown earlier. The fluorescence spectrum gives information about processes that happens when the molecules are in the excited stat.

3. Applications

In this section, the applications of fluorescence spectrophotometry as a powerful tool for quantitative analysis, characterization, and quality control in different fields will be reviewed and discussed in details. This section will include the use of fluorescence spectrophotometry as a powerful spectroscopic tool in several fields of science.

3.1 Organic analysis

In the period reviewed so many papers using fluorescence spectrophoometry for analysis, characterization, and as a tool for identification of several compounds are appeared in the literature. In this section, few recent methods will be summarized.

For example, the design and development of artificial molecular systems for sensing anions in biologically relevant conditions is a challenging task in supramolecular chemistry. In particular, sensing fluoride anion has attracted increasing interest in the molecular recognition community because of its pivotal importance in many areas of biological and chemical sciences. In recent years high levels of fluoride in drinking water have caused numerous human diseases, creating a crucial need for artificial sensors to detect fluoride anions in an aqueous environment. Recently, highly sensitive fluorescence "Turn-On" indicators, triisopropylsilyl-protected coumarin derivatives, for fluoride anion with remarkable selectivity in organic and aqueous media have been developed (Sokkalingam & Lee C-H., 2011). This developed method exhibited new fluorescent sensors systems for fluoride anion detection that proved to be simple, inexpensive, and highly selective and, achieve accurate determination with a low detection limit. In addition, the results of this study showed that the system can detect inorganic fluorides as quickly as organic ones by simply introducing chelating agents such as crown ethers. The easily prepared indicator system synthesized here could be an ideal chemodosimeter for detecting and determining fluoride anion in both organic and aqueous solutions and could lead to development of a convenient and reliable detection method for fluoride anion in practical and commercial applications.

Fluorescence is a powerful tool for structural and functional studies of a diversity of molecules. Among the various fluorophores, pyrene derivatives are attractive fluorescent probes. Therefore, pyrene "click" conjugates of 7-deazapurine and 8-aza-7-deazapurine nucleosides and a basic pyrene compound has been synthesized (Ingale et al., 2011). The influence of the nucleobase on fluorescence quenching was studied on nucleoside and oligonucleotide level. This study showed that the favorable photophysical properties of 8-aza-7-deazapurine pyrene conjugates improve the utility of pyrene fluorescence reporters in oligonucleotide sensing as these nucleoside conjugates are not affected by nucleobase induced quenching. This improves the utility of pyrene fluorescence reporters for detection of oligonucleotides.

Selective recognition of Ag^+ ions and amino acids is an important area of research due to their involvement in chemical, biological, and environmental applications. Silver compounds are used as antimicrobial agents, and the activity is closely related with the interaction of Ag^+ with sulfurdryl (−SH) groups. There are several chemosensors reported in the literature for the recognition of Ag^+ ion in nonaqueous and aqueous systems. However, the molecular receptors which can recognize Ag^+ followed by amino acids are indeed limited in the literature. Recently, a new 1,1′-thiobis(2-naphthoxy)-based receptor molecule (L) containing a benzimidazole moiety has been synthesized (Dessingou et al., 2011). The selectivity of L has been explored in aqueous methanol, resulting in selective (7.5 ± 0.5)-fold *switch-on* fluorescence response toward Ag^+ among 14 different transition, alkali, and alkaline earth metal ions studied.

There is a growing interest in the development of molecular sensors that can detect selectively metal ions even in low concentrations. Among the various techniques used for this purpose fluorescence-based methods have gained in importance because of their sensitivity. These methods depend upon the change of fluorescence intensity and/or a shift in the fluorescence band of the sensor upon interaction with the metal ion. Although such a methodology has been successful for diamagnetic metal ions, its application to paramagnetic metal ions is fraught with difficulties in view of the fact that the latter quench fluorescence either via energy or electron transfer and only in some instances a fluorescence enhancement has been observed. In this regard, development of fluorescence-based sensors for Cu(II) has assumed importance in view of the fact that it is an essential trace element and yet at slightly increased concentrations, it is toxic, being implicated in gastrointestinal, liver, and kidney diseases as well as in neurological diseases such as Alzheimer's or Parkinson's. Therefore, Chandrasekhar et al. have demonstrated a simple approach for the design of fluorescence-based sensors (Chandrasekhar et al., 2009). This methodology consists of assembly of phosphorus-supported coordinating platforms whose fluorescence properties are modulated by binding with Cu(II) as well as by the number of coordinating arms that the ligand possesses. We believe that this design is quite general and can be applied for selective detection of other type of metal ions also.

In recent decades, colorimetric and fluorometric sensors have been used in various scientific fields. In biology, for instance, such sensors are useful reagents for living cell imaging. It is important to design novel sensors because they have the potential to overcome many technical limitations in experiments. Imidazo[1,2-a]pyrazin- 3(7H)-one (*imidazopyrazinone*) often is used as a bioluminescent substrate, and it is an attractive core structure for useful sensors. In this regard, a new series of imidazopyrazinones [7-benzylimidazo[1,2 a]pyrazin-3(7H)-one derivatives] have been prepared and their fluorescent properties in the presence of various metal ions (M^{n+}) have been studied (Hirano, et al., 2010).

Copper amine oxidases (CAOs) are a large family of copper-containing quinone-dependent amine oxidases that can be found in all living organisms including bacteria, yeast, plants, and mammals. Human CAOs have been implicated in a number of diseases, including atherosclerosis, cardiovascular diseases, diabetes, Alzheimer's disease, and cancer. To study the kinetic behavior of highly potent inhibitors of the CAO bovine plasma amine oxidase (BPAO), it has been sought a sensitive and real time assay to monitor low levels of enzyme activity. The most sensitive assay for CAOs is the fluorometric coupled assay, which monitors generation of hydrogenperoxide during substrate turnover using horseradish peroxidase (HRP) as a secondary detecting enzyme. In this regard, a novel fluorogenic substrate of bovine plasma amine oxidase (BPAO), namely, (2-(6-(aminomethyl)naphthalen-2-yloxy)ethyl)trimethylammonium (ANETA) has been reported, which displays extremely tight binding to BPAO (K_m 183 (14 nM) and yet is metabolized fairly quickly (k_{cat} 0.690 (0.010 s^{-1}), with the aldehyde turnover product (2-(6-formylnaphthalen-2 yloxy)ethyl)trimethyl-ammonium serving as a real time reporting fluorophore of the enzyme activity (Ling et al., 2009). This allowed for the development of a fluorometric noncoupled assay that is two orders of magnitude more sensitive than the spectrophotometric benzylamine assay. ANETA represents the first highly sensitive, selective, and tight binding fluorogenic substrate of a copper amine oxidase that is able to respond *directly* to the enzyme activity *in real time*.

Organic molecules containing a fluorophoric unit combined with site(s) for guest binding purposes have found application in building up florescent signaling systems for biomedical research and chemical logics. Metal ions can act effectively as guests for these molecules because of their ability to enhance, shift or quench luminescent emissions of these organic ligands by coordination. The changes brought about by metal binding are mechanistically of four types, photoinduced electron transfer (PET), photoinduced charge transfer (PCT), formation of monomer/excimer, energy transfer and proton transfer. Transition metal ions with partially filled d-orbitals are known to induce fluorescence quenching by oxidative or reductive PET and energy transfer processes. In this regard, dioxomolybdenum(VI) complexes [$MoO_2(B^2)H_2O$], [$MoO_2(B^2)EtOH$], [$MoO_2(B^3)EtOH$] and [$MoO_2(B^4)EtOH$] were synthesized using several Schiff base ligands (B^1, B^2, B^3, and B^4) (Gupta et al., 2009). These ligands (B^1, B^2, B^3, and B^4) were prepared by condensation of 1-(2-pyridyl)-5-methyl-3-pyrazole carbohydrazide with salicylaldehyde, o-hydroxy acetophenone, 5-bromo salicylaldehyde and 5-nitro salicylaldehyde, respectively. Due to the presence of a substituted 1-(2-pyridyl) pyrazole unit, these ligands exhibit fluorescent emissions. As the ligands are capable of using different binding modes, according to the demands of the guest metal ions, their emission properties also change accordingly.

3.2 Inorganic analysis

The development of sensors for metal ions in solution has always been of particular importance for cations with biological and environmental interest. The molecular devices converting metal ions recognition in physical recordable signal are continuously growing. In the last few years great attention has been paid to fluorescent chemosensors and many new systems were synthesized. In particular, an effective fluorescent sensor for metal ions is a system able to interact with the metal ion in solution signaling its presence by changing fluorescence properties, as the wavelength or emission intensity, as well as by the appearance of a new fluorescence band different from those of the free sensor. Recently, paper reviews ligand molecules containing fluorophores synthesized and employed in metal ions sensing in solution in the last few years has been published (Formica et al. 2012). The large number of references reported in the review highlights the synthesis of new fluorescent chemosensors able to sense and signal metal ions in solution. This is a prosperous and still emerging field. In this review the authors concluded that the discovery of newer and more efficient emitting units is desirable. In this case, research is now focused on the synthesis of fluorescent units able to shift the emission in an optical range where the biological noise is reduced in the near infrared region (NIR) thus to increase the sensitivity of the chemosensor for in vivo analysis.

Metal–polypyridine complexes are extensively used in photochemical applications, such as solar energy conversion (Akasheh & AI-Rawashdeh, N.A.F, 1990), due to their peculiar excited-state dynamics. These complexes, of which ruthenium tris-bipyridine ($[Ru(bpy)_3]^{2+}$) is considered the prototype, exhibit a visible absorption band due to the singlet metal-to-ligand charge transfer state. The principle of the dye sensitized solar cells is based on the use of such metal-based molecular systems, of which the RuN3 ($[Ru(dcbpyH_2)_2(NCS)_2]$) dye is the most popular, adsorbed onto a semiconductor substrate (usually TiO_2). In this regards, femtosecond-resolved broadband fluorescence studies are recently reported for $[M(bpy)_3]^{2+}$ (M=Fe, Ru), RuN3 and RuN719 ($(Bu_4N)_2[Ru(dcbpyH)_2(NCS)_2]^{2-}$) complexes in solution (Bram et al. 2011). In this study, the pump wavelength dependence of the fluorescence of aqueous $[Fe(bpy)_3]^{2+}$ and the solvent and ligand dependence of the fluorescence of Ru-complexes excited at 400 nm have been investigated. The RuN3 and RuN719 are asymmetric complexes contrary to $[Ru(bpy)_3]^{2+}$, which allows us to explore the effects of molecular geometry on the ultrafast relaxation dynamics of this class of molecules.

Ru(II) polypyridyl complexes have been widely used as DNA, cation, and anion sensors, because their outstanding photophysical and electrochemical properties are quite sensitive to external stimuli (Schmittel, 2007, as cited in Cheng, 2010). Transfer of protons can be regarded as one of the simplest external stimuli and can induce the switching of properties such as fluorescence and UV–Vis absorption for pH sensors, so some Ru(II) polypyridyl complexes containing imidazole fragment have been synthesized. Imidazole-containing ligands are poor π-acceptors and good π-donors and have the appreciable ability to control orbital energies by proton transfer compared with pyridine-, pyrazine-, and pyrimidine-containing ligands, but in most cases these complexes with imidazole rings coordinated to the metal center are nonemissive or weakly emissive and only display deprotonating process [Ayers, 2002, as cited in Cheng, 2010). In this regards, tripodal ligands 1,3,5-tris{4-((1,10-phenanthroline-[5,6-d]imidazol-2-yl)phenoxy)methyl}-2,4,6-trimethylbenzene (L1), 1,1,1 tris{4-((1,10-phenanthroline-[5,6-d]imidazol-2-yl)phenoxy)methyl}propane (L2), 2,2',2''-tris{4-((1,10-phenanthroline-[5,6-d]imidazol-2-yl)phenoxy)ethyl}amine (L3), and corresponding Ru(II) complexes $[(bpy)_6L(Ru(II))_3](PF_6)_6$, have been synthesized (Cheng et al., 2010) . This study showed that the fluorescence spectra of these complexes are strongly dependent on the pH of the buffer solution. These complexes act as pH-induced off–on–off fluorescence switch through protonation and deprotonation of the imidazole-containing ligands. For the same purpose, two novel tetrapodal ligands tetrakis{4-((1,10-phenanthroline-[5,6-d]imidazol-2-yl)phenoxy)methyl}methane (L1), tetrakis {3-((1,10-phenanthroline-[5,6-d]imidazol-2-yl)phenoxy)methyl}methane (L2), and corresponding Ru(II) complexes $[(bpy)_8L(Ru(II))_4](PF_6)_8$ have been synthesized (Cheng et al., 2011). The two complexes act as off–on–off fluorescence pH switches with a maximum on–off ratio of 5. This on–off ratio is moderate compared with those reported for imidazole-containing Ru(II) complexes (Cheng et al., 2010). They have potential utility to detect pH variations of external environment due to their interesting photon-dependent photophysical properties.

Crystal engineering of coordination polymers and supramolecules have attracted lot of attention due to their potential applications as functional materials, as well as their fascinating architectures and topologies (Moulton & Zaworotko, 2010, as cited in Leong, 2009). A successful strategy in the construction of such networks is to utilize appropriate multidentate ligands with flexible backbone that are capable of binding metal ions in different modes. In this regards, a series of metal (Cu(II), Ni(II), Mn(II), Zn(II), Mg(II), Ca(II), and Al(III)) complexes containing the 4-methylumbelliferone-8-methyleneiminodiacetic acid (H_3muia, also named as Calcein Blue) has been synthesized (Leong et al., 2009). In this study, the fluorescence spectroscopy has been used to investigate the hydrogen bonding interactions along with π-π stacking in the synthesized complexes solid-state structures. Solid-state fluorescence studies indicate that complexes of muia have the similar emission properties as in the solution state. Transition metal ions quench the fluorescence of muia while alkali earth and post-transition metal complexes of muia exhibit strong blue emission.

A considerable number of papers have focused on the use of anthracene containing compounds as protein photo cleavers (Hasewage et al., 2006, as cited in Oliveria, 2007), organic light-emitting diodes and materials (Jou, et al., 2006, as cited in Oliveria, 2007), crystal engineering , molecular imprinted polymers, sensors and chemosensors (Magri, et al., 2005, as cited in Oliveria, 2007). Anthracene is one of the most employed chromophores due to its ability to induce PET (photoinduced electron transfer) processes. In this regard, a new scorpionate system (L) containing an emissive anthracene pendant arm, derived of O^1,O^7-bis(2-formylphenyl)-1,4,7-trioxaheptane and tren, has been synthesized (Oliveira et al., 2007). The fluorescence spectroscopy studies conducted suggest that this ligand is an effective complexation molecular device for several divalent metal ions of biological importance as well as for Al(III) and Cr(III), both metals of great relevance in medicine and environmental chemistry. The results of this study could be used as a starting point to develop a more efficiently fluorescence chemosensor based on macrocyclic ligands for these metals.

Since some of lanthanide ions, especially Eu^{3+} and Tb^{3+}, posses good luminescence characteristics (high color purity) based on the 4 f electronic transitions, a variety of rare earth compounds activated by Eu^{3+} and Tb^{3+} have been studied for practical applications (Zhang, et al., 1999, as cited in Xu, et al., 2004). Polydimethylsiloxane (PDMS) materials have broad application in a variety of industrial area so their well-established surface modifying properties. In this regard, Eu(III)-containing polymer complex was synthesized, in which polydimethylsiloxane was used as a polymer ligand (Xu, et al., 2004). The result of this study showed that the fluorescence intensity change of Eu(III)-PDMS complex displays typical fluorescent concentration quenching behavior, in which the emission intensity of the complex was enhanced with increasing content of Eu (III) ion and reaches a maximum at 2 wt.%, and then decreased with further increasing content of Eu(III) ion. These results indicate that the complex contains ionic aggregates in which Eu(III) ions are located close together. Eu(III)-PDMS could transform harmful ultraviolet radiation to blue luminescence that is needed by plants effectively. So Eu(III)-PDMS has a great application foreground.

Schiff base ligands have been extensively studied in coordination chemistry mainly due to their facile syntheses, easily tunable steric, electronic properties and good solubility in common solvents. Transition metal complexes with oxygen and nitrogen donor Schiff bases are of particular interest (You et al., 2004) because of their ability to possess unusual configurations, be structurally labile and their sensitivity to molecular environments. In this regard, the preparation and structures of nickel(II) zinc(II) and cadmium(II) complexes with the related Schiff base ligand N-2-pyridylmethylidene-2- hydroxy-phenylamine have been investigated (Majumder et al. 2006). The results of this study showed that the complexes with zinc(II) and cadmium(II) metals can serve as potential photoactive materials as indicated from their characteristic fluorescence properties. Since the complexes with zinc(II) and cadmium(II) metals possess an intense fluorescence property at room temperature, which is not observed for complexes with Ni(II). It is suggested that complexes with Zn(II) and Cd(II) exhibit potential applications as photoactive materials.

3.3 Pharmaceutical analysis

Over the last three decades, wide ranges of thermodynamic data concerning the guest host complexation have been reported. Cyclodextrins (CDs) are cyclic oligosaccharides composed of glucopyranose units linked together via oxygen bridges at the 1 and 4 positions (a-(l,4)-glycoside bonds) (Bender& Komiyama, 1978). This class of organized media possesses a hydrophilic upper and lower rims lined with hydroxyl groups and a hydrophobic cavity due to C^3H, C^5H, and C^6H hydrogen's and O^4 ether oxygen. This structure gives CDs the ability to extract a variety of organic guest molecules of appropriate size and hydrophobicity from the bulk aqueous solution (Szejtli, 1988). Complexation of various guest compounds with CDs generally results in the improvement of some physical properties of the guest molecules, such as stability, bioavailability, membrane permeability, and solubility. In luminescence studies, CDs have been employed to enhance fluorescence emission of different luminophors (Bender & Komiyama, 1978; Szejtli, 1988) and to induce room temperature phosphorescence under appropriate conditions. In this regards, there is a lot of literature reported in several journals of interest, as an example, the inclusion of the anti-inflammatory drug, Nabumetone (NAB), in γ-cyclodextrin (γ-CD) was studied by fluorescence measurements (Al-Rawashdeh. N.A.F., 2005). Nabumetone is poorly soluble in water and exhibits intrinsic fluorescence. The results of this study showed that the emission fluorescence spectrum, of NAB reveals a maximum whose intensity increases with the different γ-CD's growing concentrations. It is noteworthy mentioning that significant alterations in the physicochemical properties of the included molecule (NAB) are observed upon forming the inclusion complex with γ-cyclodextrin, such as stability and solubility in aqueous media.

Imidazoline-derived drugs are a family of drugs that is structurally distinguished by the existence of the heterocyclic ring of imidazoline that enables these drugs to interact with adrenergic receptors via stimulating presynaptic and postsynaptic a-adrenoceptors (Parini et al., 1996). The majority of the imidazoline-derived drugs are frequently used for their agonist activity, whereas others are used for either antihypertensive, antihistaminic, or agonistic activity (Kaliszan et al., 2006). However, the principle pharmaceutical applicability of these drugs is due to their vascoconstrictive effects. Three typical imidazoline-derived drugs were selected for this study; Naphazoline (NP) (4,5-Dihydro-2-(1-naphthalenylmethyl-1H-imidazole), Antazoline (AN) (4,5-Dihydro-Nphenyl-N-(phenylmethyl)-1H-imidazole-2-methanamine, 2-[(N-benzyl anilino) methyl]-2-imidazole, and Xylometazoline (XM) (2-(4-tert-Butyl-2,6-dimethylbenzyl)-2-imidazoline). While CDs have a wide range of applications, using CDs as additives in various separation and pharmaceutical sciences is still the foremost application of them. Hence, adding the CDs to the separation media can significantly enhance the separation process, whereas employing CDs as additives to drugs formulation can promote the bioavailability through enhancing the stability and solubility of selected drugs. In this regard, the inclusion complexes of selected imidazoline-derived drugs, namely Antazoline (AN), Naphazoline (NP) and Xylometazoline (XM) with β-cyclodextrin (β-CD) were investigated using steady-state fluorescence (Dawoud & Al-Rawashdeh, N. 2008)). Their results confirmed the formation of the inclusion complexes between the three studied drugs and β-cyclodextrin using steady-state fluorescence spectroscopy. Importantly, the results of their study showed that the geometrical size and polarity of various substituents, such as phenyl and naphthyl groups, have dramatically altered the stability and geometrical configuration of the inclusion complexes. These studies present the state-of-the-art of macromolecular binders and provide detailed illustrative examples of recent developments bearing much promise for future pharmaceutical applications.

Propranolol is a beta-adrenergic blocking drug widely prescribed for the treatment of cardiac arrhythmia, sinus tachycardia, angina pectoris and hypertension (Parfitt, 1990). It has also been suggested for use in a number of other conditions including dysfunctional labour and anxiety. When administered over a long period of time it reduces mortality caused by hypertension and lengthens survival in patients with coronary heart disease. It is also used in low activity sports, reducing cardiac frequency, contraction force and coronary flow. Therefore, it has been included in the list of forbidden substances by the International Olympic Committee. The Spanish Olympic Committee has decided that only a qualitative determination of propranolol in urine is necessary. In this regards, a sensitive fluorescence optosensor for the drug propranolol to use in the analysis of pharmaceutical preparations and as a doping test for the qualitative analysis of propranolol in human urine without lengthy preliminary procedures have been developed (Fernandez-Sanchez et al. 2003).The effect of proteins presents in urine samples were evaluated using the developed flow-through fluorescence optosensor. The results of this study showed that the proposed methods for analysis were satisfactorily applied to commercial formulations and urine samples, which offers excellent analytical parameters, such as sensitivity, selectivity, versatility, and ease of use. The development of optosensing techniques has led to a shorter turnaround analysis time and reduced costs for doping controls. As a large part of the samples prove to be non-doped, rapid analytical methods such as doping tests that provide reliable 'yes/no' responses are of increasing interest. These tests can usually be described as systems that 'filter' samples to select those with analyte content levels 'similar to' or 'higher than' a previously established threshold. These 'probably doped' samples must then be examined with more exact instrumental methods. Doping tests can significantly cut costs and save time.

Anthraquinones are known to be present in many different families such as Polygonaceae, Leguminosae, Rubiaceae, Liliaceae and Rhamnaceae. Recently, a number of pharmacological tests revealed that the anthraquinone derivatives present various biological activities including antifungal (S K. Agarwal, 2000, as cited in He, 2009), antimicrobial (Y. W. Wu, as cited in He, 2009), anticancer (J. Koyamma, as cited in He, 2009) , antioxidant (G. C. Yen, as cited in He, 2009), and antihuman cytomegalovirus activity (D. I. Barnard. As cited in He, 2009). In general, fluorescence detection is sensitive and selective. The five anthraquinones are known to possess natural fluorescence, but it is difficult to analyse and determine their contents by conventional fluorimetry due to their similar molecule structures. Therefore, recently a simple, rapid and sensitive reversed-phase high-performance liquid chromatography (RP-HPLC) method, using fluorescence detection, to simultaneously quantify aloe-emodin, emodin, rhein, chrysophanol and physcion in medicinal plants and their pharmaceutical preparations was developed by He et al. (He et al., 2009). Their method was suitable for use as a tool for routine quality assurance and standardization of the anthraquinone from the raw material and commercially available pharmaceutical preparations containing rhubarb.

Fluorimetry from the early fifties has been among the most frequently used techniques for determining both therapeutically and abuse drugs; probably due to its excellent selectivity and the low detection limits. Also, the fluorimetric technique is the recommended choice for quantifying the purity of active principles. The research on analytical fluorescence applications is in continuous expansion to new automated or semi-automate processes like classic or emergent methodologies on the continuous-flow field. In this way, fluorescence-based methods have found a wide range of analytical applications (Calatayud & Zamora, 1995)). In this regard, a new strategic tool to predict the native fluorescence of organic molecules has been proposed (Albert-Garcia et al. 2009). For this purpose, the molecular connectivity indices of different organic substances (pharmaceuticals and pesticides) were calculated. The work presented in this paper was focused to present a new tool for enhancing the research yield on new analytical applications of fluorescence.

3.4 Biological and biomedical analysis

Cancer has overtaken heart disease as the world's top killer by 2011, part of a trend that should be more than double global cancer cases and deaths by 2030. Cisplatin (cis diamminedichloroplatinum(II)) is one of the most effective anticancer drugs in the treatment of a variety of tumors and it has been clinically used widely. However, its limited usefulness in the development of resistance in tumor cells and the significant side effects have generated new areas of research, which mainly focused on searching for new metal-based complexes with low toxicity and improved therapeutic properties. In this regards, 3-Carbaldehyde chromone thiosemicarbazone and its transition metal (Cu(II), Zn(II) and Ni(II)) complexes were synthesized and characterized systematically (Li et al., 2010). The results of this study showed that the Zn(II) complex can emit blue fluorescence under UV light in solid state and may be used as an advanced material for blue light emitting dioxide devices. However, almost the solid fluorescence of Cu(II) and Ni(II) complexes could not be observed. Thus, the fluorescence property of these complexes was used as a helpful tool to understand the interaction mechanism of small molecule compounds binding to DNA. It was believed that the information obtained from the present work would be useful to develop new potential antioxidants and therapeutic agents for some diseases.

The quantitative determination of micro-amounts of nucleic acid has attracted a great deal of attention in the fields of medicine and molecular biology. Many methods have been developed, such as direct determination, including ultraviolet absorption and determination of ribose or deoxyribose in nucleic acid, spectrophotometry, chemiluminescence, electrochemical chromatography, including high-performance liquid chromatography and paper chromatography, capillary electrophoresis, and resonance light scattering. However, low sensitivity and easy interruption by protein and other biomolecules existed in these methods in common. However, the fluorometric methods make predominant concern because of their high sensitivity and selectivity. Generally, the fluorescence intensity of DNA must be enhanced by fluorescent probes because it emits weak fluorescence itself. In this regard, method for the determination of DNA based on the fluorescence intensity of the gatifloxacin-europium(III) (GFLX-Eu^{3+}) complex that could be enhanced by DNA was developed (Wang et al. 2011). The GFLX-Eu^{3+} complex showed an up to 6-fold enhancement of luminescence intensity after adding DNA. On the basis of the above findings, the fluorescence enhancement effect of the GFLX-Eu^{3+} complex by DNA was investigated in this study in detail.

The green fluorescent proteins (GFPs) originated from the bioluminescent jellyfish Aequorea victoria, were discovered by Shimomura in the early 1960s (Shimomura et al., 1962). In the last few years, green fluorescent protein (GFP) has become one of the most widely used tools in molecular and cell biology. As a noninvasive fluorescent marker in living cells, GFP allows for numerous applications where it functions as a probe of gene expression, intercellular tracer or as a measure of protein-protein interactions. The green fluorescent protein (GFP) has emerged as a powerful reporter molecule for monitoring gene expression, protein localization, and protein–protein interaction. However, the detection of low concentrations of GFPs is limited by the weakness of the fluorescent signal and the low photostability. Recently, the proximity of single GFPs to metallic silver nanoparticles increases its fluorescence intensity approximately 6-fold and the decrease in decay time has been observed by Fu et al. (Fu et al., 2008). Furthermore, single protein molecules on the silvered surfaces emitted 10-fold more photons as compared to glass prior to photobleaching. The photostability of single GFP has increased to some extent. Accordingly, longer duration time and suppressed blinking were observed. The single-molecule lifetime histograms indicate the relatively heterogeneous distributions of protein mutants inside the structure.

Detection of DNA in solution is an important problem in a large variety of biochemical assays. The most popular agents for DNA quantitation are fluorescent dyes that strongly interact with nucleic acids and significantly increase their emission intensity in the DNA complex. Fluorescent dyes are used in real-time PCR, DNA-based cell quantitation, gel staining, chromatin and other DNA-based approaches (Glazer and Rye, 1992, Jing et al., 2003, Lakowicz, 2006, Le Pecq and Paoletti, 1967, Lim et al., 1997, Szpechcinski et al., 2008, as cited in Dragan, 2010). In this regard, both a theoretical and experimental analysis of the sensitivity of a DNA quantitation assay using a fluorescent chromophore which non-covalently binds dsDNA were investigated (Dragan et al. 2010). It is well-known that the range of DNA concentrations available for fluorescence quantitation depends on the concentration of the chromophore, its affinity for nucleic acids, the binding site size on DNA and the ratio between the fluorescence intensity of the chromophore when bound to DNA compared to free chromophore in solution. In this study an experimental data obtained for a PicoGreen® (PG)/DNA quantitation assay was presented, which is in complete agreement with the results of our theoretical analysis. Furthermore, it has been shown, both theoretically and experimentally, that DNA assays based on the MEF of PG demonstrate sensitivity to DNA concentration of ≈1 pg/ml, which is several orders of magnitude more sensitive than without the silver nanoparticles, suggesting the broader practical use of this approach (metal-enhanced PicoGreen® fluorescence) for the ultra-sensitive detection of double stranded nucleic acids.

Identification of living organisms (eukaryotes, bacteria, viruses etc.) by means of quantitative analysis of their specific DNA sequences, which represent different genomes, is a challenging aim, which faces many scientists today. It also concerns the search and detection of different microorganism mutations and strains of pathogenic bacteria and causes of severe diseases in humans. In recent short communication, further development of the "Catch and Signal" technology—principles of a 2-color DNA assay for the simultaneous detection/quantification of two genome-specific DNAs in one well was presented (Dragan et al., 2011) In this method a combination of the Metal-Enhanced Fluorescence (MEF) effect and microwave-accelerated DNA hybridization has been utilized. Furthermore, it is shown that fluorescent labels (Alexa 488 and Alexa 594), covalently attached to ssDNA fragments, play the role of biosensor recognition probes, demonstrating strong response upon DNA hybridization, locating fluorophores in close proximity to silver nanoparticles, which is ideal for MEF. The 2-color "Catch and Signal" DNA assay platform can radically expedite quantitative analysis of genome DNA sequences, creating a simple and fast bio-medical platform for nucleic acid analysis. Their results clearly showed that the 2-color DNA assay can effectively be employed as a new "Rapid Catch and Signal" technological platform in the creation of an ultra-sensitive, sequence-specific approach for the fast analysis of genetic material from different organisms, for potential analysis of bacteria and virus pathogens, and search for possible mutations and sequence variations. This technology being fast, ultra-sensitive and inexpensive can effectively compete with the PCR technique, especially for the routine and rapid analysis in Point-of-Care settings and bio-medical laboratories.

3.5 Environmental analysis

For the past 20 years there has been continued growth in the applications of fluorescence spectroscopy in physical and biological sciences. Because of the sensitivity of fluorescence detection, the fluorometeric method has been one of the selected techniques to determine compounds at low concentrations (Lakowicz, 2006). Examples of such compounds are agrochemicals (pesticides, fungicides, insecticides, etc), which have attracted attention worldwide due to their usage in agriculture. Therefore, the potential increase in fluorescence of a benzimidazole-type fungicide (carbendazim) due to complexation with cucurbit[6]uril was reported (Saleh & Al-Rawashdeh, N.A.F., 2006). The fluorescence enhancement of the fungicide carbendazim by cucurbit[6]uril has been observed in solution due to formation of a host-guest inclusion complex. The enhancement of the carbendazim fluorescence (maximum factor of 10) is accompanied by a significant blue shift in the spectrum (11 nm). In general, Saleh and Al-Rawashdeh work (Saleh & Al-Rawashdeh, N.A.F., 2006) demonstrates the potential usefulness of cucurbit[6]uril in fluorometric analysis of fungicide for agricultural and environmental applications.

Recycling in agriculture of organic wastes produced by various animal breeding, such as pig slurry (PS), is a common practice throughout the world, which has recently raised serious environmental concerns. In particular, Cu(II) and Zn(II) ions, which are used abundantly as pig feed additives, may be introduced in relatively large amounts into PS-amended soils, thus representing an actual risk of phytotoxicity and/or leaching downward the soil with potential endangering groundwater quality (L'Herroux et al., 1997, Saviozzi et al., 1997, Giusquiani et al., 1998, Nicholson et al., 1999, Aldrich et al., 2002, Taboada-Castro et al., 2002, as cited in Hernandez, at al. 2006).

Bioavailability, mobility and transport of metal ions in soils are strongly influenced by binding reactions with soil organic matter, and especially its humified fractions, i.e., humic substances (HS), of which humic acid (HA) is the major component (Aldrich et al., 2002, as cited in Hernandez, at al. 2006). For these reasons, the effects of PS application on the compositional, structural and functional properties of native soil HA, and especially on their Cu(II) and Zn(II) binding behavior, need to be accurately evaluated. Therefore, The effect of the consecutive annual additions of pig slurry on the Cu(II) and Zn(II) binding behavior of soil HAs was investigated in a field experiment by a fluorescence titration method (Hernandez et al. 2006). In Hernandez et al. study (Hernandez et al. 2006), a fluorescence titration method was used for determining metal ion complexing capacities and stability constants of metal ion complexes of humic acid isolated from pig slurry and unamended and amended soils. The results of this study is expected to have a large impact on bioavailability, mobilization, and transport of Cu(II) and Zn(II) ions in pig slurry-amended soils (Hernandez et al. 2006).

Moreover, a luminescent sensor utilizing the substrate 2,6pyridine-dicarbox-aldehydebis-(*o*-hydroxyphenylimine) has been developed for low concentration detection of the environmental mercuric ion pollutants (Kanan et. al 2009). The sensor selectively detects mercury in the presence of several ions that are commonly found in aquatic environments. The sensor was found to be highly selective with strongest binding was observed between the mercuric ions and the substrate at a pH range of 6.5-7.5 which makes the substrate a distinctive fluorescent sensor for detecting mercury under normal environmental conditions. No effect was demonstrated with the addition of any other metal ions commonly found in water, making this compound selective to mercuric ions.

Dissolved organic carbon (DOC) refers to the hundreds of dissolved compounds found in water that derive from organic materials, and is composed of 'organic acids', 'organic bases', and 'neutral groups'. The amount of DOC in the hydrosphere (700 gigatons) is almost the same as the amount of carbon in the atmosphere (750 gigatons). However, a large proportion of rainwater DOC is still uncharacterized. A novel method for qualitatively characterizing DOC compounds in rainfall is to use fluorescence. A large proportion of DOC is fluorescent and this fraction can be used to characterize DOC compounds (Baker & Spencer, 2004, as cited in Muller, 2008), yet its full potential for analyzing rainwater DOC compounds has yet to be achieved. Fluorescence can be used to fingerprint fluorophores, identify DOC compounds, and detect small concentration levels which may have otherwise gone undetected, with just 0.04 ml samples. Given the importance of better characterization of rainwater DOC, and the presence of fluorescent DOC compounds, using fluorescence spectrophotometry, primarily sought to examine the fluorescent DOC compounds present in precipitation in Birmingham, UK has been investigated (Catherine et al., 2008). Furthermore, how the fluorescence of the identified DOC compounds varies with meteorological parameters, by assessing the variations with stratiform/convective storm types; examining the variations with air mass type and source area using back-trajectory analysis; and investigating variations with wind speed and wind direction have been investigated (). The results of this study demonstrate the utility of fluorescence analysis for identifying and characterizing rainwater DOC compounds. This research has revealed information regarding fluorescent rainwater DOC compounds in Birmingham, UK and provides evidence in support of using fluorescence spectrophotometry as a means of qualitatively characterizing rainfall DOC.

Organic pollution originating from industrial, agricultural and municipal wastewater discharge has become a common environmental problem in many lakes, rivers, estuaries and coastal waters. The commonly-used methods to study the composition of dissolved organic matter (DOM) (nuclear magnetic resonance, gas chromatography-mass spectrometry, liquid chromatography-mass spectrometry, size exclusion chromatography, etc.) need complicated sample pretreatment procedures and are not suitable for on-line or real-time determination. Excitation-emission matrix (EEM) fluorescence spectroscopy has been suggested as a powerful tool to characterize aquatic DOM which could detect the specific fluorescent fractions in organic matter from different sources with much higher sensitivity (Coble et al., 1996, as cited in Guo, 2010). Therefore, wastewater dissolved organic matter (DOM) from different processing stages of a sewage treatment plant in Xiamen was characterized using fluorescence spectroscopy by Gue at al. (Guo et al., 2010). The results of this study showed that excitation emission matrix fluorescence of wastewater DOM from a sewage treatment plant revealed high concentration of degradable protein-like fluorescence and the occurrence of xenobiotic-like component. This technique provides a fast and sensitive way to monitor the qualitative and quantitative variation of DOM during the whole sewage treatment process.

Perylene-3,4:9,10-tetracarboxylic bisimide, so-called perylene bisimide (PBI) is known as red vat dyes (*fluorescent dyes*), and has been applied to pigments such as automotive finishes due to its light-fastness, thermal stability, and chemical inertness . Perylene bisimide (PBI) is a fluorescent dye which has strong emission and high photostability. Although PBI has been widely used for industrial materials, the application of PBI in analytical fields was limited mainly due to its high hydrophobicity. In recent years, however, unique and useful analytical methods based on PBI platform are being successfully developed by utilizing the characteristic features of this compound including its high hydrophobicity. In this regard, the recent trend of environmental and biological analysis using PBI was reviewed by Soh &

Ueda (Soh & Ueda 2011). Furthermore, the analytical methods presented in this study are classified based on the detection mechanisms.

3.6 Food analysis

Fluorescence spectroscopy is an instrumental technique whose theory and methodology has been widely exploited for studies of molecular structure and function. It is therefore applicable to address molecular problems in food. Since many foods exhibit auto fluorescence it is considered highly relevant for characterization purposes and high-throughput screening. The main advantages of molecular fluorescence spectroscopy are its sensitivity and selectivity, in addition to its ease of use, instrumental versatility, speed of analysis and its non-destructive character.

Recently, it has been shown that excitation-emission matrix (EEM) fluorescence spectroscopy and three-way statistical methods, concretely Parallel Factor Analysis (PARAFAC) can be used for distinguishing between wine samples of different appellations, or between wine samples obtained with different ageing procedure (Airado-Rodrıguez et al., 2011). At first, excitation-emission matrices (EEMs) were obtained and examined with the aim of identifying the main families of fluorescent compounds. Then PARAFAC was applied for exploratory analysis and the scores of the four selected components were obtained. The set of sample score values obtained in the PARAFAC decomposition were plotted against each other, to visualize clustering trends of samples belonging to different appellations. The potential of the autofluorescence of wine, through its Excitation-emission matrices (EEMs), combined with the three-way method PARAFAC for the purpose of discrimination of wine according to the appellation of origin, is shown in this paper. Wine samples were monitored with rapid and non-destructive front face fluorescence spectroscopy, yielding information about the wine throughout its appellation and ageing conditions. PARAFAC revealed four groups of fluorophores that were responsible for the main fluorescence of Spanish wines. Two of them were assigned to benzoic-like phenolic acids and phenolic aldehydes, and to monomeric catequins and polymeric proanthocyanidin dimers, respectively. The exploration of the matrix of score values obtained by PARAFAC reveals some distribution trends of the samples according with their appellation.

Rapid measurements of milk properties and discrimination of milk origin are necessary techniques for quality control of milk products. A study was undertaken to evaluate the potential of using front face fluorescence spectroscopy (FFFS) and synchronous fluorescence spectroscopy (SFS) for monitoring the quality of forty-five ewe's milk samples originating from different feeding systems by Hammamia et al. (Hammamia et al., 2010). Whereas, the physicochemical analyses and fluorescence spectra were conducted on samples during lactation periods (the first 11 weeks). Results obtained showed a good discrimination among milk samples with regard to feeding systems given to the ewes throughout the lactation periods. In addition, a better discrimination was observed with front face fluorescence spectroscopy than with synchronous fluorescence spectroscopy.

Fumonisins (FBs) are worldwide distributed and produced by Fusarium verticillioides and Fusarium proliferatum, mainly in corn and corn-based products (Soriano & Dragacci, 2007, as cited in Silva et al., 2009). Although several other fumonisin analogues have been characterized, fumonisin B1 (FB1) remains the most abundant in naturally contaminated

Corn-based foods followed by fumonisin B2 (FB2).The problems and risks associated with fumonisin contamination have resulted in the development of precise, reliable and sensitive methods for its determination in corn and corn-based foods (Magan & Olsen, 2004, as cited in Silva et al., 2009). Therefore, the quality parameters in the analysis of FB1 and FB2 in corn-based products obtained with LC with fluorescence detector have been investigated (Silva et al., 2009). Furthermore, a comparison study between fluorescence detector (FD), mass spectrometry, and tandem mass spectrometry with a triple quadrupole (QqQ) analyzer using an electrospray ionization interface for the determination of fumonisin B1 and B2 in corn-based products has been performed. A comparative study of the three LC detectors, FD, single quadrupole, QqQ for the analysis of fumonisins in corn samples has been performed. The response achieved by the three detectors was sensitive enough to study the maximum contents established by the EU legislation. These LC detectors would be appropriate for quantification purposes but the acquisition of at least two transitions achieved with QqQ provided a univocal identification.

Low-molecular-weight compounds have been used widely as animal drugs, food additives, and pesticides, to achieve maximum productivity and profits directly or indirectly through food products. However, residues of such low-molecular-weight compounds in food products have been proven to be detrimental to human health. Therefore, development of a microsphere-based competitive fluorescence immunoassay for the determination of hazardous low-molecular-weight compounds in food has been described (Zou et al. 2008). In this method, antigens are covalently bound to carboxy-modified microspheres to compete monoclonal antibody with low-molecular- weight compounds in food samples; mouse IgG/fluorescein isothiocyanate conjugate is used as the fluorescent molecular probe. Thus, the hazardous low-molecular-weight compounds are quantified using a multiparameter flow cytometer. This method has been evaluated using clenbuterol as a model compound. It has a sensitivity of 0.01 ng/mL with dynamic range of 0.01–100 ng/ mL, and the concentration of clenbuterol providing 50% inhibition (IC50) is 1.1 ng/mL. The main advantages of this method are its high efficiency, biocompatibility, and selectivity, as well as ultralow trace sample consumption and low cost.

The aspects of fluorescence spectroscopy that may have value for solving problems in food science and technology have been summarized in a review article by Strasburg & Ludescher (Strasburg et al., 1995). In this review article, the techniques described, which depend on the measurement of the intensity, energy and polarization of fluorescence emission, have been illustrated by examples taken from the food science and related literature.

3.7 Optical sensors

Glucose is considered as a major component of animal and plant carbohydrates in biological systems. It acts not only as a source of energy of the living cells but also as metabolic intermediate in the synthesis of other complex molecules. Furthermore, blood glucose levels are also an indicator of human health conditions: the abnormal amount of glucose provides significant information of many diseases such as diabetes or hypoglycemia. Accurate determination of glucose is very important in clinical diagnosing as well as in food analysis. To date, various sensors for glucose analysis have been reported, and among them, fluorophotometry was used widely owing to its operational simplicity and high sensitivity (Shang et al., 2008, Li et al., 2009, Shiang et al., 2009, as cited in Jin et al., 2011). Recent advances in the noble metal clusters open a promising field toward the development of a satisfying fluorescence probe. In this regard, recently biomolecule-stabilized Au nanoclusters were demonstrated as a novel fluorescence probe for sensitive and selective detection of glucose (Jin et al., 2011). The fluorescence of Au nanoclusters was found to be quenched effectively by the enzymatically generated hydrogen peroxide (H_2O_2). By virtue of the specific response, the present assay allowed for the selective determination of glucose in the range of 1.0×10^{-5} M to 0.5×10^{-3} M with a detection limit of 5.0×10^{-6} M. In addition, it has been demonstrated the application of the present approach in real serum samples, which suggested its great potential for diagnostic purposes. In comparison with previous approaches for glucose detection, this method required no complicated preparation procedure, and used only commercially available materials. It also exhibited environmentally friendly feature and good sensitivity. Furthermore, the present nanosensor possessed red emission and excellent biocompatibility, which presage more opportunities for studying the biological systems in future applications.

Recently, metalloprotein design and semiconductor nanoparticles have been combined to generate a reagent for selective fluorescence imaging of Pb^{2+} ions in the presence red blood cells (Shete et al., 2009). A biosensor system based on semiconductor nanoparticles provides the photonic properties for small molecule measurement in and around red blood cells. Metalloprotein design was used to generate a Pb^{2+} ion selective receptor from a protein that is structurally homologous to a protein used previously in this biosensing system. This designed protein demonstrates a highly sensitive and selective biosensor that can reversibly detect Pb^{2+} ions in aqueous solutions. The modularity of semiconductor nanoparticle-based biosensors has allowed metalloprotein design and different semiconductor materials to be combined to significantly improve the detection of soluble, exchangeable Pb^{2+} ion concentrations to address inefficient Pb^{2+} ion chelation therapy.

The development of artificial receptors for molecular recognition studies of zwitterion amino acids under the physiological conditions is a very important research area since it can help to understand the important roles of free amino acids in biological systems. In this regard, a new fluorescence macrocyclic receptor based on the Zn(II) complex of a C_2-terpyridine and a crown ether has been developed for molecular recognition of zwitterion amino acids in water/DMF solution with remarkable selectivity towards L-aspartate ($K = 4.5x10^4\ M^{-1}$) and L-cysteine ($K = 2.5x10^4\ M^{-1}$) (Kwong et al., 2009).

Copper is an essential trace element, its deficiency is one of the causes of anemia, but it is toxic at higher concentration levels. The uptake of copper by human beings above a certain level is known to cause gastrointestinal catarrh, Wilson's disease, hypoglycemia, and dyslexia. Increases in copper concentration in water and plants have resulted from industrial and domestic waste discharge, refineries, disposal of mining washings, and the use of copper as a base compound for antifouling paints. Therefore, the trace copper content in water and food must be monitored on a daily basis. In this regard, a highly sensitive and selective optical sensor for the determination of trace amounts of Cu^{2+} based on fluorescence quenching has been developed (Aksuner et al., 2009). The sensing membrane was prepared by immobilization of a novel fluorescent Schiff base ligand 4-(1-phenyl-1-methylcyclobutane 3-yl)-2-(2-hydroxy-5-romobenzylidene) aminothiazole, on polyvinlyl chloride. The accuracy of the proposed sensor was confirmed by analyzing standard reference materials of natural water and peach leaves. The sensor was successfully applied for the determination of copper in tap water and tea samples. This study showed the application of a PCT dye for preparation of a new Cu^{2+} sensitive optical chemical sensor for the first time. The sensor shows a high selectivity and quick response for Cu^{2+} over other common metal ions.

Heavy metal pollution is a global problem and it causes threat to the environment and human beings. Among the different heavy metal ions, mercury has received considerable attention due to its highly toxic and bioaccumulative properties. It is released from coal burning power plants, oceanic and volcanic emissions, gold mining, and solid waste incineration. Mercury vapor lamps, fluorescent lamps, electrical switches, batteries, thermometers and electrodes are the second largest sources of mercury discharge to the environment. In this regard, an ultrasensitive and selective spectrofluorimetric determination of Hg(II) using 2,5-dimercaptothiadiazole (DMT) as a fluorophore was developed (Vasimalai & John, 2011). In this study, the practical application of the present method was demonstrated by determining Hg(II) in tap water, river water and industrial waste water samples. The obtained results have a good agreement with inductively coupled plasma atomic emission spectrometric (ICP-AES) and atomic absorption spectrometric (AAS) methods. According to the literature, this is the first report for the lowest detection with the highest selectivity for Hg(II) in a water medium by the fluorimetric method.

3.8 Nanomaterials

Noble metal nanocrystals exhibit extraordinary plasmonic properties. The excitation of their localized surface Plasmon resonance modes results in the confinement of electromagnetic waves in regions below the diffraction limit near the metal surface. On the other hand, the localized plasmon modes of elongated metal nanocrystals are inherently anisotropic. The optical signal amplification is therefore expected to be strongly dependent on the excitation polarization direction. Recently, the strong polarization dependence of the plasmon-enhanced fluorescence on single gold nanorods has been reported (Ming et al., 2009). In this study, it has been observed that the fluorescence from the organic fluorophores that are embedded in a mesostructured silica shell around individual gold nanorods was enhanced by the longitudinal Plasmon resonance of the nanorods. The polarization dependence is ascribed to the dependence of the averaged electric field intensity enhancement around the individual gold nanorods on the excitation polarization direction. The maximum fluorescence enhancement occurs when the longitudinal Plasmon wavelength of the Au nanorods is nearly equal to the excitation laser wavelength.

Organic dyes, such as fluorescein, rhodamine, and cyanine, and fluorescent proteins, such as green fluorescent protein (GFP) and its variants, are popular probes for cellular organelles and lipid and protein dynamics, due to their small sizes (<5 nm), high specificity, and aqueous solubility. However, because these fluorescent markers rapidly photobleach, have low quantum yield, and exhibit blinking, it is difficult to obtain quantitative spatial and temporal data on the cellular structures they probe (Yao et al., 2005, as cited in Muddana et al., 2009). Recently, the principle photophysical properties of calcium phosphate nanoparticles (CPNPs) using steady-state and time resolved fluorescence spectroscopy, to demonstrate the potential of these particles for biological imaging and drug delivery and to understand the underlying photophysical mechanisms of encapsulation-mediated fluorescence enhancement have been characterized (Muddana et al., 2009). The enhanced photophysical properties together with excellent biocompatibility make CPNPs ideal for bioimaging applications ranging from single-molecule tracking to in vivo tumor detection while pH-dependent dissolvability of calcium phosphate offers the possibility of timed co-delivery of drugs to control cell function.

Fluorescent nanoparticles have attracted increasing research attention due to their promising applications covering electro-optics to bio-nanotechnology. In this regard, monodispersed water-soluble fluorescent carbon nanoparticles (CNPs) were synthesized directly from glucose by a one-step alkali or acid assisted ultrasonic treatment (Li et al., 2011). The results showed that the particle surfaces were rich in hydroxyl groups, giving them high hydrophilicity. The CNPs could emit bright and colorful photoluminescence covering the entire visible-to-near infrared (NIR) spectral range. In this study they conclude that combining free dispersion in water (without any surface modifications) and attractive photoluminescent properties, CNPs should serve as a promising candidate for a new type fluorescence marker, bio-sensors, biomedical imaging, and drug delivery for applications in bioscience and nanobiotechnology.

Recent advances in ultrasensitive protein biosensors have brought significant impacts to proteomics, biomedical diagnostics, and drug discovery (Zhu et al. 2001, as cited in Huang & Chen, 2008). Advanced nanoscale biosensors based on nanoparticles, nanowires, and other nanomaterials have been developed to detect various proteins with improved sensitivity, specificity, and reliability (Fu et al., 2007, as cited in Huang, 2008). Ultrasensitive fluorescence nanosensors can detect the fluorescence signal from a fluorescence tag bound specifically with a single target molecule, but the plain fluorescence intensity measurement can hardly discriminate against a nonspecifically bound tag. Therefore, an electrically modulated fluorescence protein assay that can detect specific fluorescence from a single molecule assembled on an Au nanowire by manipulating the molecule with an electrical potential applied on the nanowire have been developed (Suxian et al., 2008). In their study, they conclude that the simple electrically modulated fluorescence detection method can be generally applied to various bioassays. The essential requirement of the method is to selectively modulate the specific fluorescence from the target molecules by an external reference field, which can be achieved by electrical, optical, magnetic, mechanical, or biochemical interactions etc.

Inorganic nanomaterials have been widely used in biological and environmental fields such as bio-labeling, imaging, drug delivery, separation processes and optical sensing. In these applications, the size and the shape of nanomaterials have very important effects on their properties. Recently, much attention has been given to one- dimensional nanomaterials for building various sensors, due to its high activity, high surface-to-volume ratio, easy assembly in an array for the device and especial suitability for intracellular detection by inserting it into cell (Zhang et al., 2004; Park et al., 2007, as cited in Xu et al., 2011). Therefore, a fluorescence sensor for selective detection of Cu(II) realized by covalently immobilizing derivatives of rhodamine6G (R6G) on the surface of silicon nanowires (SiNWs) has been designed and fabricated (Xu et al., 2011) The fabricated SiNWs-based chemosensor can be electively used for detection of Cu(II) with Cu(II)-special fluorescence enhancement over other metal ions. The Cu(II) sensor exhibits a good selectivity and sensitivity.

Drug delivery systems (DDS) are intensively investigated in the recent years for their great potential to improve the therapeutic index of small molecular drugs. In this regard, recently, new types of fluorescent nanoparticles (FNPs) were prepared through ionic self-assembly of anthracene derivative and chitosan for applications as drug delivery carriers with real-time monitoring of the process of drug release (Wei et al., 2011). In this study, the potential practical applications as drug delivery carriers for real-time detection of the drug release process were demonstrated using Nicardipine as a model drug. Upon loading the drug, the strong blue fluorescence of FNPs was quenched due to electron transfer and fluorescence resonance energy transfer (FRET). With release of drug in vitro, the fluorescence was recovered again. The relationship between the accumulative drug release of FNPs and the recovered fluorescence intensity has been established. Therefore, they conclude that such FNPs may open up new perspectives for designing a new class of detection system for monitoring drug release.

Finally, the current state-of-the-art of gold nanoparticles in biomedical applications targeting cancer has been revised (Cai et al., 2008). In which, gold nanospheres, nanorods, nanoshells, nanocages, and surface enhanced Raman scattering nanoparticles have been discussed in detail regarding their uses in *in vitro* assays, *ex vivo* and *in vivo* imaging, cancer therapy, and drug delivery.

3.9 In vivo fluorescence spectroscopy

Monitoring phytoplankton classes and their abundance is a routine task in marine scientific research. With the frequent occurrence of red tide (a colloquial term used to refer to one of a variety of natural phenomena known as a harmful algal blooms), there is an urgent need for rapid analytical methods that can provide qualitative and quantitative information. Therefore, in vivo synchronous fluorescence spectra (SFS) of phytoplankton samples for determining the relative abundance of specific classes of phytoplankton (plant-like organisms that can form dense, visible patches near the water's surface) was investigated (Li et al., 2008). This study demonstrates the potential for determining phytoplankton class abundance by in vivo SFS. The database could be expanded to include more phytoplankton species grown under different nutrient, temperature, and photon flux densities. This work brings an in situ method for determining major phytoplankton class abundance by fluorescence measurement of seawater and deserves further studies.

Magnetic nanoparticles have been widely used in biomedical research such as magnetic carriers for bioseparation (Dolye et al., 2002, as cited in Yoo at al., 2007), enzyme and protein immobilization (Cao et al., 2003, as cited in Yoo at al., 2007), and contrast-enhancing media (Wu et al., as cited in Yoo at al., 2007). In this regard, in vivo fluorescence imaging method of novel threadlike tissues (Bonghan ducts) inside the lymphatic vessels of rats with fluorescent magnetic nanoparticles has been investigated (Yoo et al., 2007). The results of this study showed new applications of nanoparticles for in vivo imaging of hardly detectable tissues using fluorescence reflectance imaging and magnetophoretic control.

Chlorophyll fluorescence has been used as an accurate and nondestructive probe of photosynthetic efficiency, which can directly or indirectly reflect the impacts of environmental factors and changes in the physiological state of the plants (Baker, 2008, as cited in Falco et al., 2011). Recently, the first in vivo observation of chlorophyll fluorescence quenching induced by gold nanoparticles has been observed (Falco et al., 2011). The results showed that laser-induced fluorescence spectroscopy can be used to investigate the alterations in the physiological response of plants induced by gold nanoparticles, and that both excitation wavelengths, 405 nm and 532 nm, were able to detect the presence of the gold nanoparticles inside the plants. Even though, further investigations must be conducted to clarify the processes of penetration, translocation, and accumulation of nanoparticles in plants.

With the increasing nutrient pollution of freshwater ecosystems, lots of reservoirs, including those used as drinking water resources, suffer from extensive cyanobacterial blooms in the summer. Most of cyanobacteria produce a broad range of compounds with various chemical and toxicological properties. Therefore, their occurrence in recreational or drinking water reservoirs represents high health risk for humans and other organisms. In vivo fluorescence methods have been accepted as a quick, simple, and useful tool for quantification of phytoplankton organisms. In this regard, a case study in which fluorescence methods were employed for the selective detection of potentially toxic cyanobacteria in raw water at the drinking water treatment plant has been presented (Gregor et al., 2007). In this study the author demonstrated that presence of cyanobacteria in raw water can be easily monitored by phycocyanin fluorescence. Measured values were in good correlation with the other parameters of cyanobacterial biomass (chlorophyll a, cell counts). However, the blue light-excited fluorescence of eukaryotic algae should be also monitored to avoid false positive signals.

Photodynamic therapy (PDT) is a treatment modality that may in some cases replace invasive, more harmful or more expensive therapies for certain disorders such as cancer (Dougherty et al., 1998; Ochsner et al., 1997, as cited in Fischer at al., 2002). The PDT treatment process uses photosensitizing compounds that are selectively retained in abnormal tissue some hours to days after administration. At an optimal time after the administration, the photosensitizer is activated with visible light, causing the formation of reactive oxygen species that can initiate a destruction process of the tissue in which they are located. In addition to this photosensitization process, however, most photosensitizers used in PDT show characteristic fluorescence properties (Wagnieres at al., 1998, as cited in Fischer at al., 2002). In this regard, in vivo fluorescence imaging using two excitation and/or emission wavelengths for image contrast enhancement has been described (Fischer et al., 2002). In their study, in vivo measurements in mice where the second image, usually the background signal only, contains new unwanted image data were described. This simple method can successfully resolve the desired image, thus demonstrating the versatility of the image processing procedure.

Currently, the use of 5-aminolevulinic acid (5-ALA) (Zaak, et al., 2001, as cited in Bulgakova et al., 2009) offers the most promising outlook for a fluorescence diagnosis method to reveal superficial bladder cancer. In this regard, a methodological approach combining fluorescence imaging with in vivo local fluorescence spectroscopy (LFS) was clinically tested in order to improve the specificity of photodynamic diagnosis of superficial bladder cancer after intravesical instillation of Alasense (Bulgakova et al., 2009). This preliminary study, which included 62 patients, suggests that in vivo LFS in the course of fluorescence cystoscopy examinations could possibly minimize false positive fluorescence cases and reduce the required number of biopsies from 5-aminolevulinic acid (5-ALA)-based agent induced red fluorescence zones.

4. Comparison of fluorescence spectroscopy with other spectroscopic methods as an analytical technique

Compare to other analysis techniques, it must suffice here to add that fluorescence measurements are rapid, accurate and require only very small quantities of sample (nanomole or less). Fluorescence instrumentation is also relatively inexpensive and easy to use. In general, fluorescence experiments are relatively easy to perform; as in many fields, it is the planning of appropriate experiments, the analysis, and accurate interpretation of the data that require more extensive experience.

Fluorescent methods have three significant advantages over absorption spectroscopy and other typical optical spectroscopy. First, two wavelengths are used in fluorimetry, but only one in absorption spectroscopy. Emitted light from each fluorescent color can be easily separated because each color has unique and narrow excitation spectra. This selectivity can be further enhanced by narrowing the slit width of the emission monochromator so that only emitted light within a narrow spectral range is measured. Multiple fluorescent colors within a single sample can be quantified by sequential measurement of emitted intensity using a set of excitation and emission wavelength pairs specific for each color. The second advantage of fluorescence over absorption spectroscopy is low signal to noise, since emitted light is read at right angles to the exciting light. For absorption spectrophotometry, the excitation source, sample and transmitted light are configured in line, so that the absorption signal is the small difference between the exciting light and the transmitted light, both of which are quite intense. The third advantage is that fluorescent methods have a greater range of linearity. Because of these differences, the sensitivity of fluorescence is approximately 1,000 times greater than absorption spectrophotometric methods (Guilbault, 1990). However, a major disadvantage of fluorescence is the sensitivity of fluorescence intensity to fluctuations in pH and temperature. However, pH effects can be eliminated by using nonaqueous solvents, and normal room temperature fluctuations do not significantly affect the fluorescence intensities of commercial dye solutions.

In addition there are other useful fluorescent techniques that have an advantages over the other typical optical spectroscopy, that not have been mentioned in this chapter, these are: (1) polarization (anisotropy), this technique gives information about the rotation of a fluorophore, and allow to infer protein shape, membrane fluidity (order parameters), and binding analysis; (2) quenching fluorescence, these processes can occur during the excited state lifetime -for example collisional quenching, energy transfer, charge transfer reactions or photochemistry -or they may occur due to formation of complexes in the ground state, thus this technique can give a useful kinetic information of the tested system, and give information about accessibility of fluorophoresis obtained allowing to correlate it with changes in protein or membrane structure; (3) Förster resonance energy transfer (FRET), this technique can be considered as a molecular ruler, which allow to determine distance from 10 to 80 Å; (4) fluorescence microscopy for image analysis, this technique gives magnification (in order to see small parts), resolution (in order to distinguish details of the small parts), and contrast (in order to magnify and resolve details, fluorescence emission provides contrast); (5) multiphoton excitation fluorescence microscopy, as an example for two photon excitation a molecule can be excited by using simultaneous photons and get fluorescence as happens with one photon excitation, this techniques has so many advantages such as sectioning effect without pinholes, low photobleaching and photodamage rate, separation of excitation and emission, no Raman from the solvent, deep penetration in tissues, single excitation wavelength for many dyes, avoid chromatic aberrations, and no expensive UV optics (for UV excited fluorophores) needed. Even though it has some disadvantages such as only is suitable for fluorescence images (reflected light images is not currently available), the technique is not suitable for imaging highly pigmented cells and tissues which absorb near infrared light, and laser source is expensive. It is worth mentioning that all analysis that can be done using fluorescence spectroscopy can be done using a multi-photon excitation fluorescence microscopy. However, multi-photon excitation fluorescence microscopy has a unique analysis applications over the fluorescence spectroscopy such as: analysis of deep tissue imaging (Brain, skin, etc), can prime photochemical reaction within subfemtoliter volumes inside solutions, cells and tissues (photolabile "caged" compounds), can be used for imaging of living specimen for longer period of time, and for live animal imaging (intrinsic fluorophores).

5. Conclusion

It could be concluded at the end of this chapter, that the fluorescence spectroscopy is intensely employed as a powerful spectroscopic tool for quantitative analysis, characterization, and quality control in different fields such as inorganic, organic, pharmaceutical, biological and biomedical, food, and environmental analysis. Furthermore, fluorescence spectroscopy has so many applications in optical sensors, and nanamaterials.

Compare with other analytical techniques, it must suffice here to add that fluorescence measurements are rapid, accurate and require only very small quantities of sample (nanomole or less) with high selectivity. The principal advantage of fluorescence over radioactivity and absorption spectroscopy is the ability to separate compounds on the basis of either their excitation or emission spectra, as opposed to a single spectra. This advantage is further enhanced by commercial fluorescent dyes that have narrow and distinctly separated excitation and emission spectra.

6. References

Airado-Rodrıguez, D., Duran-Meras, I., Galeano-Dıaz, T., Wold, J. P. (2011). Front-face fluorescence spectroscopy: A new tool for control in the wine industry. *J. Food Composition Analysis*, 24, pp.257-264, ISSN: 0889-1575.

Akasheh, T.S., Al-Rawashdeh, N.A. F. (1990). Sodium Lauryl Sulfate-Ruthenium(II) Interactions: Photogalvanic and Photophysical Behavior of Ru(II)-Mimine Complexes. *J. Phys. Chem.*, 94, 23, (Nov. 1990), pp. 8594-8598, ISSN: 0022-3654.

Aksuner, N., Henden, E., Yilmaz, I., Cukurovali, A. (2009). A highly sensitive and selective fluorescent sensor for the determination of copper(II) based on a schiff base. *Dyes and Pigments*. 83, pp. 211-217, ISSN: 0143-7208.

Albert-Garcia, J.R., Antoón-Fos, G.M., Duartb, M.J., Zamoraa, L. L., Calatayuda, J. M. (2009). Theoretical prediction of the native fluorescence of pharmaceuticals. *Talanta*, 79, 2, (July 2009), pp. 412-418, ISSN: 0039-9140.

Al-Rawashdeh. N.A.F. (2005). Interactions of Nabumetone with γ-Cyclodextrin Studied by Fluorescence Measurements. *J. Incl. Phenom. Macrocyclic Chem.*, 51, 1-2, (Feb. 2005), pp. 27-32, ISSN: 0923-0750.

Bender, M. L., Komiyama, M. (1978). *Cyclodextrin Chemistry* (1st Ed.), Springer-Verlag, ISBN: 0387085777, Berlin, Germany.

Bram, O., Messina, F., El-Zohry, A. M., Cannizzo, A., Chergui, M. (available online Dec. 3, 2011). Polychromatic femtosecond fluorescence studies of metal-polypyridine complexes. *Chem. Phys.*, DOI: 10.1016/j.chemphys.2011.11.022, ISSN: 0301-0104.

Bulgakova, N., Ulijanov, R., Vereschagin, K., Sokolov, V., Teplov, A., Rusakov, I., Chissov, V. (2009). In vivo local fluorescence spectroscopy in PDD of superficial bladder cancer. *Med. Laser Appl.*, 24, pp. 247-255, ISSN: 1615-1615.

Cai, W., Gao, T., Hong, H., Sun, J. (2008). Applications of gold nanoparticles in cancer nanotechnology. *Nanotechnology, Science and Applications*, 1 (Sept. 2008), pp. 17-32, ISSN: 1177-8903.

Calatayud, J. M., Zamora, L. L. (1995), in A. Townshend (Ed.), *Encyclopedia of Analytical Science*, Academic Press, ISBN 0-12-226700-1, Oxford.

Catherine L. Muller, C. L., Baker, A., Hutchinson, R., Fairchild, I. J., Kidd, C. (2008). Analysis of rainwater dissolved organic carbon compounds using fluorescence spectrophotometry. *Atmospheric Environment*, 42, pp. 8036-8045, ISSN: 1352-2310

Casado-Terrones, S., Ferna´ndez-Sa´nchez, J. F., Segura-Carretero, A., Ferna´ndez-Gutie´rrez, A. (2007). Simple luminescence detectors using a light-emitting diode or a Xe lamp, optical fiber and charge-coupled device, or photomultiplier for determining proteins in capillary electrophoresis: A critical comparison. *Anal. Biochem.* 365, (February 2007), pp. 82-90, ISSN: 10.1016.

Chandrasekhar, V., Bag, P., Pandey, M. D. (2009). Phosphorus-supported multidentate coumarin-containing fluorescence sensors for Cu^{+2}. *Tetrahedron*, 65, 47, (Nov. 2009), pp. 9876-9883, ISSN: 0040-4020

Cheng, F., Tang, N., Chen , J., Wang, F., Chen, L. (2010). A new family of trinuclear Ru(II) polypyridyl complexes acting as pH-induced fluorescence switch. *Inorg. Chem. Commun.*, 13, pp. 757-761, ISSN: 1387-7003

Cheng, F., Tang, N., Chen , J., Wang, F., Chen, L. (2011). pH-induced fluorescence switch of two novel tetranuclear Ru(II) polypyridyl complexes. *Inorg. Chem. Commun.*, 14, pp. 852-855, ISSN: 1387-7003

Dawoud, A. A., Al-Rawashdeh, N. (2008). Spectrofluorometric, thermal, and molecular mechanics studies of the inclusion complexation of selected imidazoline-derived drugs with *β*-cyclodextrin in aqueous media. *J. Incl. Phenom. Macrocyclic Chem.*, 60, 3-4, (April 2008), pp. 293-301, ISSN: 0923-0750

Dragan, A.I., Bishop, E.S., Casas-Finet , J.R., Strouse, R.J., Schenerman, M.A., Geddes, C.D. (2010). Metal-enhanced PicoGreen fluorescence: Application to fast and ultra-sensitive pg/ml DNA quantitation. *J. Immunolog. Meth.*, 362, pp. 95-100, ISSN: 0022-1759

Dessingou, J., Mitra, A., Tabbasum, K., Baghel, G. S., Rao, C. P. *J. org. Chem.*, (Nov. 21, 2011), in press, ISSN 10.1021/jo201926q

Dragan, A. I., Golberg, K., Elbaz, A., Marks, R., Zhang, Y., Geddes, C. D. (2011). Two-color, 30 second microwave-accelerated Metal-Enhanced Fluorescence DNA assays: A new Rapid Catch and Signal (RCS) technology. *J. Immunolog. Meth.*, 366, pp. 1-7, ISSN: 0022-1759

Falco, W. F., Falcao, E. A., Santiago, E. F., Bagnato, V. S., Caires, A. R. L. (2011). In vivo observation of chlorophyll fluorescence quenching induced by gold nanoparticles. *J. Photochem Photobio. A: Chem.*, 225, pp. 65-71, ISSN: 1010-6030

Fernandez-Sanchez, J.F., Carretero, A. S., Cruces-Blanco, C., Fernandez-Gutierrez, A. (2003). A sensitive fluorescence optosensor for analysing propranolol in pharmaceutical preparations and a test for its control in urine in sport. *J. Pharm. Biomed. Anal.*, 31, pp.859-865, ISSN: 0731-7085

Fischer, F., Dickson, E. F. G., Pottier, R. H. (2002). In vivo fluorescence imaging using two excitation and/or emission wavelengths for image contrast enhancement. *Vibrational Spectroscopy*, 30, pp. 131-137, ISSN: 0924-2031

Fu, Y., Zhang, J., Lakowicz, J. R. (2008). Metal-enhanced fluorescence of single green fluorescent protein (GFP). *Biochem. Biophys. Res. Commun.*, 376, pp. 712-717.

Formica, M., Fusi, V., Giorgi, L., Micheloni, M. (2012). New fluorescent chemosensors for metal ions in solution. *Coord. Chem. Rev.*, 256, pp. 170-192, ISSN: 0010-8545.

Gregor, J., Marsalek, B., Sipkova, H. (2007). Detection and estimation of potentially toxic cyanobacteria in raw water at the drinking water treatment plant by in vivo fluorescence method. *Water Research*, 41, pp. 228-234, ISSN: 0043-1354.

Gupta, S., Paul, B. K., Barik, A. K., Mandal, T. N., Roy, S., Guchhait, N., Butcher, R. J., Kar, S. K. (2009). Modulation of fluorescence emission of 1-(2-pyridyl) pyrazole derived Schiff base ligands by exploiting their metal ion sensitive binding modes. *Polyhedron*, 28, pp.3577-3585, ISSN: 0277-5387

Guo, W., Xu, J., Wang, J., Wen, Y., Zhuo, J., Yan, Y. (2010). Characterization of dissolved organic matter in urban sewage using excitation emission matrix fluorescence spectroscopy and parallel factor analysis. *J. Environm. Sci.*, 22, 11, (Nov. 2010), pp. 1728-1734, ISSN: 1001-0742

Guilbault, G. G. (1990). *Practical Fluorescence (Modern Monographs in Analytical Chemistry)* (2nd Ed.), Marcel Dekker, INC, ISBN-13: 978-0824783501, New York.

Hammamia, M., Rouissi, H., Salah, N., Selmi, H., Al-Otaibi, M., Blecker, C., Karoui, R. (2010). Fluorescence spectroscopy coupled with factorial discriminant analysis technique to identify sheep milk from different feeding systems. *Food Chemistry*, 122, pp. 1344-1350, ISSN: 0308-8146

He, D., Chena, B., Tiana, Q., Yaoa, S. (2009). Simultaneous determination of five anthraquinones in medicinal plants and pharmaceutical preparations by HPLC with fluorescence detection. *J. Pharm. Biomed. Anal.*, 49, pp. 1123-1127, ISSN: 0731-7085

Hernandez, D., Plaza, C., Senesi, N., Polo, A. (2006). Detection of Cupper(II) and zinc(II) binding to humic acids from pig slurry and amended soils by fluorescence spectroscopy. *Environmental Pollutution*, 143, 2, (Sept. 2006), pp. 212-220, ISSN: 0269-7491

Hirano, T., Sekiguchi, T., Hashizume, D., Ikeda, H., Maki, S., Niwa, H. (2010). Colorimetric and fluorometric sensing of the Lewis acidity of a metal ion by metal-ion complexation of imidazo[1,2-a]pyrazin-3(7H)-ones .*Tetrahedron*, 66, 21, (May 2010), pp. 3842-3848, ISSN 0040-4020.

Ingale, S. A., Pujari, S. S., Sirivolu, V. R., Ding, P., Xiong, H., Mei, H., Seela, F. (2011). 7-Deazapurine and 8-Aza-7-deazapurine Nucleoside and Oligonucleotide Pyrene "Click" Conjugates: Synthesis, Nucleobase Controlled Fluorescence Quenching, and Duplex Stability. *J. Org. Chem.*, (Nov. 30, 2011), in press, ISSN 10.1021/jo202103q

Jin,L., Shang, L., Guo, S., Fang, Y., Wen, D., Wang, L., Yin, J., Dong, S. (2011). Biomolecule-stabilized Au nanoclusters as a fluorescence probe for sensitive detection of glucose. *Biosen. Bioelect.*, 26, pp. 1965-1969, ISSN: 0956-5663.

Kaliszan, W., Petrusewicz, J., Kaliszan, R. (2006). Imidazoline receptors in relaxation of acetylcholine constricted isolated rat jejunum. *Pharm. Rep.* 58, pp. 700.

Kanan, S.M., Abu-Yousef, I.A., Hassouneh, N., Malkawi, A., Abdo, N., Kanan, M.C. (2009). A highly Selective Luminescent Sensor for Detecting Environmental Mercuric Ions. *Aust. J. Chem.*, 62, pp. 1593-1599, ISSN: 0004-942.

Kwong, H.-L., Wonga, W.-L., Lee, Ch-.S., Yeung , Ch.-T., Teng, P.-F. (2009). Zinc(II) complex of terpyridine-crown macrocycle: A new motif in fluorescence sensing of zwitterionic amino acids. *Inorg. Chem. Commun.*, 12, pp. 815-818, ISSN: 1387-7003

Leong, W. L., Vittal. J. J. (2009). Synthesis and characterization of metal complexes of Calcein Blue: Formation of monomeric, ion pair and coordination polymeric structures. *Inorg. Chem. Acta*, 362, pp. 2189-2199, ISSN: 0020-1693

Li, H., Zhang, Q., Zhu, Ch., Wang, X. (2008). Assessment of phytoplankton class abundance using in vivo synchronous fluorescence spectra. *Anal. Biochem.*, 377, pp. 40-45, ISSN: 0003-2697

Li, Y., Yang, Z-Y., Wu, J-C. (2010). Synthesis, crystal structures, biological activities and fluorescence studies oftransition metal complexes with 3-carbaldehyde chromone thiosemicarbazone. *Europ. J. Med. Chem.*, 45, pp. 5692-5701, ISSN: 0223-5234

Li, H., He, X., Liu, Y., Huang, H., Lian, S. Lee, S.-T., Kang, Z. (2011). One-step ultrasonic synthesis of water-soluble carbon nanoparticles with excellent photoluminescent properties. *Carbon*, 49, pp.605-609, ISSN: 0008-6223

Ling, K-Q., Sayre, L. M.. (2009). Discovery of a Sensitive, Selective, and Tightly Binding Fluorogenic Substrate of Bovine Plasma Amine Oxidase. *J. Org. Chem.*, 74, pp. 339-350, ISSN: 10.1021/jo8018945

Lakowicz, J.R. (2006), *Principles of Fluorescence Spectroscopy* (3rd Ed.), Springer, ISBN: 13-978-0387-31278-1, New York.

Majumder, A., Rosair, G. M., Mallick, A., Chattopadhyay, N., Mitra, S. (2006). Synthesis, structures and fluorescence of nickel, zinc and cadmium complexes with the N,N,O-tridentate Schiff base N-2-pyridylmethylidene-2-hydroxy-phenylamine. *Polyhedron*, 25, pp. 1753-1762, ISSN: 0277-5387

Ming, T., Zhao, L., Yang, Z., Chen, H., Sun, L., Wang, J., Yan, C. (2009). Strong Polarization Dependence of Plasmon-Enhanced Fluorescence on Single Gold Nanorods. *Nano lett.*, 9, 11, pp.3896-3903, ISSN: 10.1021

Muddana, H. S., Morgan, T. T., Adair, J. H., Butler. P. J. (2009). Photophysics of Cy3-EncapsulatedCalcium Phosphate Nanoparticles. *Nano let..*, 9, 4, pp. 1559-1566, ISSN: 10.1021

Oliveira, E., Vicente, M., Valencia, L., Macıas, A., Bertolo, E., Bastida, R., Lodeiro, C. (2007). Metal ion interaction with a novel anthracene pendant-armed fluorescent molecular probe. Synthesis, characterization, and fluorescence studies. *Inorg. Chem. Acta*, 360, pp. 2734-2743, ISSN: 0020-1693

Parini, A., Moudanos, C. G., Pizzinat, N., Lanier, S. M. (1996). The elusive family of imidazoline binding sites. *Trends Pharmacol. Sci.*, 17, pp. 13.

Parfitt, K. (Ed.), Martindale: The Complete Drug Reference, Pharmaceutical Press, London, 1990

Rendell, D. (1987), *Fluorescence and Phosphorescence*, Wiley, ISBN-13: 978-0471913801, Chichester, England.

Saleh, N., Al-Rawashdeh, N. A. F. (2006). Fluorescence Enhancement of Carbendazim Fungicide in Cucurbit[6]uril. *J. Fluoresc.*, 16, 4, (July 2006), pp. 487-493, ISSN: 1053-0509

Shimomura, O., Johnson, F.H., Saga, Y. (1962). Extraction, purification and properties of aequorin, a bioluminescent protein from the luminous hydromedusan, Aequorea. *Journal of Cell and Comparative Physiology*, 59pp. 223–229

Shete, V. S., Benson, D. E. (2009). Protein Design Provides Lead(II) Ion Biosensors for Imaging Molecular Fluxes around Red Blood Cells. Biochem., 48, 2, pp. 462-470

Silva, L., Fernndez-Franzon, M., Font, G., Pena, A., Silveira, I., Lino, C., Maӧes, J. (2009). Analysis of fumonisins in corn-based food by liquid chromatography withfluorescence and mass spectrometry detectors. *Food Chemistry*, 112, pp. 1031-1037, ISSN: : 0308-8146

Sokkalingam, P., Lee C-H. (2011). Highly Sensitive Fluorescence "Turn-On" Indicator for Fluoride Anion with Remarkable Selectivity in Organic and Aqueous Media. *J. Org. Chem.*, 76, 10, (May 2011), pp. 3820-3828, ISSN 10.1021/jo200138t

Soh, N., Ueda, T. (2011). Perylene bisimide as a versatile fluorescent tool for environmental and biological analysis: A review. *Talanta*, 85, 3, (Sept. 2011), pp. 1233-1237, ISSN: 0039-9140

Strasburg, G. M., Ludescher, R. D. (1995). Heory and applications of fluorescence spectroscopy in food research. *Trends in Food Science & Technology*, 6, 3 (March 1995), pp. 69-75, ISSN: 0924-2244

Suxian Huang, s., Chen, Y. (2008). Ultrasensitive Fluorescence Detection of Single Protein Molecules Manipulated Electrically on Au Nanowire. *Nano lett.*, 8, 9, pp. 2829-2833, ISSN: 10.1021

Szejtli, J. (1988). *Cyclodextrins Technology*, Kluwer Academic Publishers, ISBN: 90-277-2314-1 , Dordrecht, Netherlands.

Vasimalai, John, S. A. (2011). Ultrasensitive and selective spectrofluorimetric determination of Hg(II) using a dimercaptothiadiazole fluorophore. *J. Luminesc.*, 131, pp. 2636-2641, ISSN: 0022-2313

Wang, L., Guo, C., Fu,B., Wang, L. (2011). Fluorescence Determination of DNA Using the Gatifloxacin-Europium(III) Complex. *J. Agric. Food Chem.*, 59, 5, (Feb. 2011), pp. 1607-1611, ISSN: 10.1021

Wei Cui, W., Lu, Z., Cui, K., Wu, J., Wei, Y., Lu, Q. (2011). Fluorescent Nanoparticles of Chitosan Complex for Real-Time Monitoring Drug Release., *Langmuir*, 27, 13, (June 2001), pp. 8384-8390, ISSN: 10-1022

Xu, W., Mun, L., Miao, R., Zhang, T., Shi, W. (2011). Fluorescence sensor for Cu(II) based on R6G derivatives modified silicon nanowires. *J. Luminesc.*, 131, pp. 2626-2620, ISSN: 0022-2313

Xu, J., Huang, X.H., Zhou, N.L., Zhang, J.S., Bao, J.Ch., Lu, T.H., Li, C. (2004). Synthesis, XPS and fluorescence properties of Eu^{3+} complex with polydimethylsiloxane. *Mater. Lett.*, 58, pp. 1938-1942, ISSN: 0167-577X

Yoo, J. S., Johng, H.-M., Yoon, T.-J., Shin, H.-S., Lee, B.-Ch., Lee, Ch., Ahn. B. S., Kang, D.-I., Lee, J.-K., Soh, K.-S. (2007). In vivo fluorescence imaging of threadlike tissues (Bonghan ducts) inside lymphatic vessels with nanoparticles. *Curr. Appl. Phys.*, 7, pp. 342-348, ISSN: 1567-1739

You, Z.-L., Zhu, H.-L., Liu, W.-S. (2004). Solvolthermal Syntheses and Crystal Structures of Three Linear Trinuclear Schiff Base Complexes of Zinc(II) and Cadmium(II). *Z. Anorg. Allg. Chem.*, 630, 11, (Sept. 2004), pp. 1617-1622, ISSN: 1521-3749

Zou, M., Gao, H., Li, J., Xu, F., Wang, L., Jiang, J. (2008). Rapid determination of hazardous compounds in food based on a competitive fluorescence microsphere immunoassay. *Anal. Biochem.*, 374, pp. 318-324, ISSN: 0003-2697

12

Laser Fluorescence Spectroscopy: Application in Determining the Individual Photophysical Parameters of Proteins

Alexander A. Banishev*

Department of Physics and Astronomy, University of California, Riverside, CA, USA

1. Introduction

This work was initiated by the problem of investigating the photophysical properties of complex protein molecules and performing the diagnostics of such molecules in water environment. At the present time, fluorescence spectroscopy (fluorimetry) is widely used to study complex organic compounds (COC) (Lakowicz, 1999). Together with spectrophotometry these methods form the basis for fast and nondestructive diagnostics of COC in the natural environment, i.e. they present the diagnostic methods *in vivo* and *in situ*. However, the conventional (linear) fluorescence spectroscopy methods can not provide complete information on fluorescent objects under study because of insufficient selectivity (fluorescence bands of most COC are broad and structureless at room temperature).

The capabilities of fluorescence spectroscopy can be enhanced by using the methods of laser fluorimetry, in particular nonlinear laser fluorimetry (Fadeev et al., 1999). This method allows one to get information on the molecular level and determine the photophysical parameters of molecules (absorption cross section, lifetime in excitation state, intersystem crossing and energy transfer rates, etc.). Furthermore, the parameters can be measured *in vivo* and *in situ* in the absence of *a priori* information, which is necessary for conventional spectroscopic methods (for example, molecular concentration (Banishev et al., 2009)).

The diagnostics of protein complexes is an intricate problem if a molecule contains more than one absorption/fluorescent center (Permyakov, 1992). The problem becomes much more complex if, in addition, the protein specimen (ensemble of molecules) is a mixture of several chemically nonidentical types of molecules (subensembles) which cannot be separated, i.e. their partial concentrations are unknown. The second situation is typical for the special kind of proteins, namely, fluorescent proteins (FPs) (Piatkevich et al., 2010). The solutions of FPs are usually mixtures of several types of molecules, which are chemically different and have their own set of photophysical properties (Verkhusha et al., 2004). In this case for unambiguous interpretation of experimental data it is necessary to make simultaneous measurements of a large number of parameters, i.e. to simultaneously apply (or, better, synthesize) several spectroscopic methods.

---

* Corresponding Author

In this chapter, a new approach based on the simultaneous use of nonlinear laser fluorimetry, spectrophotometry and conventional fluorimetry methods is presented. The approach allows us to *in vivo* determine the individual photophysical parameters of fluorophores in multi-fluorophore protein complexes. The approach has been applied for investigation of the photophysical properties of the protein molecules of different complexity. Two classes of proteins have been chosen, namely, serum albumins (by the examples of human and bovine serum albumins) and fluorescent proteins (by the example of monomeric red FP mRFP1). The following new results are presented.

i. The photophysical parameters such as (a) true absorption cross section (at 266 nm) of tryptophan and intersystem crossing rate in single-tryptophan-containing protein human serum albumin and (b) true absorption cross section (at 266 nm) of tryptophans and rate of energy transfer between them in two-tryptophan-containing protein bovine serum albumin have been determined.
ii. The complete solution of the task of determining the photophysical parameters of all mRFP1 spectral forms is given. The mechanism of photophysical processes in the spectral forms under their excitation by UV radiation (at 266 nm) has been clarified.
iii. The study of the influence of a single amino acid substitution in mRFP1 protein on individual photophysical parameters of the chromophore (a heterogroup[1] responsible for light absorption and fluorescence in the visible wavelength range) of fluorescent spectral form is performed. The 66th amino acid residue (glutamine 66) has been chosen as a position to be replaced. This residue participates in formation of the chromophore and, as was shown in (Banishev et al., 2009), its substitution by polar serine or cysteine changes the spectral and photophysical properties of the resultant mutant of the mRFP1. In the present work this study has been extended. The optical properties of new variants of mRFP1 with polar (asparagine, histidine) and non-polar (alanine, leucine, phenylalanine) substitution have been investigated. It was found that the individual extinction coefficient of the chromophore and the position of the steady-state spectra of the proteins with polar substitution correlate with the volume of the substituted amino acid residue at position 66. The explanation of this effect is given.

Except for this key target, the methodological task has been put, namely, to demonstrate the unique capabilities of the nonlinear laser fluorimetry method (which is not, so far, well known in a wide circle of opticians) on the specific object.

2. The method of nonlinear laser fluorescence spectroscopy

The fluorescence signal from fluorophores of complex organic compound (COC) under powerful laser excitation is represented as the nonlinear function of the number of detected fluorescence photons N_{Fl} (or fluorescence intensity I_{Fl}) on the photon fluxes F of pumping radiation (Filipova et al., 2001). The dependence $N_{Fl}(F)$ is called fluorescence saturation curve, its typical view is represented in the Fig. 1(a). There are several reasons for that nonlinear dependence: the non-zero lifetime of organic molecules in excited state; intercombination conversion; intermolecular interactions including singlet–singlet annihilation, etc.

[1] The heterogroup is called a chromophore, even when it produces fluorescence, in other words, it is a fluorophore.

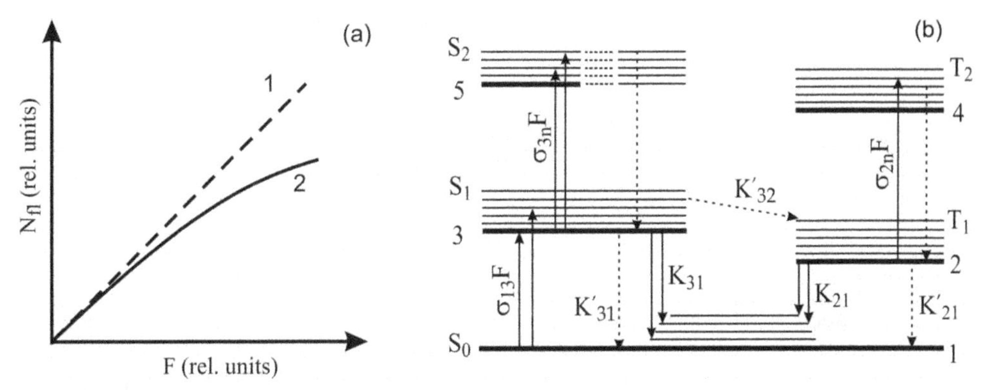

Fig. 1. (a) The $N_{Fl}(F)$ dependencies (see text): (1) in the absence of fluorescence saturation, and (1b) when saturation appears (for most COC at $F>10^{23}$ cm^{-2}s^{-1} (Fadeev et al., 1999)). (b) Photophysical processes in COC (Lakowicz, 1999), without accounting for intermolecular interactions. Solid and dotted vertical lines are radiation and radiationless transitions respectively, S_i are singlet states and T_i are triplet states.

The parameters of saturation curves depend on photophysical characteristics of molecules fluorophores, so that such characteristics can be extracted from these curves after resolving an inverse problem (Fadeev et al., 1999). This is a basement of nonlinear laser fluorimetry as a method for investigation of photophysical properties of COC. To solve the inverse problem, we should first calculate (either analytically or numerically) the theoretical saturation curves by using the fluorescence response formation model of an ensemble of fluorescent molecules under their excitation by laser radiation. In present work two models have been used: the conventional model of fluorescence response formation and the model of localized donor-acceptor (LDA) pairs.

The conventional model (Banishev et al., 2008a, 2009) in describing the fluorescence response is represented as a system of equations that describes the kinetics of concentration of COC molecules at the corresponding energy states (Fig. 1(b)). In the case of monomolecular solutions of non-interactive organic compounds, we must give priority to the following photophysical parameters when defining the saturation curve:

- the absorption cross section $\sigma \equiv \sigma_{13}$, which determines the probability of a molecule transition from the ground singlet state S_0 (level 1) to the first excited state S_1 (level 3) stimulated by a photon flux with the density F;
- the full lifetime $\tau \equiv \tau_3$ of the molecule in the S_1 state (the fluorescence decay time);
- the quantum yield $\eta_T = K'_{32}/K_3$ of the molecule transition to the lower triplet state T_1 (level 2) due to the intersystem crossing, where $K_3 \equiv \tau^{-1} = K_{31} + K'_{31} + K'_{32}$; K_{31} and K'_{31} are the rates of radiative and radiationless transitions from S_1 to S_0; K'_{32} is the rate of the transition from S_1 to T_1.

The model of the fluorescence response, which takes into account processes pointed out above, can be described by the following set of kinetic equations (Fadeev et al., 1999):

$$\begin{aligned}
\frac{\partial n_1(t,\vec{r})}{\partial t} &= -F(t,\vec{r})\cdot\sigma\cdot[n_0(t,\vec{r})-n_3(t,\vec{r})-n_2(t,\vec{r})]+(K_3-K_{32}^{'})\cdot n_3(t,\vec{r}) \\
\frac{\partial n_3(t,\vec{r})}{\partial t} &= F(t,\vec{r})\cdot\sigma\cdot[n_0(t,\vec{r})-n_3(t,\vec{r})-n_2(t,\vec{r})]-K_3\cdot n_3(t,\vec{r}) \\
\frac{\partial n_2(t,\vec{r})}{\partial t} &= K_{32}^{'}\cdot n_3(t,\vec{r}) \\
n_0 &= n_1+n_2+n_3,
\end{aligned} \tag{1a}$$

where n_0 is the total concentration of molecules; n_3, n_2, and n_1 are concentrations of molecules in the S_1, T_1 and S_0 states, respectively; $F(t,\vec{r})$ is the photon flux density of exciting radiation at the coordinate point $\vec{r}$ at instant of time t. The rest of parameters are defined above. In model (1a), the transition from T_1 to S_0 is neglected. This assumption is valid if the light pulse duration (t_p) is much less than the lifetime in the T_1 state, i.e. t_p is much less then $(K_{21}+K'_{21})^{-1}$. For pulse lasers often used in laser fluorescence spectroscopy, the t_p is ~10 ns, and this condition is fulfilled.

The conventional model (1a) describes the processes in a system in the absence of interaction between molecules. If there is the interaction and an ensemble of fluorophores generating the fluorescence response consists of subensembles of the donor and acceptor molecules, then the conventional approach is reduced to two systems of kinetic equations, i.e. separately for each subensemble. The term describing the energy transfer is in this case a "cross term" that connects these two systems of equations. Such model, based on separate mathematical description of two subensembles, is able to describe the fluorescence response when each molecule of the donor is surrounded by a large number of the acceptor molecules onto which the energy transfer can occur (Agranovich & Galanin, 1982) (i.e. the possibility that the donor molecule and all the locally surrounding it acceptor molecules simultaneously stay in the excitation state is excludes). The situation like this is typical for a concentrated binary solution of single-fluorophore molecules (for example, dye solutions with the concentration higher than 10^{-4} M) or for complexes with high local concentration (Fadeev et al., 1999), such as phytoplankton. The energy transfer process in that case is called the intermolecular one.

If there is a donor–acceptor pair within a single molecule (i.e. we have a molecule with a LDA pair), the situation is possible when the donor and the acceptor are simultaneously in an excited state. Therefore, the description of the energy transfer in the framework of a conventional scheme is impossible and the model (1a) should be modified. Let us note that the molecular objects with LDA pair are finding more and more wide applications at present time. Commonly, systems of this kind are constructed artificially from pairs of organic compounds, for example, from dye molecules (Srinivas et al., 2001) or FP macromolecules (Truong & Ikura , 2001). In (Banishev et al., 2008b) a fluorescence response formation model of an ensemble of LDA pairs has been suggested by the author. The model makes it possible to describe the energy transfer inside a LDA pair, disregarding the energy transfer between the pairs. The main idea of this approach consists in the following. Let us introduce a notion of the *collective states* of a LDA pair (Fig. 2); each of these states simultaneously describes both the donor state and the acceptor state:

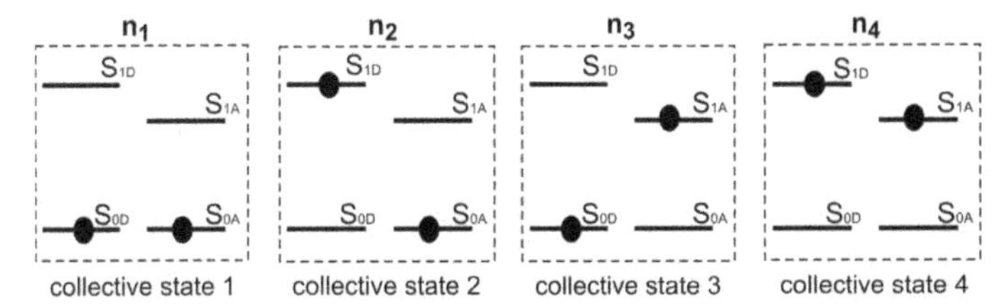

Fig. 2. The collective states of the LDA pair (nonmetering the process of singlet-singlet annihilation and intersystem crossing). The S_{0D}, S_{1D} and S_{0A}, S_{1A} are the energy levels of the donor and acceptor respectively.

- Collective state 1: the donor and the acceptor are in the ground state S_{0D} and S_{0A}, respectively; the concentration of such molecules is denoted as $n_1 \equiv n_1(t,\vec{r})$.
- Collective state 2: the donor is in the first excited singlet state S_{1D}, and the acceptor is in the state S_{0A}; the concentration of such molecules is denoted as $n_2 \equiv n_2(t,\vec{r})$.
- Collective state 3: the donor is in the state S_{0D}, and the acceptor is in the first excited singlet state S_{1A}; the concentration of such molecules is denoted as $n_3 \equiv n_3(t,\vec{r})$.
- Collective state 4: the donor and the acceptor are in the first excited singlet state S_{1D} and S_{1A}, respectively; the concentration of such molecules is denoted as $n_4 \equiv n_4(t,\vec{r})$.

As a result, there is no need to describe the fluorescence response from a sub-ensemble of donor and acceptor molecules separately, i.e. it is unnecessary to create two systems of equations (one for a donor sub-ensemble and one for an acceptor sub-ensemble) similar to (1a), as it takes place in the conventional approach. Instead of this, the system of equations that describes the populations of the collective states can be written. The dynamics of variation in the concentrations of these four collective states of the LDA pair (nonmetering the process of intersystem crossing in molecules of the donor and acceptor) is mathematically described by the following system of kinetic equations:

$$\begin{aligned}
&\frac{\partial n_1}{\partial t} = -F(t,\vec{r})\cdot(\sigma_D + \sigma_A)\cdot n_1 + \frac{n_2}{\tau_D} + \frac{n_3}{\tau_A} \\
&\frac{\partial n_2}{\partial t} = -F(t,\vec{r})\cdot\sigma_A\cdot n_2 - \frac{n_2}{\tau_D} - K_{DA}\cdot n_2 + F(t,\vec{r})\cdot\sigma_D\cdot n_1 + \frac{n_4}{\tau_A} \\
&\frac{\partial n_3}{\partial t} = -F(t,\vec{r})\cdot\sigma_D\cdot n_3 - \frac{n_3}{\tau_A} + F(t,\vec{r})\cdot\sigma_A\cdot n_1 + \frac{n_4}{\tau_D} + K_{DA}\cdot n_2 + K_{SS}\cdot n_4 \\
&\frac{\partial n_4}{\partial t} = F(t,\vec{r})\cdot\sigma_A\cdot n_2 + F(t,\vec{r})\cdot\sigma_D\cdot n_3 - \frac{n_4}{\tau_D} - \frac{n_4}{\tau_A} - K_{SS}\cdot n_4 \\
&n_1 + n_2 + n_3 + n_4 = n_0,
\end{aligned} \tag{1b}$$

where τ_D, τ_A and σ_D, σ_A are the lifetime and absorption cross section of the donor (denoted by D) and acceptor (denoted by A) as it defined above; K_{DA} is the rate of the energy transfer from the excited donor to the unexcited acceptor; K_{SS} is the rate of energy transfer from the excited donor to the excited acceptor (singlet–singlet annihilation (Fadeev et al., 1999); $F(t,\vec{r})$ is the photon flux density (see Eqs. (1a)); and n_0 is the total concentration of molecules containing a LDA pair. In this model, the following photophysical parameters are presented: the absorption cross section and the excited state lifetime of the donor and acceptor, the energy transfer rates K_{DA} and K_{SS}.

By solving systems (1a) and (1b) numerically, one can find the concentration of the fluorescent molecules in the excited state and calculate the number of fluorescence photons N_{Fl}, emitted from the volume V after the action of the laser pulse (Filipova et al., 2001; Fadeev et al., 1999). The theoretical saturation curve for the model (1a) can be calculated from following equation:

$$N_{Fl}(\lambda) = K_{31} \cdot \int_V d\vec{r} \int_{-\infty}^{+\infty} n_3(t,\vec{r})dt \tag{2a}$$

For the model (1b), for the donor (2b) and the acceptor (2c) curves, respectively:

$$N_{Fl}^D(\lambda) = \tau_D^{-1} \cdot \eta_D \cdot \int_V d\vec{r} \int_{-\infty}^{+\infty} (n_2(t,\vec{r}) + n_4(t,\vec{r}))dt \tag{2b}$$

$$N_{Fl}^A(\lambda) = \tau_A^{-1} \cdot \eta_A \cdot \int_V d\vec{r} \int_{-\infty}^{+\infty} (n_3(t,\vec{r}) + n_4(t,\vec{r}))\, dt \tag{2c}$$

where η_D and η_A are the fluorescence quantum yield, which is defined as the ratio of the radiation decay rate of S_1 state to the sum of all rates of S_1 state decay (i.e. $\eta \equiv K_{31}/K_3$), of the donor and acceptor, correspondingly; λ is the fluorescence registration wavelength. Other symbols are defined in Eqs. (1a).

In considered model (1a), the fluorescence saturation is caused by a finite lifetime τ and by intercombination conversion. In model (1b), due to the finite fluorescence lifetime and due to the saturation of the energy transfer channels. Let us note that the model (1b) could be also supplemented with the intersystem crossing mechanisms, but preliminary experiments showed that at the given parameters of the laser radiation the process for albumins and mRFP1 is small compared to the mechanisms under study and contributes little to fluorescence saturation. Therefore, this mechanism has been excluded to increase the stability of the inverse problem solution (details and mathematical basement of inverse problem solution of nonlinear laser fluorimetry can be found elsewhere (Boychuk at al., 2000). For the same reason the induced processes from the excited states (two-photon absorption or photoizomerization, etc) have been excluded.

As was mention above, the photophysical parameters of fluorophores (σ, K'_{32} and τ in the model (1a) and τ_D, τ_A, σ_D, σ_A, K_{DA} and K_{SS} in the model (1b)) can be determined from the dependence $N_{Fl}(F)$, by solving the inverse problem. However, in experiments, it is convenient to normalize the number of detected fluorescence photons N_{Fl} to the reference signal (will denote as N_{Ref}), which can represent a part of exciting radiation directed to the reference channel of the detection system by a beamsplitter or a Raman scattering signal from water molecules (Fadeev et al., 1999). In this case, one has to deal with the dependence $[\Phi(F)]^{-1}=N_{Ref}/N_{Fl}$ (which is also called a saturation curve, $\Phi(F)$ is the fluorescence parameter) rather than $N_{Fl}(F)$. According to the practical experience such normalization also helps to increase the stability of the inverse problem solution. In the absence of saturation, Φ stops being dependent on photon flux density F and tends to a constant which is denoted by Φ_0 ($\Phi_0 \equiv lim|_{F\to 0}[\Phi(F)]$). If in measurements of Φ_0 the Raman scattering band of water molecules as a reference signal is used, then it is possible to find the fluorescence quantum yield of a complex organic compound (Filipova et al., 2001).

For the reasons pointed out in (Banishev et al., 2008a), the same laser fluorimeter has been optimized for measuring the nanosecond fluorescence decay (the kinetic mode of the fluorimeter operation). The curve represents the dependence of the number $N_{Fl}(t_{del})$ of fluorescence photons in the detector gate (with wide t_g) on the gate delay time t_{del} with respect to a laser pulse. For the model (1a) an expression for kinetic curve can be written as:

$$N_{Fl}(t_{del}) = K_{31} \cdot \int_V d\vec{r} \int_{-t_g/2+t_{del}}^{t_g/2+t_{del}} n_3(t,\vec{r})dt \tag{3}$$

The t_{del} changes discretely and proportionally to the detector gate step: $t_{del}=i\times t_{step}$, where i is a number of the detector gate step. Similar expressions can be written for (2b, c). Gate position at which its centre coincided with the laser pulse maximum was taken as the zero delay (t_{del}=0). This was detected by the maximum of water Raman line (Banishev et al., 2006).

By solving the inverse problem, the fluorescence lifetime τ of a fluorophore can be determined independently from the dependence $N_{Fl}(t_{del})$. In the experiment the fluorescence signal is measured in relative units. For comparison of the experimental data with the theoretical ones it is necessary to normalize the obtained experimental curve to the fluorescence intensity at some fixed time delay. This procedure, the fluorimeter capabilities in the kinetic mode and the corresponding theory can be found elsewhere (Banishev et al., 2006). The difference of such variant of kinetic fluorimetry from the conventional time-resolved fluorimetry is that the fluorescence is excited by a pulse with rather long duration (~10 ns), and for fluorescence registration an optical gated multichannel analyser is used. Whereas, the conventional time-resolved fluorimetry (Lakowicz, 1999) is based on the analysis of fluorescence decay curves after the excitation pulse, whose duration is much shorter than the lifetime of a fluorophore in the excitation state (picoseconds).

For determination of the photophysical parameters from experimental curves (solution of the inverse problem) the variation algorithm (Banishev et al., 2008a) was used. It is based on the procedure of minimizing the functional of the residue between the experimental curves and curves calculated from models (1a) or (1b) by varying the photophysical parameters (Boychuk at al., 2000).

It is necessary to point out two distinctive features of nonlinear laser fluorimetry: (i) as the method implies detection of fluorescence photons (see Eqs. (2)), the photophysical parameters derived from the saturation curve relate only to a fluorescent molecule of COC; (ii) information on the concentration of fluorescent molecules is not used in deriving the photophysical parameters from the saturation curve (Banishev et al., 2009; Filipova et al., 2001). Thus, the method allows one to determine individual photophysical parameters of a molecule in the case when the following complex situation takes place: (i) the sample under study is a multicomponent ensemble of molecules, the absorption bands of its subensembles overlapping (i.e. when the sample is excited, all the subensembles absorb light); (ii) the concentrations of molecules from the subensembles in the mixture are *a priori* unknown.

3. Experimental

3.1 Nanosecond laser fluorescence spectrometer

A homemade laser fluorimeter (Fig. 3) has been built for the experiments. The fluorimeter consist of a laser source, optical elements for light conversion, a fiber-optic cable, a cuvette and an optical multichannel analyser.

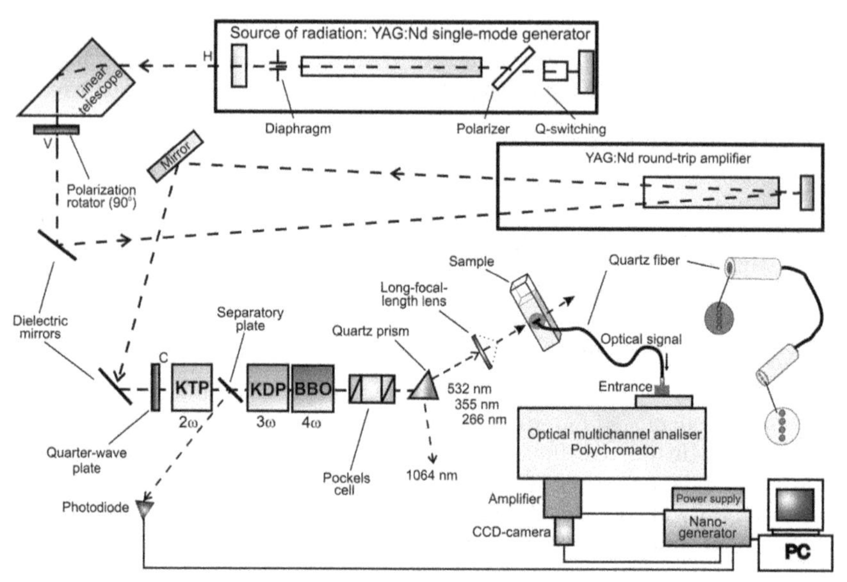

Fig. 3. Nanosecond laser fluorescence spectrometer scheme (for details see text). The letters H, V and C denote the horizontal, vertical and circular light polarization.

3.1.1 Laser excitation system

The pulsed single mode Nd:YAG laser (the fundamental wavelength is 1064 nm) with the set of nonlinear crystals for generating the 2nd (532 nm), the 3d (355 nm) and the 4th (266 nm) harmonics of the fundamental radiation was used. At a pulse repetition frequency of 10 Hz, the average pulse energy was: 5 mJ (532 nm), 2 mJ (355 nm) and 0.7 mJ (266 nm). The 2-mm-diameter Nd:YAG crystal with the diaphragm (the diameter is 1.5 mm) on the outlet face was used to obtain the light generation in a single mode regime (single transverse mode). Q-switching was made by the electro-optical shutter, working based on the Pockels effect. Between the Nd:YAG crystal and the shutter, the polarizer, transmitting light only with horizontal polarization, was placed. Thus, at the generator output, we had a single mode beam with horizontal polarization. After the generator, the light beam passed through the linear telescope, which also worked as a rotating prism. The main purpose of this component was to decrease the beam divergence in the horizontal plane, which is necessary for effective conversion of the fundamental frequency to its harmonics. Then, the laser beam was sent to the 90 degree polarization rotator, which changed the horizontal polarization to vertical one (required for reducing energy losses on rotating mirrors). Passed through the rotator and reflected from the rotating mirror beam was directed through the round-trip amplifier, which was composed of the Nd:YAG crystal (5 mm in diameter) and the rear mirror. Then the beam arrived at the polarizer, which reflects only vertical polarization. The reflected beam was directed through the quarter-wave plate for changing the vertical polarization to the circular one. That improves the efficiency of frequency-doubling in a KTP crystal. The KDP and BBO nonlinear crystals were used for generating the 3d and 4th harmonics. After the frequency conversion, the radiation of the 4th (or the 3d), 2nd and the fundamental harmonics was transmitted through the quartz prism for their spatial divergence in the horizontal plane. The continuous adjustment of the laser intensity during the saturation curves measurement was carried out by the Pockels cell, which was placed in the beam way right after the KTP crystal. The cell consists of the electro-optical component (DKDP crystal) and two Glan prisms. Changing the voltage on the DKDP, one can adjust the radiation output intensity at the outlet from the Pockels cell.

3.1.2 Registration system

The laser radiation was focused on the cuvette with the sample solution by a long-focal-length lens (the focal length is 20 cm). For collecting the fluorescence photons, the light-guide cable (the length is 5 m), consisting of seven quartz fibers (the diameter of each fiber is 600 micron), which were laid out in a row (like a slit) at both ends, was used. The cable inlet was fixed at the cuvette side, and the outlet was clasped to the polychromator entrance slit; the sample solution was excited by transmitted laser radiation. As a detector of radiation, the optical multichannel analyser (OMA) was used. The optical chamber (DeltaTech, Scientific Park of MSU) of the analyser consists of the electro-optical converter based on a gated microchannel plate (MCP), CCD matrix, and optical device for transferring an image from the MCP to the CCD matrix (a pixel size is $11\times11\mu m^2$). The chamber was fixed to the polychromator (MUM without the output slit, reciprocal linear dispersion is 0.15 nm per channel) optical output. The multichannel analyser was connected to the PC. As a result, optical image in the polychromator output slit plane could be obtained as a 2D picture on the PC monitor. The software installed on the PC allowed the OMA to operate both in the continuous and gated modes. When working in gated mode, a part of light was sent to the silica photodiode PD-265 (the building-up time of the leading edge is less than 2 ns). The photodiode was connected to the nanogenerator triggering inlet (the trigger level is 0.6 V). The gating of the MCP was implemented by high-voltage pulses from the nanogenerator (the amplitude is 800 V). The detector gate delay time could be adjusted through the nanogenerator over the range of 50 ns (the dead time of the detector) to 1200 ns with the step t_{step}=2.5 ns. Exactly because of the dead time of the registration system, when it operates in the gated mode (i.e. in the case of kinetic curves measurements), the light-guide cable as the dead time compensator (the optical delay line) was used. The detector gate width t_g could be varied from 10 to 1200 ns; in our experiments it was set to 10 ns.

3.1.3 Laser radiation parameters

The laser pulse duration at the wavelengths of 532 and 266 nm were fitted well by Gaussian function with the full width at half maximum (t_p) of 12 and 10 ns, respectively. When measuring (i) the kinetic curves, the laser beam diameter was 3 mm and (ii) the saturation curves, the beam was focused to a spot with a diameter from 600 to 800 μm, depending on the protein under investigation. The values of the photon flux density F were determined according to the equation $F^{-1}=E^{-1}\hbar\omega\cdot S\cdot t_p$, where $\hbar\omega$ is a photon energy, E is average pulse energy, S is cross section area of a laser beam, t_p is a pulse duration. The photon flux density F_{max} at a maximum of the saturation curve was measured before each experiment. Then, the photon flux density was gradually decreased. The F value at each point was obtained by the value of the Raman scattering signal from water molecules N_{RS}: $F=F_{max}\cdot N_{RS}/N^{max}_{RS}$, where N^{max}_{RS} is the Raman scattering signal at $F=F_{max}$. In nonlinear laser fluorimetry experiments, the fluorescence saturation curves were measured in the following ranges of F: (i) $5\times10^{23}\div7\times10^{25}$ $s^{-1}cm^{-2}$ when the excitation wavelength was 532 nm and (ii) $2\times10^{24}\div5\times10^{25}$ $s^{-1}cm^{-2}$ when excited at the wavelength of 266 nm.

3.2 Picosecond time-resolved and steady-state spectral measurements

The fluorescence lifetime measurements were performed with a streak camera (Agat SF 3M, VNIIOFI, Russia). A Nd:YAG laser with excitation wavelengths of 532 and 266 nm (the second and fourth harmonic of fundamental radiation) was used as a light source. The laser radiation parameters of the fluorimeter were as follows (for 532 nm): pulse energy 160 μJ, duration 20 ps (fwhm), beam diameter 5 mm. The error in determining the fluorescence lifetimes in time intervals of several nanoseconds did not exceed 5 %. In addition to the laser equipment, the Cary 100 spectrophotometer (Varian, Inc., USA) and the Cary Eclipse spectrofluorimeter (Varian, Inc., USA; slits width was 5 nm) were used for optical density measurements and fluorescence registration, respectively.

4. The object

4.1 Albumins

In this work, the solutions of human serum albumin (HSA) (>96%, Sigma) and of bovine serum albumin (BSA) (>98%, MP Biomedicals) in a phosphate buffer (0.01 M, pH 7.4) have been used. The proteins concentrations were 10^{-5} (absorption spectra measurement) and 10^{-9} M (fluorescence measurement at the nanosecond laser fluorimeter). All of the experiments were performed at a temperature of 25±1 ^{0}C. The structure and biological functions of HSA and BSA can be found in (Peters, 1996). Tryptophan, tyrosine, and phenylalanine (with relative contents of 1:18:31 in HSA and 2:20:27 in BSA) are the absorption groups in these proteins (as in many other natural proteins). The tyrosine fluorescence in HSA and BSA (as in many other natural proteins) is quenched due to the effect of adjacent peptide bonds, polar groups (such as CO, NH_2), and other factors, and phenylalanine has a low fluorescence quantum yield (0.03) (Permyakov, 1992). Therefore, the fluorescence signal in these proteins is determined mainly by tryptophan groups. In that case the fluorescence, registered in nonlinear and kinetic laser fluorimetry measurements, correspond to tryptophan residues (this fact will be used in Section 6.1).

As described in (Peters, 1996), HSA and BSA have similar structure and amino acid sequences that differ insignificantly by location of some certain amino acids. However, HSA contains one tryptophan residue in the protein matrix (Trp-214), and BSA contains two residues (Trp-212 and Trp-134). Trp-212 in BSA and Trp-214 in HSA have a similar microenvironment and, hence, their spectral properties are similar (Eftink et al., 1977). Tryptophans of BSA are not spectrally identical due to the stronger integration of Trp-212 into the protein's structure and the more hydrophobic environment of Trp-212 in comparison with Trp-134. The distance between tryptophans in BSA is about 3.5 nm. This fact makes the intramolecular energy transfer between them using the Forster resonance energy transfer (FRET (Valeur, 2002)) mechanism possible.

4.2 Red fluorescent proteins

Fluorescent proteins are a class of proteins that have a distinguishing property of forming their chromophore without involvement of any additional cofactors and ferments (autocatalytic reaction), except for molecular oxygen. In recent years, FPs have gained enormous popularity as genetically encoded fluorescence markers that enable to visualize a broad range of biological processes in cells and tissues. The most popular for practical applications are FPs whose fluorescence is shifted to the red (red FPs) and whose molecules are monomers (Piatkevich et al, 2010). The mRFP1 protein possesses these properties (Cambel et al., 2002), which made it an object of the research.

In this work the red FP mRFP1 and its mutants at 66 amino acid residue (glutamine 66) have been used. The seven mutants with substitution of the glutamine 66 for the serine (protein mRFP1/Q66S[2]), cysteine (mRFP1/Q66C), asparagine (mRFP1/Q66N), histidine (mRFP1/Q66H), alanine (mRFP1/Q66A), leucine (mRFP1/Q66L) and phenylalanine (mRFP1/Q66F) have been created. The method for fabrication and purification of proteins is described in (Vrzheshch et al., 2008). All the experiments were performed in a 0.06 M phosphate buffer at a temperature of 25±1 °C. The Bradford method (McCluskey, 2003) was used to determine initial concentration of the proteins. The proteins concentration was: (1) 10^{-9} M when measured fluorescence saturation curves; (2) $5.5\times10^{-10} \div 3\times10^{-10}$ M (depending on protein) when measured fluorescence quantum yield; (3) 10^{-6} M when measured steady-state spectra and fluorescence life-times on a picoseconds laser fluorimeter.

The formation of a fluorescent molecule of red FPs is a complicated process (usually called maturation) consisting of several stages (Verkhusha et al., 2004; Strack et al., 2010). At some stages, intermediate protein forms are produced, which remain in the resultant specimen (solution) of protein. In other words, the solution of red FPs is an ensemble of protein molecules consisting of several chemically non-equivalent subensembles (a mixture of different spectral forms of FPs). At neutral pH, the solution of mRFP1-like proteins contains three different spectral forms, namely, the mature form (will denote below as R form), the immature form with protonated chromophore (GH form) and the immature form with deprotonated chromophore (G form). The detailed information about formation of the chromophore of red FPs and analysis of composition of the samples can be found elsewhere (Verkhusha et al., 2004; Strack et al., 2010). According to a widely used terminology (Verkhusha et al., 2004), I'll also call the mature form as a red form and the immature forms as a green form in this work. The spectral forms can be detected in an absorption spectrum as the bands with maxima in the blue (360-420 nm, GH form), green (450-540 nm, G form) and red (550-600 nm, R form) spectral ranges. It is known (Pakhomov et al., 2004) that the balance between GH and G forms of FP can be disturbed with the changing of the solution acidity or as a result of light (from the UV, blue, or green spectral ranges) influence on the

---

[2] In this work, the name of each mRFP1 mutant is designated like mRFP1/Q66#, where Q66 is the abbreviation of the glutamine (single letter code is Q) at position 66, which is substituted. In place of the symbol #, the single letter code of amino acid the glutamine is substituted for is written down. The list of the amino acids abbreviation can be found elsewhere (Zamyatin, 1972).

protein. Irradiation of the protein molecules by the light from the green spectral range provokes a conversion of G form molecules to the molecules of GH form; as a result, the protein solution will represent a mixture of only two forms, namely, GH and R. In current work, for conversion the radiation of an Ar laser (LG-106-M1) at 488 nm (300mW) was used. An increment of the protein solution pH to 9 leads to a practically total transfer of the GH form molecules to the G form molecules (in that case, the protein solution is a mixture of the G and R forms). These procedures will be used as a way to decrease the amount of simultaneously existing forms in the mRFP1 solution when the method of the nonlinear fluorimetry will be implemented (see Section 6).

5. Steady-state fluorescence and absorption spectroscopy

5.1 Human and bovine serum albumins

The absorption spectra of the proteins and (for comparison) the corresponding equimolar solutions (solution of tryptophan, tyrosine, and phenylalanine at the same ratio as they are contained in protein) are shown in Fig. 4. It is seen that the parameters of the absorption bands of proteins do not coincide with the corresponding parameters for the equimolar solutions.

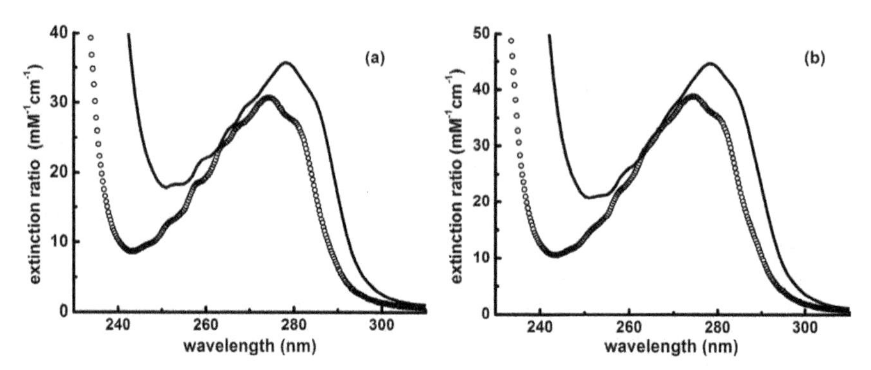

Fig. 4. Solid line is an absorption spectrum of (a) human and (b) bovine serum albumin. Circles are absorption spectrum of an equimolar solution with relative contents of tryptophan, tyrosine, and phenylalanine as (a) 1:18:31 and (b) 2:20:27.

The measurements of the fluorescence emission spectra of the proteins (data not showed) revealed that the fluorescence of both proteins is blue shifted relative to the tryptophan fluorescence (353 nm) in a buffer solution (Banishev et al., 2008a). This is due to a decrease in the tryptophan environment polarity in the proteins. The maximum of the HSA fluorescence (332 nm) is blue shifted in comparison with BSA (342 nm). Since the fluorescence spectrum of the tryptophan residues reflects the polarity of their nearest environment, and since the properties of the environments of Trp-212 in BSA and Trp-214 in HSA are similar (Eftink et al., 1977), such a shift can be related to the fact that BSA contains tryptophan Trp-134 located in the environment with a higher polarity (in comparison with Trp-212). Thus, the total fluorescence spectrum of BSA is red shifted. This result will be necessary for choosing the registration wavelength in measuring the acceptor and donor fluorescence when the nonlinear and kinetic curves will be measured (Section 6.1).

5.2 Monomeric red fluorescent protein and its mutants at residue 66

One can see from the absorption spectra of the mRFP1 (Fig. 5(a)) that in the wavelength range from 370 to 650 nm, there exist three absorption bands, which are explained by the presence of three spectral forms in the solution (Verkhusha et al., 2004), i.e. R, G and GH forms (the corresponding absorption maxima at 584, 503 and 380 nm). One can also see that the absorption bands of G and R forms overlap. Excitation of fluorescence in the absorption band of each form indicates that G and GH forms do not fluoresce, the R form fluoresces with maxima at 607 nm. The excitation of the protein solution by irradiation at a wavelength of 270 nm (tryptophan absorption band) leads to the appearance in the signal spectrum not only of an UV band (maximum at 330 nm), which corresponds to the tryptophan fluorescence in the protein matrix, but also of a band in the visible region of wavelengths (maximum at 607 nm) corresponding to fluorescence of the chromophore mRFP1 R form (Fig. 5(b)). The chromophore of GH and G forms is non-fluorescent (Campbell et al., 2002).

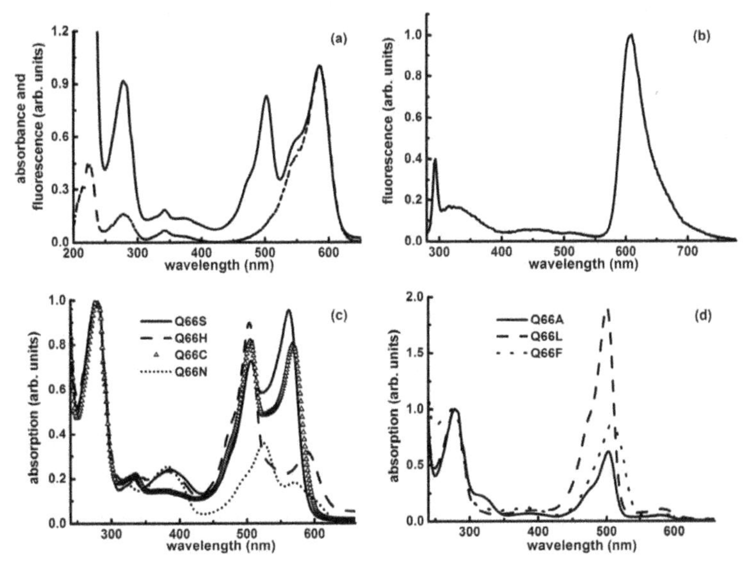

Fig. 5. (a) Absorption (solid) and fluorescence excitation (dotted, registration at 607 nm) spectra of mRFP1. The spectra are normalized to the signal at 584 nm. (b) Emission of mRFP1 (excitation by 270 nm). (c) Absorption spectra of the mutants with polar and (d) non-polar substitutions at position 66. The spectra are normalized to optical density at 270 nm.

Absorption spectra of the variants of mRFP1 with mutation at position 66 are shown in the Fig. 5(c, d).The absorption spectra structure of the mutants is similar to one for mRFP1. One can see that for all proteins in the wavelength range from 370 to 650 nm, the spectrum contains three bands with the maxima at 360-420 nm, 450-540 nm and 550-600 nm depending on the protein. These absorption peaks are corresponding to the GH, G and R forms. There only signal in the red spectral range were observed in the fluorescence emission and excitation spectra (i.e. the G form was found to be non-fluorescent) for all proteins except mRFP1/Q66F. For mRFP1/Q66F, the additional band in fluorescence excitation (maxima at 502 nm) and emission (maxima at 512 nm) spectra was detected, which attributed to the chromophore of G form. The spectral characteristics of the proteins are presented in the Table 1.

Note that the presence of the three protein forms can be qualitatively seen in the absorption spectrum (Fig. 5). However, as is described in (Banishev et al., 2009), the quantitative determination of the individual photophysical parameters of their chromophore with the help of only conventional methods are problematic. This is explained by the fact that the preparative separation of the forms is rather difficult and, as a result, it is hard to find their partial concentrations. At that rate, for example, for calculating the molar extinction coefficient (or absorption cross section) of chromophore from absorption spectra, the total protein concentration (total concentration of all forms) is used. As a result, the extinction coefficients are artificially underestimated (Kredel at al., 2008). At present, the only method that used for determining the individual extinction coefficient of chromophore of each spectral form can be found in (Ward, 2005). However, as it was pointed out in (Kredel et al., 2008), the values measured by the method are inaccurate in case of red FPs. Although, the procedure which enables to reduce the experimental errors has been proposed by (Kredel et al., 2008), the problem remains still topical.

Protein	Residue at 66	Green (G) form	Red form		
		λ^{max}_{abs}	λ^{max}_{abs}	λ^{max}_{ex}	λ^{max}_{em}
mRFP1	Glutamine (Gln)	503	584	584	607
mRFP1/Q66S	Serine (Ser)	506	562	561	579
mRFP1/Q66C	Cysteine (Cys)	505	568	568	588
mRFP1/Q66N	Asparagine (Asn)	525	570	570	604
mRFP1/Q66H	Histidine (His)	504	588	588	618
mRFP1/Q66A	Alanine (Ala)	500	582	578	605
mRFP1/Q66L	Leucine (Leu)	500	582	577	613
mRFP1/Q66F	Phenylalanine(Phe)	507	591	595	624

Table 1. The position (nm) of the maximum of the absorption, fluorescence excitation and emission spectra of the proteins.

6. Nonlinear fluorescence spectroscopy of proteins

6.1 Determination of photophysical parameters of single- and double-tryptophan-containing proteins

The emission spectra at several values of photon flux density F are shown in Fig. 6(a, b). In the figures, the band with a maximum value of F (4×10^{25} $cm^{-2}s^{-1}$) corresponds to the maximum value of the Raman scattering signal from water molecules. One can see that the maximum of the human serum albumin (HSA) fluorescence band does not change its position when F is changed. This is due to the fact that HSA contains one saturating fluorophore. However, the bovine serum albumin (BSA) fluorescence band is blue shifted (from 340 to 335 nm) when F is increased, owing to the fact that BSA contains two fluorophores in different environments (therefore, with different spectral properties), which exhibit different degrees (factors) of saturation. Taking into account the blue shift of the HSA fluorescence spectrum (in comparison with the BSA fluorescence spectrum) and the similarity of the properties of Trp-214 in HSA and Trp-212 in BSA (see Section 5.1), one can assume that, in the system of two tryptophans of BSA, Trp-212 serves as the donor of the energy (the energy transfer occurs via Forster mechanism), and Trp-214 is the acceptor (i.e., its fluorescence spectrum is presumably shifted towards long wavelengths).

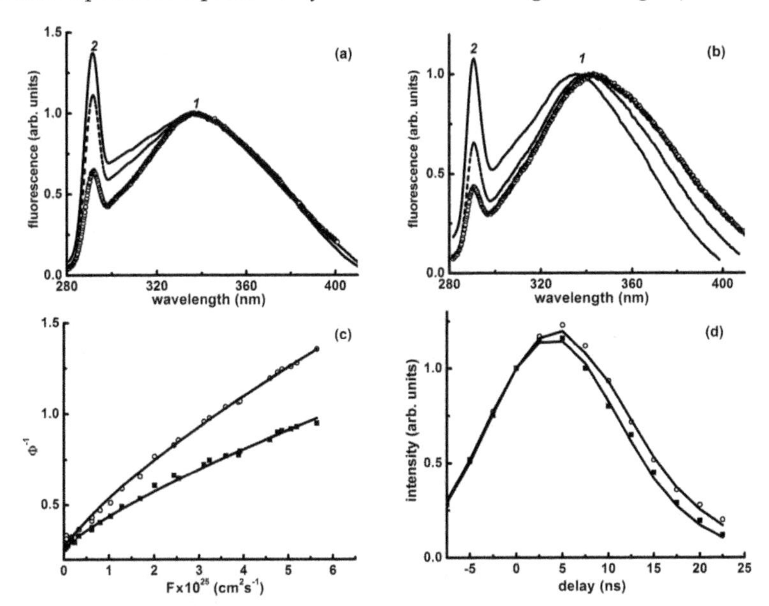

Fig. 6. (a) HSA and (b) BSA emission spectra at several values of photon flux density (see text); (1) is fluorescence and (2) is water Raman scattering band. (c) Saturation and (d) kinetic curves for BSA; fluorescence was registered at 390 (squares) and 310 (circles) nm. Lines are plotted using model (1b) and Eqs. (2b, c) for parameters from the Table 2.

The kinetic curves (see Eq. 6) and the fluorescence saturation curves of BSA are shown in Fig. 6(c, d). For BSA, the saturation curves depend on the registration wavelength in the wavelength range 310–390 nm. This is due to the fact that the BSA fluorescence band is a superposition of the bands of two tryptophans possessing different spectral properties. A similar difference in the curves for HSA is negligible. For the determination of the photophysical parameters of HSA fluorophore from fluorescence saturation and kinetic curves, the model (1a) and Eq. (2a) have been used. The same calculation procedure was done for BSA experimental curves, but with using the model (1b) and Eqs. (2b,c); the fluorescence signal was measured at 310 nm (when registered the fluorescence saturation and kinetic curves of the donor) and 390 nm (for similar curves of the acceptor). The resulting values of the parameters of protein fluorophores are presented in the Table 2.

As one can see from the Table 2 that the values of photophysical parameters σ and τ of Trp-214 in HSA and Trp-212 in BSA are similar. This result should have been expected based on a comparison of the structures of these proteins. The rates of energy transfer in BSA from excited donor to unexcited acceptor (K_{DA}) and to excited acceptor (K_{SS}) are small in comparison with the rate of intramolecular relaxation (τ^{-1}). This can be due to following reasons: (i) in BSA, the tryptophan residues in the D–A pair are located at a distance that is insufficient for a noticeable energy transfer between them. According to the data on the BSA structure (Peters, 1996), the distance between two tryptophans in the molecule is about 3.5 nm. For comparison, the Forster radius for the energy transfer between free tryptophans ranges from 0.6 to 1.2 nm (depending on the solvent). (ii) perhaps, the mutual orientation of the transition dipoles of fluorophores impedes the energy transfer (Lakowicz, 1999).

<table>
<tr><td>Protein</td><td>Parameters</td><td colspan="2">Tryptophan residues</td></tr>
<tr><td rowspan="2">HSA</td><td rowspan="2">τ, ns
$\sigma(\lambda_{ex}=266)$, cm^2
K'_{32}, s^{-1}</td><td colspan="2">Trp-214</td></tr>
<tr><td colspan="2">4.5 ±0.5
(1.3±0.1)×10^{-17}
<10^{-7}</td></tr>
<tr><td rowspan="2">BSA</td><td rowspan="2">τ, ns
$\sigma(\lambda_{ex}=266)$, cm^2
K_{DA}, s^{-1}
K_{SS}, s^{-1}</td><td>Trp-134 (donor)</td><td>Trp-212 (acceptor)</td></tr>
<tr><td>6.2±0.5
(3±0.3)×10^{-17}
<10^{-7}
<10^{-7}</td><td>5±0.5
(1±0.1)×10^{-17}
<10^{-7}
<10^{-7}</td></tr>
</table>

Table 2. Photophysical parameters of fluorophores (tryptophan residues) in HSA and BSA.

It is significant that the values of the absorption cross section determined using the method of nonlinear fluorimetry are true values for fluorophores and they are obtained without *a priori* information about the contribution of other groups into absorption at a specific wavelength and about the concentration of fluorophores. This is a unique feature of nonlinear fluorimetry. As mentioned above, there are three absorption groups in proteins (tryptophan, tyrosine and phenylalanine) and the absorption spectra of these three amino acids are overlapping. In that situation it is hard to separate the contribution of each amino acid group from total protein absorption without any preparative action on a molecule. For that reason, the absorption cross section of the tryptophan residue in protein is assumed to be equal to the absorption cross section of free tryptophan in solution (Pace et al., 1995), because it is supposed that this parameter is weakly dependent on the environment. However, a comparison of the equimolar solution and protein solution absorption spectra (Fig. 4) shows that these spectra do not coincide for HSA and BSA. Therefore, it is clear that such an assumption is just estimation and for diagnostics of the state of a protein it is necessary to be able to determine the true photophysical parameters of tryptophan merged in protein matrix. That has been done in this Section. The absorption cross section for tryptophan in aqueous solution is equal to 1.6×10^{-17} cm^2 (Banishev et al., 2008a) (at 266 nm); now, it can be compared with the true values of the absorption cross section of tryptophan residues in native proteins (see Table 2). The abilities of the approach are not limited by albumins.

6.2 Determination of photophysical parameters of mRFP1 protein under UV excitation

According to three-dimensional structure (PDB ID1g7k) of the mRFP1, there is a tryptophan at a distance of about 15Å from the protein chromophore, which could be a potential partner (the donor) for inductive FRET to the protein chromophore. Thereby, the tryptophan and the chromophore form a LDA pair inside a molecule of the FP.

As a preliminary step in determination of the full set of the mRFP1 photopysical parameters, the analysis of fluorescence decay at picoseconds excitation has been done. The excitation wavelength was 266 nm in order to match the acceptor (tryptophan) absorption band. The acceptor fluorescence decay under excitation of an ensemble of donor-acceptor pairs by δ-pulse is described (Valeur, 2002) as:

$$I_A(t) \sim B \cdot \exp(-t/\tau_A) - A \cdot \exp(-t/\tau_{D+A}) \quad (4)$$

where, $B=A+[A^*]_0$, $[A^*]_0$ is the excited acceptor molecules concentration at a time point $t=0$ (immediately after the excitation pulse is over); $A=[D^*]_0 \cdot K_{DA}/(1/\tau_{D+A}-1/\tau_A)$; $[D^*]_0$ is the excited donor molecules concentration at a time point $t=0$; $\tau_{D+A}=(1/\tau_D+K_{DA})^{-1}$ is the fluorescence lifetime of the donor in the presence of the acceptor; other designations are given above.

The fluorescence decay signal was measured at picoseconds time-resolved fluorimeter described in Section 3.2. The excitation wavelength was 266 nm and the signal registration was done in the range of the R form chromophore fluorescence (the chromophores of others forms are non-fluorescent). The experimental time dependence was fitted by function (4), as a result, the lifetimes τ_A, τ_{D+A}, and the partial contributions of B and A components of the fluorescence decay curve were obtained. The K_{DA} cannot be determined only from the fluorescence decay curve, because in this case it would be necessary to remove the acceptor and measure the fluorescence lifetime of the donor (in the absence of the acceptor). As it will be shown below, the nonlinear fluorimetry method makes it possible to resolve this problem for the native protein without any preparative action. The processing of the fluorescence decay curve of the mRFP1 allowed us to determine the lifetime values, τ_A=1.6 ns and τ_{D+A}=0.24 ns. The values of B and A in (4) were found to be practically equal. It is indicative of the absence of the direct excitation of the acceptor; therefore, in the equation system (1b) one can assume σ_A=0 (under excitation at 266 nm) and for this reason, the acceptor fluorescence (under this wavelength excitation) is a result of the energy transfer from the tryptophan to the chromophore.

The next step was to measure and analyse the fluorescence saturation curves with excitation at the wavelength of 266 nm. To improve the stability of the inverse problem solution, the process of singlet-singlet annihilation from the model (1b) was excluded. At pH 7.4 (the acidity at which the protein was initially produced), the mRFP1 solution is represented as a sum of three subensembles of molecules, each having its own set of photophysical parameters, and the dynamics of the populations of the collective states is described by its own system of equations similar to (1b). The number of the photons of the tryptophan fluorescence is calculated from Eq. (2b), where the sum under the integral is the population of the collective states n_2 and n_4 for all three forms. The fluorescence of the R form is calculated from Eq. (2c), where the populations of the collective states n_3 and n_4 of this form are present. It is difficult to resolve the inverse problem of nonlinear fluorimetry in such a situation, because the number of unknown parameters is too large. But, as was mentioned in Section 4.2, there are the techniques that allow reducing the number of the simultaneously present forms to two. In that case if the portion of the concentration of the R form is c_R and that of the second form (for example, G) is c_G, then the numbers of the fluorescence photons of the donor and acceptor (from the unit of volume) are:

$$N_{Fl}^{D}(\lambda)=\tau_{D}^{-1}\cdot\eta_{D}\cdot\int_{-\infty}^{+\infty}[c_{R}\cdot(n_{2}{}^{R}(t,r)+n_{4}{}^{R}(t,r))+c_{G}\cdot(n_{2}{}^{G}(t,r)+n_{4}{}^{G}(t,r))]dt \quad (5a)$$

$$N_{Fl}^{A}(\lambda)=\tau_{A}^{-1}\cdot\eta_{A}\cdot\int_{-\infty}^{+\infty}c_{R}[n_{3}{}^{R}(t,r)+n_{4}{}^{R}(t,r)]\,dt \quad (5b)$$

where the symbols R and G denote the R and G form molecules; n_2, n_3, and n_4 are the collective states (see (1b)); η_D and η_A are the fluorescence quantum yields (see (2b,c)); and c is the relative concentration of the forms in total molecules concentration. The amount of the R form molecules c_R can be found from the algorithm that described in the next Section.

In this connection, the following procedure has been realized:

a. The value of the protein solution pH was set near 9; in this situation, the protein solution contains only protein molecules of the R and G forms. After that, two saturation curves were taken under excitation at the wavelength of 266 nm and fluorescence registration at 330nm (the fluorescence saturation curve of the donor, i.e., the molecules of the tryptophan contained in the protein matrix of the R and G forms) and at 607nm (the fluorescence saturation curve of the acceptor, i.e., the chromophore only in the protein molecules of the R form). Resolving the inverse problem, in which fluorescence response formation is described by two systems of the kind of (1b) for the populations of the collective states of the LDA pairs in macromolecules of R and G forms, and the values of the parameters τ_{D+A}, τ_A, σ_A=0 for the R form of the protein are considered to be known (see above), the values of K_{DA} (for the R and G forms) and τ_A (for the G form) have been determined.
b. The protein sample was irradiated with the Ar laser at a wavelength of 488 nm (the pH value is near to a neutral one) and made only the R and GH forms present in the solution. After that, I used the same procedure of measurement and calculation of the saturation curves and determined the following parameters: K_{DA} (for R and GH forms of protein) and τ_A (for the GH form of the protein). The values of the K_{DA} for the R form in both cases coincided in error limits of the experiment.

The experimental dependences $\Phi^{-1}{}_D(F)$ and $\Phi^{-1}{}_A(F)$ for case (a) can be found in (Shirshin et al., 2009). When measured the saturation curves $\Phi^{-1}{}_D(F)$ and $\Phi^{-1}{}_A(F)$, the intensity in the spectra of the first and second orders of the RS valence band of water molecules (the wavelengths are 291 and 582 nm, correspondingly) were used as a reference signal; the excitation was at 266 nm. Having performed this procedure, the photophysical parameters of FP mRFP1 under UV excitation (266 nm) have been determined (see Table 3).

Let us discuss the main results of this Section. First of all, it is interesting to compare the true value of the absorption cross section obtained for tryptophan in the FP mRFP1 with the values for tryptophan in an aqueous solution (1.6×10^{-17} cm^2 (Banishev et al., 2008)), human serum albumin (1.3×10^{-17} cm^2), and bovine serum albumin ($\sigma_D=1\times10^{-17}$ cm^2 and $\sigma_A=3\times10^{-17}$ cm^2), which were determined in previous Section. One can see that the values for tryptophans in proteins are different and do not equal to the value for a free tryptophan, as is often assumed. I want to emphasize that the lifetime of the excited state of the donor (tryptophan) τ_D in the absence of the acceptor (chromophore) has been obtained without the removal of the acceptor (as it is often supposed when determine the energy transfer efficiency in LDA pairs by conventional methods (Valuer, 2002)). For mRFP1, this value can be compared with ones for the free tryptophan (τ_D=2.8 ns (Banishev et al., 2008a)), HSA (τ_D=4.5 ns) and BSA (τ_D=5 and τ_A=6.2 ns). Simultaneously, the excited state lifetime values of the chromophores of the GH and G form have been obtained (τ_A in the Table 3), although the chromophores of these forms are non-fluorescent. The obtained results show that the high volume (E=0.89) of the energy transfer efficiency from the tryptophan to the chromophores in all three forms of the protein is of special scientific and practical interest. This permits employing mRFP1 as a promising fluorescence indicator that makes use of its own inner LDA pair (an alternative is the preparation of such pairs of two proteins (Srinivas et al., 2001; Truong & Ikura, 2001).

Parameter	Values for R form	Values for GH form	Values for G form
$\sigma_D(\lambda_{ex}=266)$, cm^2	$(1\pm0.2)\times10^{-16}$	$(1\pm0.2)\times10^{-16}$	$(1\pm0.2)\times10^{-16}$
$\sigma_A(\lambda_{ex}=266)$, cm^2	0	not defined	not defined
K_{DA}, s^{-1}	$(3.7\pm0.7)\times10^9$	$(7.8\pm1)\times10^9$	$(2.5\pm0.7)\times10^9$
E	0.89	0.94	0.84
τ_A, ns	3±0.15	1.9±0.4	1.7±0.4
τ_D, ns	2.1±0.5	2.1±0.5	2.1±0.5

Table 3. The photophysical parameters of the LDA pairs in mRFP1 by UV excitation. In the table: (1) $\sigma_D(\lambda_{ex}=266)$ and $\sigma_A(\lambda_{ex}=266)$ are the absorption cross section of the donor (tryptophan) and the acceptor (chromophore) at the wavelength of 266 nm; (2) K_{DA} and $E\equiv K_{DA}/(K_{DA}+1/\tau_D)$ are the rate and efficiency of the energy transfer from the excited donor to the unexcited acceptor; (3) τ_D and τ_A are the excited state lifetimes of the donor (in the absence of the acceptor) and the acceptor.

6.3 Influence of a single amino-acid substitution on the individual photophysical parameters of the fluorescent form of the mRFP1 protein

In this Section, the influence of a single amino acid substitution in mRFP1 at position 66 on optical characteristics of the chromophore of fluorescent spectral form (R form) is performed. For that purpose, the method of nonlinear laser fluorimetry was realized in the version when the protein fluorescence is excited by the wavelength of 532 nm (i.e. the only protein chromophore was excited). All technical details of the procedure can be found in (Banishev et al., 2009).

At first, the photophysical parameters of R form chromophore of the proteins were determined. The σ and K'_{32}, defined in Section 2, have been determined from fluorescence saturation curve for each of the eight protein samples. Because the solution of each of the eight proteins contains mature (red) and immature (green) form, the only fluorescence in the red spectral range (from 550 nm) was detected (to obtain the parameters only for R form chromophore). The typical view of the measured fluorescence saturation curves can be found in (Banishev et al., 2009). To simplify the inverse problem solution, the fluorescence lifetime τ was measured independently with the picosecond laser fluorimeter (excitation at 532 nm). It was found that for all protein samples the fluorescence decay best fit by a single-exponential dependence (Banishev et al., 2009). Solving the inverse problem of nonlinear fluorimetry for each saturation curve at given τ, the σ and K'_{32} for each protein sample have been defined. Note that in this scheme of nonlinear laser fluorimetry the 532-nm laser pulses were used for exciting fluorescence, and, hence, σ is the absorption cross section of the protein R form at 532 nm, i.e., $\sigma \equiv \sigma_R^{(532)}$.

At the second stage, the partial concentration of the mature and immature species in the resultant solution of each mutant has been determined. As was said above, the equilibrium between the GH and G forms of FPs can be shifted under the action of external factors. Using this property of red FPs, it is possible to find the ratio of concentrations of all forms. However, the only the red fluorescence of R form is useful for practical application (Piatkevich et al., 2010). The green form is the by-products of maturation and supposed to be absent in ideal case. For that reason the measurement procedure has been simplified and the concentration of the R form and the total concentration of the GH and G forms were determined.

Indeed, given above-mentioned assumptions, one can write the following system of equations:

$$\begin{aligned} &\frac{\Phi_0^{(570)}}{\Phi_0^{(532)}} \cdot \frac{\sigma_{RS}^{(570)}}{\sigma_{RS}^{(532)}} = \frac{\sigma_R^{(570)}}{\sigma_R^{(532)}} \\ &C_R \cdot \sigma_R^{(570)} = 2.3D^{(570)} l^{-1} \\ &C_R \cdot \sigma_R^{(532)} + C_G \cdot \sigma_G^{(532)} = 2.3D^{(532)} l^{-1} \\ &C_R + C_G = C_0, \end{aligned} \tag{6}$$

where C_R, C_G are the concentrations of the R form and the total concentration of the G and GH forms in the solution (in cm^{-3}); C_0 is the total concentration of protein molecules determined by conventional methods (McCluskey, 2003); $\sigma_R^{(570)}$ and $\sigma_R^{(532)}$ are the individual absorption cross section of the chromophore of fluorescent form at 570 and 532 nm; $\sigma_R^{(570)}$ is integral absorption cross section of the chromophore of green form; $D^{(570)}$, $D^{(532)}$and $\sigma_{RS}^{(570)}$, $\sigma_{RS}^{(532)}$ are the optical density of the protein solution and Raman scattering cross section of water (Filipova et al., 2001) at 570 and 532 nm, respectively.

The first equality in system (6) reflects the fact that the quantum yields (expressed in terms of the fluorescence parameter Φ_0 (Filipova et al., 2001)) upon excitation of the protein solution at 532 and 570 nm are the same. The second and third equalities are the optical density (determined from the absorption spectrum of the proteins) written in terms of the concentration of protein molecules absorbing light at 570 and 532 nm and in terms of their absorption cross section. The wavelengths of 570 and 532 nm were chosen for reason mentioned in (Banishev et al., 2009). In system (6) the sought-for quantities are C_R, C_G, $\sigma_R^{(570)}$, $\sigma_G^{(532)}$, while experimentally measured values are $\Phi_0^{(570)}/\Phi_0^{(532)}$, $D^{(570)}$, $D^{(532)}$, l, C_0, $\sigma_R^{(532)}$ (the latter found by means of nonlinear laser fluorimetry). After the $\sigma_R^{(570)}$ was found, one can find the maximum value of the individual absorption cross section $\sigma_R^{(max)}$ of the R form chomophore, or the extinction coefficient $\varepsilon_R^{(max)}$, which is more convenient for comparison with data from literature. This value can be calculated using the absorption spectrum and relation $D^{(max)}/D^{(570)} = \varepsilon_R^{(max)}/\varepsilon_R^{(570)}$, where $D^{(max)}$ is the optical density at the maximum of the absorption band of the R form. The results for the eight samples are given in the Table 4.

One can see from the Table 4 that at the absorption maximum of the R form of the mRFP1 $\varepsilon_R^{(max)}$=(215±40) mM^{-1}cm^{-1}, which is drastically (four times) larger than the value published in (Campbell et al., 2002). This difference is due to the fact that the (Campbell et al., 2002) calculated the extinction coefficient using the total protein concentration (and, therefore, found the integral extinction coefficient) rather than the partial concentration (as in our case). As a result, the determination of the partial concentration of fluorescent molecules allowed us to find the individual extinction coefficient of the chromophore of the R form.

Protein	$\varepsilon_R^{(max)}$ (mM^{-1}cm^{-1})	C_R/C_0**, %	η	η_T*
mRFP1	215±40	26±6	0.24±0.03	0.01±0.01
mRFP1/Q66S	85±13	34±6	0.20±0.04	0.05±0.02
mRFP1/Q66C	135±20	17±6	0.19±0.04	0.01±0.01
mRFP1/Q66N	133±15	9±4	0.17±0.03	0.02±0.02
mRFP1/Q66H	230±27	8±4	0.13±0.03	0.07±0.02
mRFP1/Q66A	171±16	2±2	0.19±0.04	0.05±0.02
mRFP1/Q66L	240±40	2±2	0.12±0.03	0.06±0.02
mRFP1/Q66F	142±18	2±2	0.04±0.03	0.12±0.03

Table 4. Individual photophysical parameters of the R form and its fraction in the protein sample. Note: τ, η and η_T are the fluorescence lifetime, fluorescence quantum yield and quantum yield to the triplet state (converted from K'_{32}), respectively. Other parameters are defined after the system of Eqs. (6). * Determined from the fluorescence saturation curve; ** determined by solving system of Eqs. (6).

In the general case, the determination of photophysical parameters of FPs with the help of integral characteristics of the sample is incorrect, which can be proved by several examples. By using the dynamic difference method, (Kredel et al., 2008) obtained the individual extinction coefficient 143 mM^{-1}cm^{-1} for the chromophore of the R form of mPlum, which is also drastically larger than other published values ranging from 22 to 41 mM^{-1}cm^{-1} (Shcherbo et al., 2007) for this protein. It is interesting to note that (Gross et al., 2000) have earlier reported a similar value of 150 mM^{-1}cm^{-1} for the R form chromophore of red FP DsRed (the table value for this protein is assumed to be 75 mM^{-1}cm^{-1}). In their approach, the amount of immature species was deduced from mass spectroscopic analysis. Another example can be found in (Strack et al., 2010). Using the method described in (Ward, 2005), (Strack et al., 2010) got an assessed value of 123 mM^{-1}cm^{-1} for DsRed.T7, which is close to that obtained by (Kredel et al., 2008). From these examples one can see that the values of the individual extinction coefficients of the R form chromophore are close, as it is expected to be, because the chromophores of these proteins are considered to be chemically identical. One can assume a minor disagreement due to chromophore orientation change relative to protein matrix or composition of its closest environment (this can explain the difference in the extinction values for mPlum, DsRed and DsRed.T7 determined by (Kredel et al., 2008; Gross et al., 2000) and (Strack et al., 2010)). However, the published values of the extinction of red FPs with the same chromophore drastically vary depending on the protein: 75 mM^{-1}cm^{-1} per a polypeptide chain for the DsRed, 120 mM^{-1}cm^{-1} per a polypeptide chain for tdimer2(12) (Campbell at al., 2002), 22 mM^{-1}cm^{-1} for mPlum (Shcherbo et al., 2007), 50 mM^{-1}cm^{-1} for mRFP1 (Campbell at al., 2002) and 90 mM^{-1}cm^{-1} for mStrawberry (Shu et al., 2006). Discrepancies follow directly from the content of immature form in the protein samples.

In the measurements I obtained 215 $mM^{-1}cm^{-1}$ for mRFP1, which is larger than the value for DsRed and mPlum. However, it should be noted that I did not take into account the photochemical processes (photoionisation, photobleaching, etc. (Banishev et al., 2008a) in the model of fluorescence response generation (1a). The efficiency of these processes in the protein samples under study may be different. When efficient enough, the photochemical processes may contribute noticeably to fluorescence saturation. In this case their emission can result in the saturation curve giving an overstated value of the absorption cross section and, therefore, overstated quantity of $\varepsilon_R^{(max)}$. On the other hand, as it was mentioned in Section 5.2, the method applied by (Kredel et al., 2008) for determining the individual extinction coefficient of mPlum is not accurate enough in the case of red FPs (the method is well adapted only for GFP-like FPs). Therefore, the obtained value of 143 $mM^{-1}cm^{-1}$ can be underestimated and the precise value of the chromophore extinction is greater.

As it was shown earlier (Banishev et al., 2009), for the R form of the proteins mRFP1, mRFP1/Q66S and mRFP1/Q66C, the position of the maximum of absorption, fluorescence excitation and emission bands depends on the substituted amino-acid residue at position 66 and positively correlates with the volume of this residue: the maximum moves to the red range with increasing the volume of the residue. A similar correlation was described for the individual extinction coefficient of the R form chromophore, i.e. a higher extinction coefficient corresponds to a larger volume of the residue. The results for the new mutants mRFP1/Q66N, mRFP1/Q66H, mRFP1/Q66A, mRFP1/Q66L and mRFP1/Q66F are presented in Fig. 7(a, b). In the same figure the results obtained in (Banishev et al., 2009) for mRFP1, mRFP1/Q66S and mRFP1/Q66C are plotted. The values of the amino acids volume were taken from (Zamyatin, 1972). One can see that the dependence of the position of steady-state spectra (at maximum) of two new mutants (mRFP1/Q66N and mRFP1/Q66H) on the volume of amino-acid residue at position 66 has the same trend as described in (Banishev et al., 2009). The same can be observed for the individual extinction coefficient (but not for the integral one). There are no such dependences for characteristics of the proteins mRFP1/Q66A, mRFP1/Q66L and mRFP1/Q66F.

The results can be explained in the following way. It is known that formation of R form chromophore of red FPs goes through formation of a double bond between the Cα and N atoms of the 66th amino acid residue. Since dehydrogenation of a bond between Cα and N atoms involves the carbonaceous framework of the 66th amino acid residue in the system of conjugation, then the changes in the side radical of this residue can lead to the changes in the spectral and photophysical properties of the new mutants.

In the case of a polar amino acid at position 66 (serine, cysteine, asparagines and histidine), its side radical can form hydrogen bonds with the side radicals of glutamine-42, glutamine-213 and glutamate-215 (see Fig. 7 (c, d)). These radicals, in turn, belong to the protein shell (the β-barrel) and are rather rigidly bonded to it (Khrameeva et al., 2008). A change in the geometry of the side radical at position 66 will in this case cause a change in the geometry of the chromophore imidazolidine ring, because the interaction of the chromophore with the glutamine-42 and glutamine-213 through the hydrogen bond network can be distorted and a new bond with glutamate-215 can be formed. As a result, the chromophore tilt- and twist-angles (the pictures with the explanation of the angles can be found in (Piatkevich et al., 2010)) will change and the chromophore coplanarity will be distorted. In (Piatkevich et al., 2010) it was shown on the basis on x-ray diffraction data that the chromophore planarity is directly connected with the optical properties (the steady-state spectra positions, fluorescence quantum yield, etc.) of red FPs. The deviations from chromophore coplanarity are responsible for the changes in the optical characteristics for mCherry and mStrawberry (Shu et al., 2006). The interrelation between the optical properties of the monomeric red FPs and the geometry of their chromophores was also confirmed by the molecular dynamics simulations conducted for mRFP1 mutants with single polar amino acid substitutions at position 66 (Khrameeva et al., 2008). The simulations have shown that the substitutions have an influence on the torsion angles in the phenolic and imidazolidine rings of the chromophore as well as on the torsion angles in the regions of connection between these rings and chromophore attachment to β-barrel. It was predicted that the volume of the amino acid residue at position 66 can correlate with the optical characteristics of the mutants. The experimental results presented in this Section are consistent with the results of simulations performed by (Khrameeva et al., 2008).

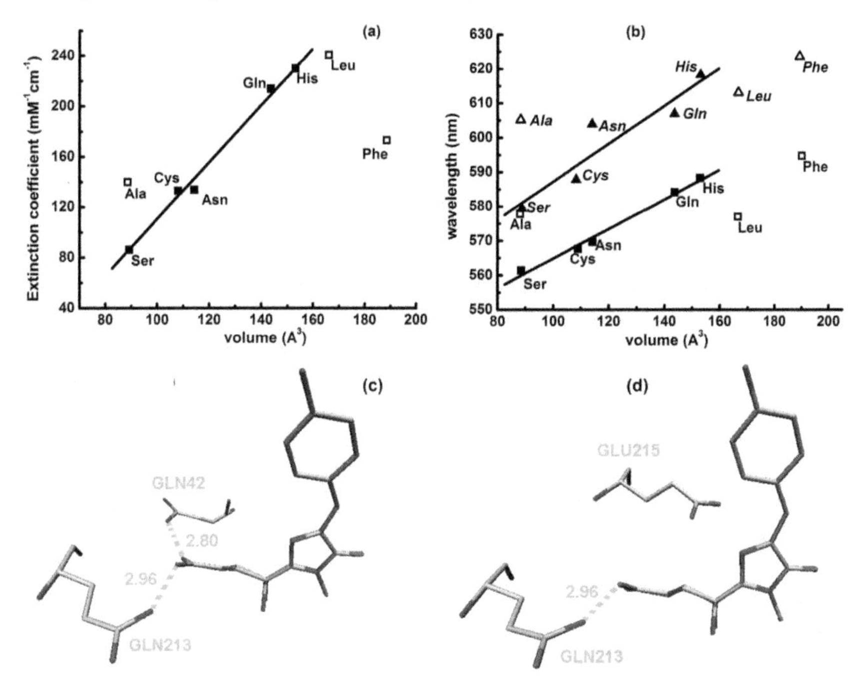

Fig. 7. The dependence of the (a) extinction coefficient, (b) position of absorption/ excitation (squares) and fluorescence emission (triangles) maximum of the protein R form on the volume of 66th amino acid residue. The solid and hollow scatters are polar and non-polar amino acids, respectively. (c) and (d) are schematic diagrams of chrmophore environment, showing the residues location at positions 213, 42 and 213 , 215, respectively. Hydrogen bonds are shown in dashed lines, labeled with lengths in angstroms. Glu is glutamate.

For the mutants with non-polar groups at residue 66 (alanine, leucine and phenylalanine) there is no correlation effect. The non-polar substitutions lead to breakage of hydrogen bonds between the 66th residue of the chromophore and glutamine-42 and glutamine-213. Formation of a new bond between the chromophore and the glutamate-215 is unlikely because of non-polarity of substituted residues. Therefore, there is no defined correlation between the chromophore geometry (consequently, the volume of the substituted amino acid residue) and the optical properties of the proteins.

At the present time the red FPs whose molecules are monomers are of particular interest (Piatkevich et al., 2010) as fluorescent markers. Attempts to find new variants of red FPs in order to improve their properties (higher brightness, photo and pH stability, etc.) are performed. However, the interrelation between optical or photophysical parameters and structural properties of FPs, which is necessary for development of these studies, is rather unclear. A method for prediction of properties of FPs based on their structure is still not developed. This problem might be solved by analysis of properties of mutant proteins with point mutations. Therefore, the results obtained in this Section can be used to tackle the general problem of the development of an algorithm, which could provide the prediction of the spectral properties of FPs based on their structures. The data will also be useful for revealing promising positions for directed mutagenesis.

7. Conclusion

In current work the approach based on the simultaneous use of nonlinear laser fluorimetry, spectrophotometry and conventional fluorimetry methods has been applied for investigation of the photophysical properties of the protein molecules of different complexity. The full set of photophysical parameters of the fluorophores (tryptophan residues) of human and bovine serum albumins has been determined. The photophysical processes in the spectral forms of the red FP mRFP1 under UV (266 nm) and visible (532 nm) irradiation are described quantitatively. The individual photophysical parameters of the new mutants of the mRFP1 protein (a single substitution at the 66 amino acid position) were determined. It was shown that the individual extinction coefficient of the red chromopore of the proteins correlate positively with the volume of the substituted amino acid residue at position 66 (for polar substitution). A similar correlation has been described for the position of the maximum of the absorption, fluorescence excitation and emission spectra: the position of the maximum moves to the red with increasing the volume of the residue. In addition, the partial concentration of the fluorescent spectral form in the resultant solution of each FP variant has been determined.

8. Acknowledgment

The author is grateful to Prof. Victor Fadeev for providing the ability to work in his group and for great help in mastering the fluorescence spectroscopy methods. The author also thanks Evgeny Vrzheshch for the samples of the red FPs and for valuable discussions.

9. References

Agranovich, V.M. & Galanin, M.D. (1982). *Electronic Excitation Energy Transfer in Condensed Matter*, Elsevier Science Ltd, ISBN 978-0444863355, North-Holland

Banishev, A.A.; Vrzhechsh, E.P. & Shirshin, E.A. (2009). Application of laser fluorimetry for determining the influence of a single amino-acid substitution on the individual photophysical parameters of a fluorescent form of a fluorescent protein mRFP1. *IEEE J. Quantum Electron.*, Vol. 39, No. 3, pp. 273-278, ISSN 10637818

[a]Banishev, A.A.; Shirshin, E.A. & Fadeev, V.V. (2008). Determination of photophysical parameters of tryptophan molecules by methods of laser fluorimetry. *IEEE J. Quantum Electron*, Vol. 38, No. 1, pp. 77-81, ISSN 10637818

[b]Banishev, A.A.; Shirshin, E.A. & Fadeev, V.V. (2008). Laser fluorimetry of proteins containing one and two tryptophan residues. *Laser Physics*, Vol. 18, No. 7, pp. 861–867, ISSN 1054-660X

Banishev, A.A.; Maslov, D.V. & Fadeev, V.V. (2006). A nanosecond laser fluorimeter. *Instrum. Exp. Tech.*, Vol. 49, No. 3, pp. 430–434, ISSN 0020-4412

Boychuk, I.V.; Dolenko, T.A.; Sabirov, A.R.; Fadeev, V.V.; Filippova, E.M. (2000). Study of the uniqueness and stability of the solutions of inverse problem in saturation fluorimetry. *IEEE J. Quantum Electron*, Vol. 30 , No. 7, pp. 611-616, ISSN 10637818

Campbell, R.E.; Tour, O.; Palmer, A.E.; Steinbach, P.A.; Baird, G.S., Zacharias, D.A. & Tsien, R.Y. (2002). A monomeric red fluorescent protein. *Proc. Natl. Acad. Sci. USA*, Vol. 99, pp. 7877–7882, ISSN 1091-6490

Eftink, M.R.; Zajicek, J.L. & Ghiron, C.A. (1977). A hydrophobic quencher of protein fluorescence: 2,2,2-trichloroethanol. *Biochim Biophys Acta.*, Vol. 491, No. 2, pp. 473–481, ISSN 0006-3002

Fadeev, V.V.; Dolenko, T.A.; Filippova, E.M. & Chubarov, V.V. (1999). Saturation spectroscopy as a method for determining the photophysical parameters of complicated organic compounds. *Opt. Commun.*, Vol. 166, pp. 25-33, ISSN 0030-4018

Filipova, E.M.; Fadeev, V.F.; Chubarov, V.V.; Dolenko, T.A.& Glushkov, S.M. (2001). Laser fluorescence spectroscopy as a method for determining humic substance. *Appl. Spectrosc.*, Vol. 36, No. 1, pp. 87-117, ISSN 0003-7028

Gross, L.A.; Baird, G.S.; Hoffman, R.C.; Baldridge, K.K. & Tsien, R.Y. (2000). The structure of the chromophore within DsRed, a red fluorescent protein from coral. *Proc. Natl. Acad. Sci. USA*, Vol. 97, pp. 11990–11995, ISSN 1091-6490

Khrameeva, E.E.; Drutsa, V.L.; Vrzheshch, E.P.; Dmitrienko, D.V. & Vrzheshch, P.V. (2008). Mutants of monomeric red fluorescent protein mRFP1 at residue 66: structure modeling by molecular dynamics and search for correlations with spectral properties. *Biochemistry*, Vol. 73, No. 10, pp. 1082-1095, ISSN 0006-2979

Kredel, S.; Nienhaus, K.; Oswald, F.; Wolff, M.; Ivanchenko, S.; Cymer, F.; Jeromin, A.; Michel, F. J.; Spindler, K.D.; Heilker, R.; Nienhaus, G.U & Wiedenmann, J. (2008). Optimized and far-red-emitting variants of fluorescent protein eqFP611. *Chem. Biol.*, Vol. 15, pp. 224–233, ISSN 1074-5521

Lakowicz, Joseph R. (1999). *Principles of Fluorescence Spectroscopy (2nd edition)*, Kluwer/Plenum Publishers, ISBN 0-306-46093-9, New York

McCluskey, K. (2003). The fungal genetics stock center: from molds to molecules. *Adv. Appl. Microbiol.*, Vol. 52, pp. 245–262, ISSN 1365-2672

Pace, C.N.; Vajdos, F. & Fee, L. (1995). How to measure and predict the molar absorption coefficient of a protein. *Protein Sci.*, Vol. 4, pp. 2411, ISSN 1469-896x

Pakhomov, A.A.; Martynova, N.Y.; Gurskaya, N.G.; Balashova, T.A. & Martynov, V.I. (2004). Photoconversion of the chromophore of a fluorescent protein from Dendronephthya sp. *Biochemistry*, Vol. 69, pp. 901–908, ISSN 0006-2979

Permyakov, E.A. (1992). *Luminescent Spectroscopy of Proteins,* CRC Press Inc, ISBN 978-0849345531, Boca Raton, USA

Peters, Jr., T. (1996). All *About Albumin: Biochemistry, Genetics, and Medical Applications,* Academic Press, ISBN 978-0125521109, San Diego, USA

Piatkevich, K.D.; Efremenko, E.N.; Verkhusha, V.V. & Varfolomeev S.D. (2010). Red fluorescent proteins and their properties. *Russ. Chem. Rev.*, Vol. 79, No. 3, pp. 243-258, ISSN 1468-4837

Shcherbo, D.; Merzlyak, E.M.; Chepurnykh, T.V.; Fradkov, A.F.; Ermakova, G.V.; Solovieva, E.A.; Lukyanov, K.A.; Bogdanova, E.A.; Zaraisky, A.G.; Lukyanov, S. & Chudakov, D.M. (2007). Bright far-red fluorescent protein for whole-body imaging. *Nat. Methods*, Vol. 4, pp. 741–746, ISSN 1548-7091

Shirshin, E.A.; Banishev, A.A. & Fadeev, V.V. (2009). Localized donor–acceptor pairs of fluorophores: determination of the energy transfer rate by nonlinear fluorimetry. *JETP Letters*, Vol. 89, No. 10, pp. 475–478, ISSN 0021-3640

Shu, X.; Shaner, N.C.; Yarbrough, C.A.; Tsien, R.Y. & Remington, S.J. (2006). Novel chromophores and buried charges control color in mFruits. *Biochemistry*, Vol. 45, No. 32, pp. 9639–47, ISSN 1520-4995

Srinivas, G.; Yethiraj, A. & Bagchi, B. (2001). FRET by FET and dynamics of polymer folding. *J. Phys. Chem. B*, Vol. 105, pp. 2475–2478, ISSN 1520-5207

Strack, L.R.; Strongin, D.E.; Benjamin, L.M.; Glick, S. & Keenan, R.J. (2010). Chromophore formation in DsRed occurs by a branched pathway. *J. Am. Chem. Soc.*, Vol. 132, pp. 8496–8505, ISSN 0002-7863

Truong, K. & Ikura, M. (2001). The use of FRET imaging microscopy to detect protein—protein interactions and protein conformational changes in vivo. *Curr. Opin. Struct. Biol.* Vol. 11, pp. 573–578, ISSN 0959-440X

Valeur, B. (2002). *Molecular Fluorescence: Principles and Applications*, Wiley-VCH, ISBN 978-3527299195, Weinheim, Germany

Verkhusha, V.V.; Chudakov, D. M.; Gurskaya, N. G.; Lukyanov, S.; Lukyanov K.A. (2004) Common pathway for the red chromophore formation in fluorescent proteins and chromoproteins. *Chem. Biol.*, Vol. 11, pp. 845–854, ISSN 1074-5521

Vrzheshch, E.P.; Dmitrienko, D.V.; Rudanov, G. S.; Zagidullin, V.E.; Paschenko, V. Z.; Razzhivin, A.P.; Saletsky, A.M. & Vrzheshch, P. V. (2006). Optical properties of the monomeric red fluorescent protein mRFP1. *Moscow Univ. Phys. Bull.*, Vol. 63, No. 3, pp. 109-112, ISSN 0096-3925

Ward, W.W. (2005). *Biochemical and Physical Properties of Green Fluorescent Protein. In Green Fluorescent Protein: Properties, Applications and Protocols (2nd Edition)*, Wiley John & Sons , ISBN 9780471736820, New Jersey, USA

Zamyatin, A.A. (1972). Protein Volume in Solution. *Prog. Biophys. Mol. Biol.*, Vol. 24, pp. 107-123, ISSN 0079-6107.

Protein structure available from Protein Data Bank, www.pdb.org, ID 2vad

Permissions

The contributors of this book come from diverse backgrounds, making this book a truly international effort. This book will bring forth new frontiers with its revolutionizing research information and detailed analysis of the nascent developments around the world.

We would like to thank Jamal Uddin, for lending his expertise to make the book truly unique. He has played a crucial role in the development of this book. Without his invaluable contribution this book wouldn't have been possible. He has made vital efforts to compile up to date information on the varied aspects of this subject to make this book a valuable addition to the collection of many professionals and students.

This book was conceptualized with the vision of imparting up-to-date information and advanced data in this field. To ensure the same, a matchless editorial board was set up. Every individual on the board went through rigorous rounds of assessment to prove their worth. After which they invested a large part of their time researching and compiling the most relevant data for our readers. Conferences and sessions were held from time to time between the editorial board and the contributing authors to present the data in the most comprehensible form. The editorial team has worked tirelessly to provide valuable and valid information to help people across the globe.

Every chapter published in this book has been scrutinized by our experts. Their significance has been extensively debated. The topics covered herein carry significant findings which will fuel the growth of the discipline. They may even be implemented as practical applications or may be referred to as a beginning point for another development. Chapters in this book were first published by InTech; hereby published with permission under the Creative Commons Attribution License or equivalent.

The editorial board has been involved in producing this book since its inception. They have spent rigorous hours researching and exploring the diverse topics which have resulted in the successful publishing of this book. They have passed on their knowledge of decades through this book. To expedite this challenging task, the publisher supported the team at every step. A small team of assistant editors was also appointed to further simplify the editing procedure and attain best results for the readers.

Our editorial team has been hand-picked from every corner of the world. Their multi-ethnicity adds dynamic inputs to the discussions which result in innovative outcomes. These outcomes are then further discussed with the researchers and contributors who give their valuable feedback and opinion regarding the same. The feedback is then collaborated with the researches and they are edited in a comprehensive manner to aid the understanding of the subject.

Apart from the editorial board, the designing team has also invested a significant amount of their time in understanding the subject and creating the most relevant covers. They scrutinized every image to scout for the most suitable representation of the subject and create an appropriate cover for the book.

The publishing team has been involved in this book since its early stages. They were actively engaged in every process, be it collecting the data, connecting with the contributors or procuring relevant information. The team has been an ardent support to the editorial, designing and production team. Their endless efforts to recruit the best for this project, has resulted in the accomplishment of this book. They are a veteran in the field of academics and their pool of knowledge is as vast as their experience in printing. Their expertise and guidance has proved useful at every step. Their uncompromising quality standards have made this book an exceptional effort. Their encouragement from time to time has been an inspiration for everyone.

The publisher and the editorial board hope that this book will prove to be a valuable piece of knowledge for researchers, students, practitioners and scholars across the globe.

List of Contributors

José Manuel González-López
Miguel Servet University Hospital, Clinical Biochemistry Service, Zaragoza, Spain

Elena María González-Romarís
Galician Health Service, Clinical Laboratory, Santiago de Compostela, Spain

Isabel Idoate-Cervantes
Navarra Hospital Complex, Clinical Laboratory, Pamplona, Spain

Jesús Fernando Escanero
University of Zaragoza, Faculty of Medicine, Department of Pharmacology and Physiology, Zaragoza, Spain

Shigeru Watanabe
Meikai University, Japan

Cynthia Ibeto, Akuzuo Ofoefule and Eunice Uzodinma
Biomass Unit, National Center for Energy Research & Development, University of Nigeria, Nsukka, Enugu State, Nigeria

Chukwuma Okoye
Department of Pure and Industrial Chemistry, Faculty of Physical Sciences, University of Nigeria, Nsukka, Enugu State, Nigeria

Raquel Borges Pinto and Themis Reverbel da Silveira
Post Graduate Program in Medicine: Pediatrics, Brazil

Pedro Eduardo Fröehlich and Tiago Muller Weber
Post Graduate Program in Pharmaceutical Sciences, UFRGS, Porto Alegre, Brazil

Ana Cláudia Reis Schneider and André Castagna Wortmann
Post Graduate Program in Medical Sciences: Gastroenterology and Hepatology, Universidade Federal do Rio Grande do Sul (UFRGS), Hospital de Clínicas de Porto Alegre, Brazil

Rita Giovannetti
University of Camerino, Chemistry Section of School of Environmental Sciences, Camerino, Italy

Petero Kwizera and Alleyne Angela
Department of Mathematics, Edward Waters College, Jacksonville, Florida, USA

Moses Wekesa and Md. Jamal Uddin
Department of Natural Sciences, Coppin State University, Baltimore, Maryland, USA

M. Mobin Shaikh
Sophisticated Instrument Centre (SIC), School of Basic Science, Indian Institute of Technology Indore, Indore, India

Judyta Cielecka-Piontek and Przemysław Zalewski
Poznan University of Medical Sciences, Department of Pharmaceutical Chemistry, Poland

Anna Krause and Marek Milewski
PozLab Contract Research Organization at Centre of Transfer of Medical Technologies, Poland

Abdelouaheb Djilani and Amadou Dicko
Université Paul Verlaine-Metz/LCME 1, Metz, France

Babakar Diop
Université de Bamako/Faculté des Sciences et Techniques, BP, Mali

Drissa Diallo
INRSP/Département de Médecine Traditionnelle, Bamako, Mali

Donatien Kone
Université Paul Verlaine-Metz/LCME 1, Metz, France
Université de Bamako/Faculté des Sciences et Techniques, BP, Mali

Zhang Li, Wang Minzhu, Zhen Jian and Zhou Jun
State Food and Drug Administration Medical Device Supervising and Testing Center of Hangzhou Zhejiang Institute for the Control of Medical Device, Hangzhou, China

Adina Elena Segneanu, Ioan Gozescu, Anamaria Dabici, Paula Sfirloaga and Zoltan Szabadai
National Institute for Research and Development in Electrochemistry and Condensed Matter, Timisoara (INCEMC-Timisoara), Romania

Nathir A. F. Al-Rawashdeh
United Arab Emirates University, Department of Chemistry, UAE
Jordan University of Science and Technology, Department of Chemistry, Jordan

Alexander A. Banishev
Department of Physics and Astronomy, University of California, Riverside, CA, USA